The
SCIENCE *and*
PHILOSOPHY *of*
MARTIAL ARTS

Exploring the Connections
Between the *Cognitive*, *Physical*,
and *Spiritual* Aspects of
Martial Arts

ALEX W. TONG

Foreword by Frederick Turner

BLUE SNAKE BOOKS
HUICHIN, UNCEDED OHLONE LAND
AKA BERKELEY, CALIFORNIA

Published by Blue Snake Books,
an imprint of North Atlantic Books
Huichin, unceded Ohlone land
aka Berkeley, California

Cover art by Caitlyn Tong
Cover design by Howie Severson
Book design by Happenstance Type-O-Rama

Printed in the United States of America

The Science and Philosophy of Martial Arts: Exploring the Connections Between the Cognitive, Physical, and Spiritual Aspects of Martial Arts is sponsored and published by North Atlantic Books, an educational nonprofit based in in the unceded Ohlone land Huichin (aka Berkeley, CA) that collaborates with partners to develop cross-cultural perspectives, nurture holistic views of art, science, the humanities, and healing, and seed personal and global transformation by publishing work on the relationship of body, spirit, and nature.

North Atlantic Books' publications are distributed to the US trade and internationally by Penguin Random House Publishers Services. For further information, visit our website at www.northatlanticbooks.com.

PLEASE NOTE: The creators and publishers of this book disclaim any liabilities for loss in connection with following any of the practices, exercises, and advice contained herein. To reduce the chance of injury or any other harm, the reader should consult a professional before undertaking this or any other martial arts, movement, meditative arts, health, or exercise program. The instructions and advice printed in this book are not in any way intended as a substitute for medical, mental, or emotional counseling with a licensed physician or health-care provider.

Library of Congress Cataloging-in-Publication Data

Names: Tong, Alex W., 1952– author.
Title: The science and philosophy of martial arts : exploring the
  connections between cognitive, physical, and spiritual aspects of
  martial arts / Alex W. Tong ; foreword by Frederick Turner.
Description: Berkeley, CA : Blue Snake Books, [2022] | Includes
  bibliographical references and index. | Summary: "An exploration of the
  physical, mental, and spiritual components of martial arts that
  integrates contemporary sports psychology, kinesiology, and
  neuroscience"— Provided by publisher.
Identifiers: LCCN 2021038681 (print) | LCCN 2021038682 (ebook) | ISBN
  9781623176655 (Trade Paperback) | ISBN 9781623176662 (eBook)
Subjects: LCSH: Martial arts. | Martial arts—Philosophy.
Classification: LCC GV1101 .T6 2022 (print) | LCC GV1101 (ebook) | DDC
  796.8—dc23
LC record available at https://lccn.loc.gov/2021038681
LC ebook record available at https://lccn.loc.gov/2021038682

1 2 3 4 5 6 7 8 9 KPC 27 26 25 24 23 22

North Atlantic Books is committed to the protection of our environment. We print on recycled paper whenever possible and partner with printers who strive to use environmentally responsible practices.

Opinions expressed in this book are solely those of the author. They do not purport to reflect the views of the American Amateur Karate Federation (AAKF) or its members. Discussions throughout the book have cited current theories and practice in human physiology and clinical psychology. They are not intended as medical advice. Readers should consider these references only as informational to the raised subject matter.

To my parents, Robert and Agnes.
For their love and sacrifice,
And for giving me the learning
opportunities that they did not have.

CONTENTS

ACKNOWLEDGMENTS

This book is very much the product of community, friends, and family. Accolades are due for the North Atlantic Books editorial and production staff, who turned loose pages into a highly professional final product. In particular, my special thanks go to Shayna Keyles for her insightful editing, Janelle Ludowise for her guidance throughout production, Christopher Church for the incredibly astute copyedits, and Jasmine Hromjak for the gorgeous cover design. I also thank Karen Davis for her meticulous proofreading.

I thank Anjali Krishnan, PhD, Suresh Lakhanpal, MD, Susan Radtke, Professor Frederick Turner, and Bob Young for their valuable comments; Nicole Tong for legal counsel; Alexander Tong as sparring partner; Caitlyn Tong for the amazing graphic illustrations; and Caitlyn and Maddie Tong for their keen eyes as photographers. My appreciation goes to Sander Neggo sensei for sharing his valuable out-of-print *Samurai* magazines. The contents provided indispensable background on the tenor of martial arts practice in the 1970s.

As the very idea of writing a book turned all-consuming, my children, Nicole, Alexander, Caitlyn, and Maddie, inspire by their youthful exuberance, self-assuredness, and sense of adventure, for which I am eternally proud and thankful.

I also thank the American Amateur Karate Federation Technical Committee members, Mahmoud Tabassi, Richard Kageyama, and Albert Cheah for their friendship and encouragements; and *yūdansha* in my *dōjō*, Tuan Huynh, Anjali Krishnan, Lanny Little, Vincent May, J. Robert Todd, Frederick Turner, Attila Vari, and Adam Blanchard, MD, who through their enthusiasm and inquisitiveness spur the process of reflection and deeper understanding.

The highest accolades are extended to the teachers and mentors throughout my life's journey as a martial artist and research scientist. These role models, in particular Hidetaka Nishiyama sensei and Dr. Marvin J. Stone, practice their craft with passion and lead by example, impressing that the circle of life is much about the passing of past generations' knowledge to the next.

FOREWORD

The Science and Philosophy of Martial Arts is an amazing work. It is surely the most comprehensive book yet written on the art of karate. But it is more than comprehensive: it does not just collect the many aspects of karate practice and theory, but points to a larger and more universal vision of the human body and spirit. Though a useful handbook on every aspect of *kihon* (basics), *kata* (form), and *kumite* (sparring) in karate, it is also an argument that dissolves ancient conflicts between spiritual and scientific knowledge, between East and West, and between the arts of the body and the arts of the mind. Throughout the book, the concept of emergence, where a new level of order arises out of the competitive interaction between the parts to generate a whole, is implicit in every chapter.

Teachers of karate need to read the careful systematic research into the biology and cultural history of the body in karate. It deepens and confirms what contemporary sensei know from their experience in learning and teaching this art. It is the chapters on the spirit that are perhaps most exciting, especially the discussion of the *dantians,* the qi focus centers of the body. Tong is a distinguished medical biologist and can translate between bioelectric and Taoist descriptions of how the body works.

This work of translation is also cultural and political. The elegant vocabulary of martial arts theory is echoed by Western bodily metaphors and vocabulary for moral, psychological, and emotional qualities in battle—heart, courage (as in the French *coeur,* "heart"), guts, bowels of compassion, having the stomach for a fight, having backbone, second wind, spirited defense, breathing space, level-headedness, having the neck (self-assurance) or kidney (inner strength) to do something, putting your shoulder to the wheel, and so on. This unanimity confirms the universal validity of the teaching. One would surely find the equivalent in Swahili and Quechua.

But Tong is after bigger game yet, perhaps. By including fundamental and classical physics, biochemistry, bioelectric theory, psychobiology, immunology, and other fields—almost the whole suite of scientific disciplines—and interleaving the discussion of these fields with a thorough grounding in Vedic, Chinese, and Japanese meditative and martial arts theory—Tong is making an important point. Martial study is a way of understanding the universe, not just a Western or Eastern art. And it is revealed partly in science, partly in religious language.

Tong's discussion of baseball and track and field expertise, with sound math behind it, points toward a grand pan-human vision of human solidarity in understanding our nature and our relationship to the universe we live in. The moral parts of his message do the same thing, noting the kinship between Western Christian-Aristotelian concepts such as honor, justice, faith, and temperance and the Confucian, Taoist, and Buddhist virtues. Tong's own origins in the great city of Hong Kong, which until recently was a brilliant synthesis of Eastern and Western traditions, may have something to do with the inclusiveness of his insights.

So he is not, as some superficial treatments of martial arts sometimes do, cheaply downgrading "superficial" and "materialistic" Western values and offering a cult of secret "Oriental" mysteries that will solve all our life's problems—arguably, a divisive and troublemaking message—but contributing a wider humanistic vision that integrates the best in all human cultures.

Tong's style, with its mixture of argument, instruction, anecdote, scientific explanation, humor, and witty contrasts, is an unorthodox but highly effective and always surprising medium. As a born teacher, he knows when to relieve a serious discussion by a story or show how a commonplace saying reveals deep truths.

On a personal note, as his student for thirty-five years, I can testify to the transformative effectiveness of his teaching both in my life and my work. As well as karate, I practice archery (kyudō in Japanese). I had gotten into a really destructive mood of self-criticism and excessive effort, attempting to force a good shot or to change some element of what I was doing, as if it were a magic trick that would suddenly fix everything. I read his section "The Psychology of Martial Arts: Mind Like Water, Mind Like Moon," and one day at practice I remembered his words and the explanation. I simply cast away all my inner conflict and striving, and suddenly my accuracy doubled.

Frederick Turner

Founders Professor Emeritus,

School of Arts and Humanities, University of Texas at Dallas

November 15, 2020

PREFACE

Today's martial arts date back to centuries of refinements. Traditional styles emerged in approximately 500 CE, having since crossed geographical boundaries, leaving indelible footprints as Northern and Southern Chinese wushu, Okinawan karate-dō, Japanese *budō*, and Korean taekwondo, to name a few. The various ethnicities' physicality and culture were integrated as distinct practices and applications at each stop. Yet the fundamental essence of duty, honor, respect, and dignity remains unchanged.

Martial arts practice aligns with our survival instincts to ward off physical harm. Our civilized society rarely requires this skill set for self-defense or in combat. The enthusiasm for martial arts continues to thrive nonetheless, also serving as the path for heightened physical performance, mental discipline, and spiritual growth. Training broadens our mental and physical reach, rendering confidence as we navigate through the uncertainties of everyday life, and in times of crisis.

Apart from increased dexterity and heightened awareness, these disciplines lower anxiety, mitigate stress, and sustain clarity of mind under duress. Deliberate practice, the process of repetition learning with focused attention and defined goals, uncovers the practitioner's strengths and shortcomings and gives a better understanding of the inner self.

The unarmed art of karate-dō became highly popular in the United States in the 1960s.[1] Its "exotic" combat techniques dazzled by their sheer power and explosiveness. The scientific community was equally enthralled and sought explanations for the remarkable feats of the discipline's practitioners. Studies now show that all of us are endowed with these seemingly mysterious gifts of strength, speed, confidence, and premonition, all of which can be unlocked through dedicated training in the ancient arts.

As a professional biomedical researcher for the past forty years, I am awed by the prescience of traditional martial arts insights that I have experienced firsthand as a practitioner of Nishiyama Shōtōkan Karate since 1972. These concepts are now validated by contemporary insights in neuropsychology, kinesiology, and sports medicine research. All point to the depths of untapped human potential. In particular, martial arts promote mind-body synergy, a common attribute among high achievers in all forms of competitive sports. As we are entertained by superhero tales spun in comic books and movies, we can take heart at the astounding breadth of our limits—mental and physical—that are well within reach.

Most robust art forms evolve by incorporating new ideas without forsaking basic tenets. Karate as a form of budō (martial way) practice was introduced to Japan in May 1922, when Okinawan Master Gichin Funakoshi demonstrated Okinawan *kara-te jutsu* (Chinese hand techniques, 唐手術) at a physical education exhibition sponsored by Japan's first Ministry of Education.[2] His students, Masatoshi Nakayama and Hidetaka Nishiyama, made the quantum leap of integrating biomechanical concepts to traditional practice, significantly adding appeal for the modern audience.

As the popularity of martial arts continues to the present day,[3] it is reasonable to ponder the future of traditional practice from the Western perspective, particularly in view of the public's ardor for mixed martial arts.

In an October 2019 opinion piece, staff sports writer Patrick Blennerhassett of the Hong Kong *South China Morning Post* commented on a well-known incident when forty-year-old mixed martial artist Xu Xiaodong challenged, then demolished, a tai chi master thirty years his senior.[4] The topic of martial art conflicts evidently ranked high among Western expatriate journalists in spite of their lack of relevant expertise.[5] "It's time to admit most traditional martial arts are fake," asserted Blennerhassett. His online video remained accessible some twelve months later, where Blennerhassett adopted even more pointed rhetoric to question the validity of traditional martial arts from karate to Wing Chun gōngfu; his opinion was backed by fights that he took part in as a teenage ice hockey player. Although Blennerhassett claimed the famed Bruce Lee as a pioneer of mixed martial arts, he skimmed over Lee's past as a teacher of traditional styles and his disciplined training for over twenty years (chapter 12).

Should budō training revert back to full contact combat and abandon long-held mores of discipline, honor, mutual respect, and humility for the sake of popular appeal? Or perhaps the traditional curriculum would be better received when dismantled into separate, consumer-driven end products, such as for self-defense skill-building, weight loss, and stress relief.[6]

This author is partial to a more sanguine tack, having committed his adult life to traditional karate-dō training. The Japan Traditional Karate Association of Dallas (formerly JKA Dallas), which I founded in 1982, has graduated more than thirty black belts and shaped champions in all categories of competition. I served as national and international judge after retiring from national competition, and hence have been party to the triumphs, setbacks, and sorrows of elite martial arts athletes for the past thirty years.

There is no question that the whole of traditional practice is more than the sum of its parts. The budō mind-set guides an individual in seeking perfection of character. As further elaborated in various chapters of this book, an attitude of humility does not preclude competitiveness, but impels the individual to give their best in all endeavors. This lifelong journey brings purpose and inner strength, enabling the person to better handle stress from day to day.[7]

Constituents of a typical dōjō (training hall) cover a broad demographic. Those who dedicate themselves can realize noticeable gains, regardless of age, training goal, or athletic talent. Many embark on the path for thrill, competition, and dominance. Yet over time, punching, kicking, or throwing come to be experiences toward a deeper understanding of personal identify. It behooves us to remind younger generations that these disciplines reinforce accountability and contribution to society, according to the martial arts morality of deed.[8]

Accordingly, martial arts are not merely idiosyncratic Asian pursuits that intrigue the Western mind. Each represents a human heritage that we can all share, a gift to celebrate our amazing presence on earth. Leaning on his scientific background, the author is partial to the idea that the what, why, and how of traditional practice can be better appreciated through the prism of contemporary science.

Researchers have long been intrigued by the interplay of mind, body, and spirit, also a treasured developmental goal in traditional martial arts. The sports sciences have played an increasingly important role to stretch mental and physical limits since the early twentieth century. An overlaying of proven scientific ideas can likewise bolster martial arts performance by elevating awareness and crystallizing training goals, and can minimize injury without diminishing traditional values.

In the book *Homo Deus: A Brief History of Tomorrow,* Yuval Noah Harari considers science and humanism to be opposing but complementary forces of modern society.[9] Science powers our way of life, whereas humanism offers meaning and ethical judgments. As we entertain the relevance of scientific insights in future practice, philosophical reflections provide a moral compass for the intended path to enlightenment. Chinese Daoist and Japanese Zen Buddhist thought has profoundly influenced martial arts since the thirteenth century. Budō arts seek *busshō* (Buddha nature) within—transcending inherent worth and dignity beyond the materialistic self.[10] Spirituality balances our physical and emotional needs. A strong spirit renders confidence and positivity.

Sānzhàn (now called Sanchin, 三戰) is the name of a historic, Okinawan training form (kata) introduced by Kanryō Higaonna and Kanbun Uechi at the turn of twentieth century.[11] The kata traces its origin to the sixteenth-century Chinese form Saam Chien of Five Ancestors Fist and Fujian White Crane gōngfu. Multiple versions are practiced today in Chinese wushu and Okinawan and Japanese karate and kobudō.[12] The literal translation of Sanchin (and Saam Chien) is "Three Battles," directed at the opponent's mind (*xīn,* 心), body (*shēn,* 身, or *tǐ,* 体), and spirit (*shén* or *kami,* 神).

Sanchin also serves as an instrument to order our constitutions. Clarity of mind and strength in spirit are part and parcel of the discipline it takes to accomplish the remarkable physical feats celebrated in the martial arts. A balanced development of mind, body, and spirit defines the character of the individual and that of the art.

Coming to terms with our three inner conflicts enriches our everyday life. It is within this context that this book has been organized. Its discussions combine technical analyses of the three domains of traditional practice with a counterpoint of philosophical and scientific perspectives.

Alex W. Tong

January 18, 2020

Dō (道, The Way) is the road of the Cosmos, not just a set of ethics for the artist or priest to live by, but the divine footprints of God pointing the Way.
—MIYAMOTO MUSASHI, *THE BOOK OF FIVE RINGS*

If you know the enemy and know yourself, you need not fear the result of a hundred battles.
—SUN TZU, *THE ART OF WAR*

Knowledge must become capability.
—CARL VON CLAUSEWITZ

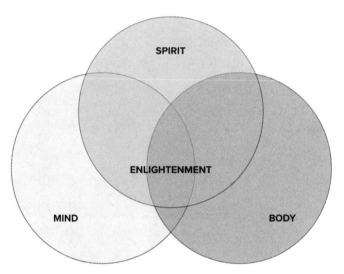

Sanchin (*sānzhàn,* 三戰) Three battles to unify the
mind, body, and spirit.

PART I

Introduction

Traditional Martial Arts: An Amalgamation of Heritage, Culture, and Philosophy

East-West Fusion: Coming to America

Hundreds of styles of martial arts beckon our attention in this culturally diverse society. About 3.5 million people participated in martial arts in the United States in 2019.[1] Once held in secrecy in their native country, wushu (China), kendō (Japan), taekwondo (Korea), muay thai (Thailand), and many other styles are now publicized as national treasures and heritage. This knowledge is accessible for a small fee and requires no more than enthusiasm, an open mind, and a sense of adventure.

Presidents Ulysses Grant and Theodore Roosevelt were known to practice judō, although popularization of Asian martial arts in the United States did not come about until the early 1960s. The arrival of this twentieth-century phenomenon can be credited to the U.S. Armed Forces over the Cold War era.[2]

In 1948 the Strategic Air Command (SAC) arranged a series of traditional martial arts demonstrations at U.S. military bases in Japan, subsequently leading to the formation of karate and judō clubs at these installations. Isao Obata* and Masatoshi Nakayama of the Japan Karate Association (JKA) gave karate demonstrations and provided instructions for the karate clubs.[3] Both were students of Gichin Funakoshi, founder of Shōtōkan (House of Pines and Waves, 松濤館) karate-dō.

Curtis E. LeMay, general of the U.S. Air Force, was then Commander of the Strategic Air Command (SAC). Mindful of the casualties of the U.S. bomber group during World War II, he felt that Air Force personnel should acquire a working knowledge of hand-to-hand combat to aid in escape and evasion.[4] SAC Combative Measures Programs

---

* Romanized names for historic Japanese figures are entered as personal name (Isao) then surname (Obata).

were implemented across U.S. SAC bases by 1951, incorporating judō, karate-dō, and aikidō techniques. The initiative was led by Airman Emilio (Mel) Bruno, a former AAU National Wrestling Champion and 5th degree judō black belt.[5] The U.S. Judo Federation was founded within the same period.[6]

In 1953, Bruno invited ten Japanese martial arts instructors to a four-month demonstration tour at U.S. and Cuban SAC bases. The team comprised seven premier *judō-ka* (practitioner of judō) from the Tokyo Kodokan Institute (Sumiyuki Kotani, Tadao Otaki, Kenji Tomiki, Kusuo Hosokawa, Tsuyoshi Sato, Takahiko Ishikawa, and Kiyoshi Kobayashi) and three renowned *karate-ka* (practitioner of karate) from JKA (Hidetaka Nishiyama and Toshio Kamada, led by Isao Obata). Hand-picked SAC airmen subsequently attended intensive eight-week Combative Activity Training Programs in Japan. The curriculum comprised judō, karate-dō, aikidō, and taiho jutsu (arresting art).[7]

Master Tsutomu Ohshima, student of Funakoshi, arrived from Japan in 1957 and set up the first university karate club at the California Institute of Technology in Pasadena, California. JKA dispatched three premier instructors in 1961. Its aim was to extend karate teaching nationwide. They were Hidetaka Nishiyama (Los Angeles), Teruyuki Okazaki (Philadelphia), and Takayuki Mikami (New Orleans), all students of Master Funakoshi.

A plan to globalize karate-dō practice soon followed. JKA instructors were seeded at major cities in Europe, the Middle East, and Central and South America. Hidetaka Nishiyama was appointed as president of JKA-International for this visionary endeavor.

Premier teachers of various karate styles also started dōjōs in the mainland United States within the same period, including Richard Kim (San Francisco), Takayuki Kubota (Los Angeles), Fumio Demura (Santa Ana, California), Masataka Mori (New York City), Shojiro Sugiyama (Chicago), Yutaka Yaguchi (Denver), and Teruo Chinen (Spokane, Washington).

Returning U.S. servicemen, having trained in Japanese and Korean martial arts, came back with their adopted art. Robert Trias, a U.S. Navy veteran, opened the first public karate school in the mainland United States in Phoenix and taught a form of American Karate that combined Western boxing with Chinese gōngfu and Okinawan *Shuri-te'* techniques.[8] Other notables include Ed Parker, a native Hawaiian, Coast Guard veteran, and founder of American kenpo; George Mattson, who studied Uechi-ryū; and Peter Urban, a Navy veteran who trained in Japan and founded American Gōjū-ryū.[9]

Robert Fusaro began his JKA karate training during the Korean War. He was awarded a black belt by Masatoshi Nakayama in 1959. Fusaro sensei opened his first dōjō in 1960 in downtown Minneapolis. Robert Graves, a Marine Corps veteran, returned after the Korean War and became the founding instructor of Shōtōkan Karate in the Pacific Northwest. Arriving Korean instructors also introduced various indigenous Korean martial arts, later developed as taekwondo. The notables included Jhoon Rhee, Henry Cho, Kum Soo, and Jack Hwang.[10]

The emergence of these famed instructors coincided with the global infatuation with James Bond movies, based on the fictional novels of Ian Fleming, a retired British spy. The lone-wolf secret agent code-named 007 alternated between high-browed demeanors and cold-blooded actions. He waded through danger, armed with nothing more than attitude, a pistol, a handful of high-tech gizmos, and mastery of Western and Eastern combat techniques—in particular the ubiquitous "karate chop." Martial arts dōjōs sprang up in every state's suburbs to cater to the public's fascinations, particularly in Hawaii and West Coast cities with a concentration of Asian Americans.

Ever the inquisitive lot, Americans were not satisfied with the traditional Asian method of learning by rote. They pestered their Asian sensei as to why techniques were performed in a particular way. JKA instructors became early proponents of applied biomechanics to highlight the explosive power and speed of karate techniques.[11]

Over time, academics also became intrigued with the philosophy and practice of Asian martial arts. Biophysicists performed sophisticated in-person studies on karate and taekwondo experts. These biomechanical discourses, alongside scientific breakthroughs in neuropsychology and professional sports training, gave context to long-held practices such as posture, stance, abdominal breathing, and mental focus. A convergence of these ideas plows fertile grounds for furthering human potential, in particular the interdependence of mind, body, and spirit. This knowledge base truly represents a fusion of Eastern and Western ideas, well before the term was popularized in Western consumerism.

Amalgamation with Culture and Philosophy

Martial arts originally referred to the combat systems of Europe as early as the 1550s.[12] Derived from Latin, meaning "the arts of Mars," the Roman god of war,[13] the term now primarily designates East Asian fighting arts, encompassing Chinese quánfǎ (拳法), also called wushu (武術) or gōngfu (功夫) in southern China; Japanese budō (武道); and indigenous and evolved fighting arts in Korea and mainland Southeast Asia. Many excellent publications have described their respective lineage and evolution, including those by Mark Bishop, Itzik Cohen, Donn Draeger, Peter Lorge, Patrick McCarthy, Jwing-Ming Yang, and Wong Kiew Kit.[14] These titles are listed in "Further Reading," below; hence, only an abbreviated version is provided here.

Chinese Quánfǎ

Historically, Chinese martial arts are categorized broadly as combat systems of kicking (*tī*, 踢), hand punching and striking (*dǎ*, 打), wrestling (*shuāi*, 摔), and joint lock (*ná*, 拿), which roughly correspond to Japanese karate-dō (striking and kicking), sumō and judō (wrestling), jujitsu (joint lock), and aikidō, which combines throwing and joint lock techniques.

"Southern fists and northern kicks" (南拳北腿) commonly alludes to a predilection for high kicks and linear striking techniques in northern wushu, and rooted, circular hand actions for their southern counterparts. Many believe that this diversification comes from the cultural roots of "south boat, north horse" (南船北馬).[15]

Ancient agricultural and commercial societies in southern China traveled by boat through vast waterways. There are speculations that southern quánfǎ emulates the upright stance and the circular rowing motions of a rower. The close-in defense movements and short-range, vital point strikes can be executed while sustaining balance in the confines of a moving vessel.

Northern China had nomadic roots, and distance travel was enabled by riding on horseback. A rider's well-developed lower body strength gave preference to high kicks and deep stances as well as the powerful full range linear striking actions that were in line with combat on horseback.

Legends attributed the genesis of Chinese martial arts to the Xia dynasty some four thousand years ago.[16] Over two hundred styles of contemporary wushu have emerged over the last fifteen hundred years. These can be traced directly or indirectly to Shaolin quan (Shaolin fist, 少林拳) in the north and Wudang quan (Wudang fist, 武當拳) in the south. Shaolin and Wudang were widely recognized as the two pillars of wushu. Their mutual influence over time has produced extensive overlaps in both training methods and techniques.

Shaolin and Wudang

The quintessential role of Shaolin quan in Chinese martial arts is without dispute,[17] and it was a companion to the Buddhist movement from India in the Eastern Han period under Emperor Ming-ti (58–76 CE). Buddhism took root in China over the next five hundred years, impelled by the religious conversions of emperors.

The Buddhist monk Bodhidharma (Chinese name Da Mo, 達摩) arrived from Persian Central Asia circa 500 CE. He introduced Chán (Zen) Buddhism to Shaolin monks, along with external techniques for physical conditioning (Luohan's Eighteen Hands, 羅漢十八手) and internal methods (Sinew Metamorphosis Classic, 易筋經) of meditation (*zazen,* 坐禪). These exercises became the foundation of Shaolin quan. The Chinese word *luohan* is synonymous to *arhat* in Sanskrit, referring to monks or bodhisattvas who have attained enlightenment (nirvana). The original Luohan's Eighteen Hands represented the eighteen martial postures of arhat synthesized into a form.[18] This relic survived as the oldest documented style of Shaolin quan, synonymous with references to Luohan quan (monk's fist).

At around the mid-thirteenth century, the Shaolin martial monk Jue Yuan traveled cross-country and invited indigenous Chinese martial artists to visit the temple.

Bai Yufeng* (白玉峰, 1260–1368) was one of many who remained and converted to the Buddhist faith. Bai transformed Luohan quan into 173 techniques and compiled the monograph called *Essence of the Five Fists* (*Wǔ Xíng Quán*, 五形拳). The incorporated movements emulate the spirit and fighting patterns of five animals (dragon, tiger, snake, panther, and crane).[19]

Shaolin martial arts continued to weave legendary tales up until the twentieth century Republic Era (post–1911), alternating throughout dynasties as the instrument of the oppressed populace or the oppressor. The powerful, linear actions of Shaolin quan is generalized as "external" (*wàijiāquán*, 外家拳), denoting focus on biomechanical power through skeletomuscular movements that is emblematic of northern Chinese martial arts.

Legends and folklore spoke of a second monastery established in the Tang Dynasty, when martial monks (monks who specialized in martial arts) were dispatched by the Emperor Taizong (626–649 CE) to combat piracy on the East China Sea. The Southern Monastery (Nan-Shaolin, 南少林), located on the Jiulan Mountain in Putian, East Fujian Province, became the root of southern martial arts.[20] The Northern Temple in Henan had been consecrated by emperors, revered throughout history as birthplace of Chán Buddhism. It was nevertheless burned down, then restored many times, the last in 1732 by edict of the Manchu Emperor Yongzheng of the Qing Dynasty in order to suppress Han uprisings. By the mid-nineteenth century, the Southern Temple was also destroyed by Qing Imperial forces to dispel revolutionary gatherings.

Rebirth of the current Northern Temple came in 1986. It was believed that on-site filming of the *Martial Arts of Shaolin* movie spurred both public and private interest in renovating the decrepit relic.[21] The temple has served as global hub of Shaolin culture since the late 2000s and mecca for the study of Shaolin gōngfu.

According to legend, Wudang quan was created in the twelfth century CE by the immortal Daoist priest Zhang Sanfeng (張三峰) while he lived in the monasteries of the Wudang Mountains. The genesis of Wudang quan was closely tied to Daoism (sometimes written as Taoism; often translated as The Way, 道) as philosophy, religion, and a way of life that traced back to the sixth century BCE. The actual existence of Zhang Sanfeng is another matter of debate, but Daoist philosophy unquestionably asserted major influence on the southern Chinese agricultural society. Daoist rituals sustain a high profile to the present day as specified by the Chinese lunar calendar. Practices extend from birth to burial, including feng shui, health-improving tai chi (taijiquan), qigong, and other indigenous martial arts. Its influence on pop culture can be found in movies such as *Crouching Tiger, Hidden Dragon*.

---

* Names for historic Chinese figures are entered by the convention of surname (Bai) then personal name (Yufeng).

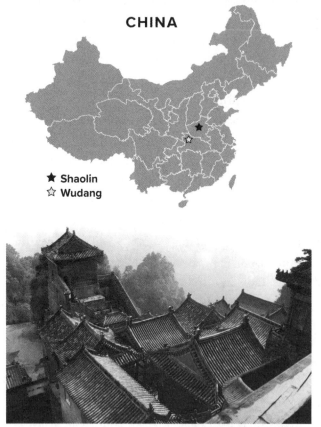

Figure 1.1 Shaolin and Wudang. Top: Shaolin Temple, Mount Songshan, Henan Province. Center: geographical locations of Shaolin and Wudang. Bottom: Purple Cloud Monastery, Wudang Mountains, Hubei Province.

The temples of Wudang and Shaolin sit on opposite sides of the Qinling-Huaihe line (Qin Mountains–Huai River, 秦岭淮河线, see fig. 1.1), the historical geographic boundary of north and south China, each entrenched in unique cultural beliefs. Wudang martial arts of the south are considered "internal" styles (*nèijiāquán*, 内家拳). Where "external" northern Shaolin styles cultivate skeletomuscular strength, "internal" *nèijiāquán* focuses on breathing and meditation, under the belief that directional channeling of universal energy (qi or chi, 氣) constitutes the ultimate source of power. Techniques gravitate toward circular, parrying, and joint-lock actions and near-mystical "light-body" (*qwīng-gong,* 輕工) skills to scale walls and skip along roofs. The three major styles attributed to Wudang quan are baguazhang (八卦掌), xingyiquan (形意拳), and taijiquan (tàiji quán or t'ai chi ch'üan, 太極拳). Taijiquan is the only style that traces back to the Wudang Mountains and the legend of Zhang Sanfeng.[22]

Wudang quan emphasizes rooted stances, hip snap, intricate hand techniques, and body power transmission by directing inner qi.[23] Consistent with Daoist philosophy, Wudang quan seeks to emulate the polarity of nature (yang and yin, hard versus soft, slow versus quick, attack versus defense) as reflected by the name tai chi (*t'ai chi, taiji,* "ultimate polarities," 太極), which signifies "polar opposites yet complementary aspects of the entire cosmos."[24] Techniques and strategies represent the alignment with nature's Five Elements according to Daoism: metal (hard, linear, rigid); wood (flexible, expansive); water (flowing, penetrating); fire (active, fast moving); and earth (heavy, slow, sedate).[25]

When non-Han Manchus invaded from far north in the mid-seventeenth century and founded the Qing dynasty, Bai's disciples and other North Shaolin quan masters fled to the Fujian province, including the famed Fan Zonggong (方種公) and his daughter Fan Qiniang (方七娘). The exiles contributed to the flowering of southern Chinese martial arts styles collectively called nanquan (Southern Fists, 南拳). Nanquan was epitomized by various versions of the Five Ancestors Fist (*wu zu quan* or *ngo-cho kun,* 五祖拳), precursors of today's Southern Shaolin Boxing (南少林拳), Monkey Fist (猴拳), Hong Qia quan (Hong family gōngfu, 洪家拳), and yǒng chūn (Wing Chun–Fujian White Crane quan, 咏春白鶴拳).[26]

The Five Ancestors Fists incorporate ritualized breathing and "iron body" practice of North Shaolin quan, the stances and dynamics of Luohan quan, the efficient movements in tai chi, soft and hard hand techniques of Fujian White Crane, and the agility and footwork of the Monkey Fist.[27] Five Ancestors Fist–based styles also profoundly influenced Okinawan and Japanese karate, as reflected by today's practice of multiple versions of Sanchin (three battles, 三戰), a conditioning kata (form) that originated as Saam Chien in many southern Chinese styles, including White Crane quan.

For over three millennia, warlords inflicted near constant strife among China's citizenry, interspersed only by brief periods of peace. Five main historic ethnicities inhabited what was known as the Middle Kingdom, parsed now into the current fifty-six ethnic groups of the People's Republic of China. The majority Han populated former China proper, which, together with the four northern ethnicities of Man (Manchus), Meng (Mongols), Hui (ethnic groups of Islamic faith in northwestern China), and Zang (Tibetans),[28] are represented by the five golden stars on the Republic's flag.

Han and non-Han feudal lords ruled as alternating dynastic eras from 1279 CE on, fomenting resentments among northern and southern ethnic groups. In spite of considerable intermingling over time, reference to "internal" versus "external" styles likely coded for indigenous ("domestic") Daoist-influenced southern Wudang versus external ("foreign") Buddhist-inspired northern Shaolin styles. Nonetheless, contemporary wushu practice abides by the common attitude of martial morality (*wudě*, 武德, see chapter 14), commanding mutual respect for all styles of practice.

Okinawa-te

From 1368 to 1644 CE, China prospered under Ming Dynasty rule. This respite of more than 250 years was known as the Golden Age of China. Commerce increased with Europeans as well as the neighboring Ryūkyū Kingdom, comprising the three former kingdoms of Okinawa. Chinese diplomats and merchants, particularly those who hailed from Fujian Province, with its long history of maritime trade, emigrated to Ryūkyū and served in government or engaged in business. Fujian's Thirty-Six Families from Min (*bin-jin san-jūroku*, 閩人三十六姓) was the first documented group that settled in the community of Kumenmura in 1392 (see table 1.1),[29] bringing with them Chinese cultural practices that included martial arts.

The weapons-ban theory has been a popular yet highly contested theory for the emergence of unarmed combat arts in sixteenth-century Okinawa.[30] King Shō Shin, first king of the Ryūkyū Kingdom, reportedly had imposed a bladed-weapon ban to suppress uprisings. This ban was extended in 1609 by the invading Satsuma samurai domain from southern Kyushu, Japan,[31] which ultimately annexed the Ryūkyū Kingdom as a prefecture of Japan. It was commonly believed that Okinawan martial artists would rely on Okinawa-te as secret unarmed combat practices along with kobudō (a fighting system with farming implements)[32] while they served as bodyguards for trade convoys and for protection of the *shizoku* (Ryūkyū ruling class).

Table 1.1 Coming to America.

ERA	CHINA	OKINAWA	JAPAN	UNITED STATES
600 BCE	Birth of Daoist philosophy			
525 CE	Bodhidarma introduced Chán Buddhism, Luo-han's Eighteen Hands, and Sinew Metamorphosis Classic to Shaolin Temple			
626–649	Genesis of Nan Shaolin in Fujian Province (according to legend)			
1200	Legends of Zhang Sanfeng and Wudang quan			
mid-13th century	Bai Yu Feng compiled *The Essence of Five Fists*			
1312–1346			Japanese monks Da Zhi and Shao Yuan studied at the Shaolin Temple and returned to teach gōngfu to the Japanese Martial Arts Society.	
1392		Emigration of Fujian's Thirty-Six Families from Min		
1600–1800	Popularization of Nanquan based on the Five Ancestors Fist			
1609		Satsuma samurai domain extended bladed weapon ban		

(continued)

ERA	CHINA	OKINAWA	JAPAN	UNITED STATES
1754		Kūsankū, martial artist and diplomat, arrived in Naha.		
1762		Kanga Sakugawa started teaching *tōde*.		
1800		Popularization of tōde as Shuri-te, Tomari-te, and Naha-te		
1871			Emergence of modern budō practice post–Meiji Restoration; abolition of the samurai caste	
1922			Gichin Funakoshi moved to Japan, changed *kara-te jitsu* to karate-dō	
1933			Karate-dō recognized as modern budō art by Dai Nippon Butokuka	
1936			Gichin Funakoshi established first karate dōjō in mainland Japan	
1948				Martial arts demonstration at U.S. military bases in Japan; formation of judō and karate-dō clubs on bases
1951				Combative Measures Programs (judō, karate-dō, aikidō) at U.S. SAC bases

ERA	CHINA	OKINAWA	JAPAN	UNITED STATES
1957–1961				Shōtōkan/JKA premier instructors (T. Ohshima, H. Nishiyama, T. Okazaki, T. Mikami) established U.S. dōjōs
1961–1970				Popularization of Japanese and Korean martial arts

The influence of Chinese quánfǎ on indigenous Okinawan fighting arts (Okinawa-te) has been documented since the mid-eighteenth or early nineteenth centuries CE. Theories abound regarding how the *Bubishi* (武备志), recognized as that era's definitive Chinese martial manual, came to be part of Okinawa-te. The *Bubishi* was founded on Shaolin quan and Wing Chun–Fujian White Crane quan, its impact reflected by parallels with contemporary karate techniques.[33]

Historical archives recognized Kūsankū (also known as Kwang Shang Fu, 公相君), Chinese ambassador to Ryuku, to have formally introduced Chinese quánfǎ to Okinawa.[34] Kūsankū learned martial arts at the Shaolin Temple and continued to practice in Fujian Province for much of his life. He arrived at his post in Naha around 1756 CE. Naha is located in the southwest quadrant of the kingdom, and the city directly faces Fujian, one of China's predominant ports of call, across the East China Sea. Shuri Castle, residence of the Ryūkyū ruler and seat of the Ryūkyū government, is no more than six miles from Naha.

During his tenure, Kūsankū gave a demonstration of Chinese boxing and grappling to the nobles at Shuri, many believed to be Chinese expatriates. Kūsankū teachings were known as *tōde'* (Chinese hand or Chinese technique, 唐手). Kanga Sakugawa,* student of the Ryūkyūan monk Peichin Takahara, became a student of Kūsankū. Sakugawa began to spread his expertise to others after six years of training, and he was given the name Tōde Sakugawa.[35]

Sakugawa developed the Kūsankū kata (form) in honor of his teacher after Kūsankū's death.[36] The form contained flowing technical elements that were consistent with Fujian White Crane quan.[37] Current renditions, known variously as Kankū (Dai and Sho), or Kong

---

* To maintain consistency with contemporary figures such as Gichin Funakoshi, romanized names for historic Okinawan masters are written as personal name (Kanga), then surname (Sakugawa).

Sang Koon, are practiced by Okinawan Shōrin-ryū, Japanese Shōtōkan, and Korean Tong Soo Do. Sakugawa's most famous student was Sōkon Matsumura, founder of the Shuri-te line of tōde. Okinawan martial artists studied in China within the same period, mostly in Fuzhou, capital of Fujian Province. They returned to Okinawa to teach tōde. It was commonly believed that tōde was used for close-in, unarmed combat in that era.[38]

Tōde styles were initially categorized according to the city of origin as Shuri-te, Naha-te, and Tomari-te. Tomari was a small village between Shuri and Naha and best known for producing bodyguards for the Shuri nobilities.

Shuri-te, Naha-te, and Tomari-te were forerunners of well-known contemporary Okinawan styles denoted with the suffix *ryū* (stream or training style, 流). Shuri-te and Tomari-te became Shōrin-ryū (Sōkon Matsumura, Kosaku Matsumura, Ankō Itosu) in deference to their roots from Shaolin stylists, then branched to Shōtōkan-ryū (Gichin Funakoshi), Shitō-ryū (Kenwa Mabuni), and Shōrinji-ryū (Chotuku Kyan). Naha-te (Kanryo Higaonna) evolved into Shōrei-ryū, then Gōjū-ryū (Chōjun Miyagi) and Uechi-ryū (Kanbun Uechi). These practices in turn evolved into contemporary Japanese karate-dō.[39]

Japanese Bujutsu and Budō

According to written history, Japan's exposure to Shaolin martial arts went back at least seven hundred years. The Buddhist monk Da Zhi studied martial arts at the Shaolin Temple for nearly thirteen years during the Yuan Dynasty (1312 CE). On his return, he taught Shaolin gōngfu to the Japanese Martial Arts Society. Another Japanese monk, Shao Yuan, traveled to the Shaolin Temple to master calligraphy, painting, and gōngfu. He was regarded as Japan's *guohuen* (Nation's Spirit) when he returned in 1346 CE.[40]

Contemporary Japanese martial arts are grouped collectively under budō (martial way, 武道). Budō is often used interchangeably with *bujutsu* (martial technique, 武術). However, budō practice strives beyond technical applications, serving as "the way" for personal improvement through mind and physical conditioning and spiritual enlightenment.

Budō practice prior to the Meiji Period was referred to as kobudō (ancient budō, 古武道), which drew heavy technical influence from *koryu bugei* (ancient martial arts school, 古流武芸) and the *bushidō* code of ethics (way of the warrior class, 武士道).

Modern budō (*gendai budō,* 現代武道) refers to philosophy and practice after the Meiji Restoration, and after abolition of feudalism and the samurai caste in 1871.[41] Alarmed by perceived Westernized excesses of Japanese society, advocates moved to formalize gendai budō practice into the Japanese education system to reinforce indigenous cultural practice.[42] While retaining kobudō's traditional aspects of spiritual and physical discipline, gendai budō is founded on the path (dō, 道) for perfection of the human character (*ningen keisei no michi*, 人間形成の道). Accordingly, self-defense-oriented jujitsu (柔術) was modified to judō (柔道) by the great Kano Jigoro, the first of this cultural movement. Other disciplines followed, giving rise to contemporary disciplines that include kendō (剣道), sumō (相撲), karate-dō (空手道), aikidō (合気道), kyudō (archery, 弓道), shōrinji kempo (Shaolin Temple Fist, 少林寺拳法), jukendō (bayonet fighting, 銃剣道), and naginata-dō (bladed pole, 薙刀).

Dentō Karate-dō

Ginchin Funakoshi introduced modern karate to Japan in 1922, for which he was recognized by many as the father of modern Japanese karate. Funakoshi changed the original connotation of *kara-te* from "Chinese hand" (唐手) to "empty hand" (空手), in consideration that 唐 ("Chinese") and 空 ("empty") are homophones in the Japanese language. Funakoshi maintained that "empty" more appropriately symbolizes this weaponless self-defense art that strives for the Buddhist belief of emptiness, the "void" that embodies all matters and is found within all creation.[43] "Empty hand" implies endless possibilities in the application of a single technique. He revised the reference of *karate-jutsu* (karate-techniques, 空手術) to karate-dō (way of the empty hand, 空手道). The assignment of *dō* aligns its practice with other contemporary budo arts. Current-day karate practices that abide by Okinawan- or Japanese-tenets of practice are also called Dentō (Traditional) Karate-dō. Funakoshi grew up in Okinawa and studied under Ankō Itosu and Ankō Azato, both students of the famed Sōkon Matsumura of Shōrin-ryū. He also learned from Shiroma (Gusukuma) of Tomari-ryū and another Chinese national who resided in the same city.

Funakoshi was the principal of an Okinawan High School when, at the request of the Okinawan Prefecture, he traveled and gave a demonstration of karate in Tokyo. He lectured on the art of karate at the Kodokan Judō Hall to a packed audience at the invitation of Jigoro Kano, the judō legend, using three long scrolls of photographs of stances, kata, and hand and foot techniques. Kano later requested another kata demonstration by Funakoshi, which was attended by well over a hundred spectators among Japan's aristocrats.[44]

With encouragement from Itosu and Azato, Funakoshi stayed in Tokyo to introduce karate-dō to the people of Japan. He supported himself as a karate teacher and founded study groups at Keio and Takushoku Universities, the firsts of their kind in Japan. In December 1933 the Dai Nippon Butoku Kai (Greater Japan Martial Virtue Society, 大日本武徳) formally recognized karate-dō as a modern Japanese budō art. Funakoshi opened his first dōjō in Japan in 1936. He named it Shoto-kan (Hall of the Pine Trees and Waves, 松濤館), corresponding to his pen name, Shoto. By then, Funakoshi was nearly seventy years old.[45]

Funakoshi brought sixteen kata to Japan: five Pinan (now known as Heian) and three Naihanchi (also called Tekki), along with Kūsankū Dai (Kankū Dai), Kūsankū Sho (Kankū Sho), Seisan (Hangetsu), Patsai (Bassai), Wanshū (Empi), Chintō (Gankaku), Jutte (Jitte), and Jion (see chapter 17). Students were required to practice different Heian and Tekki forms for at least three years before progressing to advanced kata.

In *Karate-dō, My Way of Life,* Funakoshi acknowledged the Shaolin and Wudang influence in Okinawan tōde development, in particular for Shōrin-ryū and Shōrei-ryū. Funakoshi considered each style to have its own strengths and weaknesses. He regarded the Wudang-influenced Shōrei-style techniques as better suited to a bigger person, whereas a smaller person would be more adept with the Shaolin-influenced Shōrin movements. "Shōrei ... taught a more effective form of self-defense, but it lacked the mobility of Shōrin," according to Funakoshi. He advocated that contemporary karate-dō should incorporate the strengths of the different schools. "The ultimate aim of both karate and sumō was the same," he concluded: "the training of both body and mind."[46]

To Be, or Not to Be: Contemporary American Practice

Over the last millennia, Shaolin-based martial arts that hailed from India have transformed into the physical prowess of northern wushu, the intricacy of southern wushu, the explosiveness of Okinawa-te, the dynamic precision of Japanese budō, and the agility of Korean taekwondo. Each is coveted as a national treasure despite evidence of foreign influence on indigenous practice.

The United States is well known as a melting pot of foreign cultures. Expatriate practices are adopted then modified as uniquely American. Ed Parker, founder of American kenpo, launched the first Long Beach International Karate Championship in 1964 and introduced a largely unknown Bruce Lee to the American audience. The championship showcased the talents of Chuck Norris, Tony Martinez Sr., Benny "The Jet" Rodriguez, Billy Blanks, Bill "Superfoot" Wallace, and Joe Lewis, who later founded the sport of full-contact kickboxing. Their success as competitors and movie icons built lasting impressions of how martial arts are perceived in this country.

However, American-style martial arts have increasingly gravitated toward business models of combat sports and entertainment. In 1993, the Ultimate Fighting Championship (UFC) incorporated grappling and ground fighting to rebrand kickboxing into the professional extreme sport of full-contact mixed martial arts (MMA). Also called cage fighting, professional MMA caters to the public's penchant for violence. MMA incurs an injury rate that is greater than most, if not all, commonly practiced combat sports.[47] Popular tactics include "ground and pound," when the opponent is pinned on the ground and receives a volley of head punches until submission, and "sprawl and brawl," when a fighter is pummeled with elbow and knee strikes while trapped standing in a corner or against the cage. The era likely marked an inflexion point that runs counter to the true meaning of traditional martial arts.

Human action is predicated on intent and purpose. Traditional practice specifies that qi follows *yi* (*yi yi yin qi,* 以意引氣), where energy and action are governed by decision and intent (see chapter 16). For centuries, morality of deed has imparted accountability of the martial artist to the community, and morality of mind for their spiritual growth (see chapter 14). Contemporary budō practice strives beyond technical applications, its purpose as "the way" (*dō,* 道) for perfection of character. Students are guided by a code of conduct (*dōjō kun,* 道場訓) that seeks to define personal attitudes and instill mutual respect, including refraining from "hot blooded" behaviors (see chapter 15).

Competition is necessary to spur confidence and to challenge one's skills (*shiai,* 試合, see chapter 2), but it is not an end goal in traditional practice. In a civilized society, there is little justification in decimating another human being for the sake of self-validation. Addiction to violence brings lasting damage to the psyche. On the opposite end of the spectrum, entrepreneurs consider packaging the traditional curriculum into separate consumer-driven activities for confidence building, weight loss, or stress relief.[48] These enterprising processes regrettably will forsake the spiritual aspects of martial arts.

Traditional martial arts are time-tested systems that balance physical clout with advancement of mind and spirit. Biomedical research has confirmed that the overall benefits of the entire curriculum exceed the sum of its parts. Before these national treasures are summarily dismissed by future generations, it is timely to examine the contemporary relevance of traditional thoughts and practices, alongside remarkable parallels that have been promoted in contemporary sciences.

Xīn

The brain is wider than the sky.

—EMILY DICKINSON

Between stimulus and response there is a space. In that space is our power to choose our response. In our response lies our growth and our freedom.

—VICTOR FRANKL

Mind and body are to become one in true karate.

—GICHIN FUNAKOSHI

When your sparring partner scratches or head-butts you, you don't then make a show of it, or protest, or view him with suspicion or as plotting against you.... We should give a pass to many things with our fellow train-ees. You should act this way with all things in life.... It's possible to avoid without suspicion or hate.

—MARCUS AURELIUS, *MEDITATIONS*

PART II

The Mind

2

Mind Over Matter

The *Dō* of Martial Arts

Many traditional Asian philosophies are grounded in the concept of 道, pronounced variously as "dao" (pinyin romanization), "tao" (Wade-Giles romanization), or "dō" (Japanese). This concept has been pivotal in Chinese thought since the Eastern Zhou Dynasty (circa 600 BCE). Commonly referred to as "The Way," *dao* defies definition when applied in the abstract, encompassing "course," "method," "manner," "style," "practice," "technique," or the "principle" (*dàolǐ,* 道理) of "how" or "what to do,"[1] to accommodate the Chinese pragmatist viewpoint.

Over time, Chinese Dao teachings (Dào Jià, 道家) garnered a philosophical and religious following known as Daoism (or Taoism). At its most fundamental, Daoism considers the human microcosm to have the same constitution as the cosmos. The Daoist philosophy (and religion, Dào Jià, 道家) seeks to align human dao with the dao of the universe, thereby achieving harmony of human qi with the qi of the universe—the source, pattern, and substance of everything that ever existed.[2]

Laozi, Founder of Dào Jià, considered *wu-wei* (action without deliberation, 無爲) as central thought. Wu-wei seeks a state of naturalness, effortless action, or letting one's actions flow with the simple and spontaneous course of nature. While Confucianists advocated social order and work ethics to unify humanity with heaven, Daoism peaked in popularity during the reign of the Ming Emperor Jiajing (1522–1566).[3] Daoists were the "hippies" of historic times—romanticized as carefree, nonconforming intellectuals. They aspired toward *wei wu-wei* (action from nonaction, 爲無爲), loosely interpretable as acting in such a way as to facilitate matters to take their own course. In practical terms, wu-wei is the level of mastery when daily chores are performed without intervention of conscious thoughts.

Buddhism was regarded as "foreign Daoism" when it was introduced to fifth- or sixth-century China.[4] This spiritual tradition seeks relief from pain and suffering of the material world through compassion and wisdom. According to legend, Bodhidharma brought Chán Buddhism and *dhyāna* meditation to the Shaolin Temple after having spent nine years in seated meditation (zazen, 坐禪) in a nearby cave, facing a wall. The Buddhist scriptures were eventually transcribed into Chinese with Daoist vocabulary. For millennia, the singular cultural influence of Buddhism, Daoism, and Confucianism waxed and waned, also comingling as the Harmonious Aggregate of the Three Teachings (*san jiao*, 三教, see fig. 2.1).[5] The Three Teachings ultimately coalesces around the concepts of dao, and *de* (virtue, 德).[6]

Figure 2.1 *The Three Vinegar Tasters.* The artwork portrays Confucius, Buddha, and Laozi tasting a vat of vinegar, an allegory of the "Essence of Life." Confucius viewed life as sour in light of the chaos in society. Life was bitter for Buddha, filled with longings that brought heartache and pain. Yet Laozi relished the taste, as it befitted the nature of vinegar. Their congregation favored the interpretation that all three beliefs should be considered in harmonious aggregate.

Indigenous Daoist beliefs blended with Chinese Chán (禪) Buddhism over time to emphasize introspection and meditation (zazen). Chán Buddhism in turn shares Daoist leanings for simplicity, which, along with its antiauthoritative and iconoclastic bend, foments the belief that Buddha Nature manifests within one's true nature.[7] Chán aspires for nirvana (the transcendent state of perfect happiness) through enlightenment of the mind, to be accomplished by total immersion and incessant practice of a highly cultivated skill.[8]

These beliefs gained purchase as Chán evolved into Zen Buddhism in nineteenth-century Japan, particularly among the ruling class. Zen concepts shaped the philosophy of Japanese budō, including aikidō, judō, karate-dō, and kendō. Despite differences in practice, each discipline follows "The Way" to transcend mind, body, and spirit. The attitude of seeking enlightenment through devoted practice is often referred to as "Moving Zen" (see chapter 14).

Budō arts subscribe to the practice of *mokusō* (silent contemplation, 黙想) to nurture the spirit and cultivate the mind. Mokusō aims to elevate awareness that is devoid of analysis or emotion (see chapters 14 and 15). Martial arts dōjō sessions begin and end with mokusō, when students clear their thoughts while kneeled in formal position and focus on the very moment and the task at hand (rhythmic breathing).

Mind over Matter

Researchers have been trying to understand the intriguing effects of meditation on the mind for quite some time. The first Western account was described by the intrepid Belgian-French explorer Alexandra David-Néel in her 1929 book *Magic and Mystery in Tibet.*

David-Néel recounted how neophyte Tibetan Buddhist monks draped icy water-soaked sheets over their naked bodies as they took part in winter training of Inner Fire Meditation *(g Tum-mo)*. They dried the icy sheets with body heat repeatedly without getting sick. As the monks sat in the open snow with no clothes, their meditative power was measured by the amount of snow melted by each's radiant heat.[9]

By the 1950s, the advent of electroencephalography (EEG) enabled researchers to profile brain electrical activity, soon becoming the rave of that era's brain research. Scientists turned their attention to the brain functions of yoga meditation experts, alongside their breathing pattern, blood functions, and skin temperature.

The yogis could alter their brain wave patterns as they entered deep meditation. The wave patterns replicated those found in the rapid eye movement (REM) stage of sleep, when the brain actively engages in vivid dreams as related to learning and memory functions. Further, the yogis could survive in an airtight box for eight to ten hours by going into the meditative state. Their skin temperature dropped, their heart rate was reduced, and respiration output went down to a minimum. These conditions would have been unsustainable for a normal person.[10]

While most of us can hold our breath from thirty seconds to two minutes, deep-sea free divers routinely cease breathing for over ten minutes. They sustain conscious breath control and overcome the asphyxiation reflex by entering into the *pranayama* (life force extension) yogic state. Practiced meditation also conditions their cardiovascular system. Periods of oxygen starvation are accompanied by slowed breathing rhythm, lower heart rate, and reduced metabolism.[11]

Even nonmeditating athletes can compensate to improve performance if they are given the impression that they are merely competing under normal conditions. This has been proven in a "seeing is believing" study in the United Kingdom in which three cohorts of cyclists took part in thirty-minute stationary bicycle ergometer time trials.[12] The study compared the performance of cyclists under heat and high humidity (HOT group) with a control group (NORM) who cycled under optimal conditions. HOT cyclists, as expected, covered far less distance: 0.47 miles less than NORM cyclists. Their mean power output was also markedly reduced. A third group of cyclists (DEC) was then put to the same unfavorable settings as the HOT group, but told that they were performing in merely marginally warmer conditions. Remarkably, the DEC cyclists covered the same distance as NORM cyclists. There was no marked difference in the two groups' monitored power output. DEC athletes felt far better than the HOT cyclists when quizzed on muscle fatigue, breathing exertion, and heart rate. These findings showed that subconscious expectations were a powerful incentive that can overrule sensory perceptions, enabling the deceived cyclists to overperform under challenging conditions.

These are but a few examples of brain power that can elevate physiological thresholds. Studies illustrate repeatedly that the motivated body can be impelled to work harder. By extension, athletes consistently achieve heightened performance through conditioning of the mind.[13] Psychologists talk of neuroplasticity, or adaptability of the brain to meet life's elevated challenges.[14]

Meditation represents a discipline of mindfulness practice to improve health and brain performance (see chapter 15).[15] Mindfulness exercises reduce wandering thoughts and garner structural modifications in brain areas that are charged with "default mode" operation for creative thinking,[16] the everyday process of arriving at innovative ideas through review of preconceived notions. In a study involving eight weeks of mindfulness-based stress reduction (MBSR) training, participants exhibited thickening in their hippocampus region that governs learning and memory and a reduced brain cell volume in the amygdala, which dispenses emotions of fear, anxiety, and stress (fig. 5.2).[17] The use of transcendental meditation continues to be a popular area of clinical research in seeking noninvasive relief for chronic ailments.

Mindfulness and Martial Arts

Decisiveness takes on a high priority in a compromised scenario of having to defend oneself. A strong mental constitution promotes the knack to instantly appraise impending threats and the confidence to execute countermeasures without hesitation. People who have acquired a focused mind-set tend to value hard work and patience in order to achieve their goals. For those of us who pursue martial arts as a worthy endeavor, immersion in practice is for the most part enjoyable, but also fulfills the purpose of realizing elevated mental and physical performance.

A heightened awareness, and strong grasps of one's mental and physical reach, are important aspects in martial arts practice. In *Karate-dō kyōhan: The Master Text,* Gichin Funakoshi asserted that "True karate, i.e. karate-dō, strives internally to train the mind to develop a clear conscience, enabling one to face the world truthfully while externally developing strength until one may overcome even ferocious wild animals."[18] These statements affirm that martial arts represent discipline of the mind as well as that of body.

Complex technical sequences demand intense mental concentration, requiring the practitioner to be free from errant distractions of the surroundings. Consistency in performance requires repeated practice of the same techniques at different levels of challenge and under varying scenarios.

In this regard, traditional practice is no different from competitive sports training that strives for mental, physical, and technical mastery.[19] The immersion process incorporates single and paired sparring drills (see chapter 7) and complex, predetermined movement sets (kata, 形). These different approaches bring distinct perspectives in technique execution, at the same time reinforcing the student's knowledge base, building confidence and resolve while under duress.

Over time, the student gains tenacity through experience and a level of decisiveness that overrides self-doubt and inaction. Repetitive practice streamlines their mental loop of perception–decision-making to physical action (see chapter 4). As the student advances in rank, they come to rely on their mental and physical acuity to handle affronts and deal with unpredictability.

Budō tournaments, called *shiai* (testing of one's skill, 試合), test a competitor's mettle under various settings. The two major categories of competition are kumite, or close-quarters unarmed one-on-one sparring matches, and kata, when the competitor performs a long-established and standardized series of movements and techniques in front of judges and audience, either individually or as teams.

An athlete is expected to demonstrate unfettered concentration, balance, control, precision, and correct tempo in solo kata competition. Their mind functions like that of a concert pianist as they look within to call upon their inner strengths. The performance is more than replicating correct movements; that task is now relegated to their highly practiced, autonomous mind-body continuum (see chapter 4). Rather, the competitor showcases mastery in power, speed, and rhythm as if in combat and exudes confidence that signifies determination and insights on strategy of attack and defense. Moreover, the athlete fully invests their emotions to interpret the kata's tenor (see chapter 17), conveying subdued vigilance in slow, defensive movements, and spontaneous intensity in climactic counterattacks. The judges' final scores reflect technical form, physical control, power, and the athlete's composure in conveying resolve and imagery (see chapter 17). Any loss of balance or hesitation is deducted from the final score.

Sparring competitions, called kumite (組手) in karate-dō and kendō and *randori* (乱取り) in judō, tend to be the main attractions of martial arts tournaments. Budō competitions employ *shobu ippon* rules, where both athletes strive for the outcome of scoring *ippon* (one full point, 一本). *Ippon* is awarded when judges determine that the athlete successfully executed a controlled finishing blow (*todome waza,* 留め技). The score recognizes a competitor's game-changing action that dismantles the opponent's mental composure or offense capabilities. Todome waza techniques are similar in concept to boxing's knockout punch. The competitor launches a powerful technique without hesitation, catching the adversary with no viable defense, and connects to a target area within reach. Unlike boxing, the full force of the technique is withheld right before penetration to mitigate injury. Otherwise, the competitor will be penalized for lack of control.

As a competitor steps into the competition area amidst cheers and jeers, it may well be the loneliest place over the duration of the match. There is no turning back once they face the opponent. The athlete understands subconsciously that the competition area of eight meters square signifies a raised platform, where stepping off is instant death. Accordingly, their maneuvers have to be well calculated or be penalized for stepping out of bounds.

The challenge in kumite is for the competitor to come to terms with the limits of their own skills. Their singular commitment, regardless of win, lose, draw, or forfeit due to injury or abandonment, impels the karate-ka to give their very best and to accept the consequences regardless of outcome. This sense of finality is a key component of martial arts training (see chapter 7) and is also an essential mind-set in self-defense.

As soon as the match begins, the mind intensifies and blanks out mundane thoughts other than self and adversary. Both contestants maneuver to survey, probe, and seek weaknesses of the opponent. Both realize that any action may be exploited by the other as an opening that is vulnerable to an attack.

The match calls on each athlete's decisiveness, resolve, and tenacity in the urgency of time-sensitive, close-in confrontations. Traditional competition rules contain safeguards against excessive contacts. Bruises and sprains are nonetheless unavoidable from incidental impact or miscalculated techniques, which, depending on severity, exact proportionate tolls to the psyche. The competitor's brain now borders on hyperdrive. They attend to their adrenalized state with elevated heartbeat and blood pressure, and a mental hair trigger. Solace rests on past trainings that built confidence and skill, sustaining the athlete's stable emotions and physical precision (see chapter 5).

Todome waza requires technical finesse, utmost composure, and mental commitment. These actions are predicated on steeliness of the mind that peers into the opponent's mental breach. Spontaneous resolution by ippon is a sight to behold amid the volley of exchanges, invariably eliciting visceral approval from the audience. In international traditional karate championships, effective techniques are scored in the hundreds, yet awarded todome waza are counted on one hand. The high intensity of these events makes them unforgettable to competitor and audience alike. Win, lose, or draw, participants will be awarded with a good night's sleep thereafter.

The Psychology of Martial Arts: Mind Like Water, Mind Like Moon

Budō arts aspire for two mental states, shown over time to sustain mental stability and physical performance when under pressure. *Tsuki no kokoro* (mind like the moon, 月の心), and *mizu no kokoro* (mind like water, 水の心) are the backbones of Japanese martial arts as well as the Zen Buddhist mind-set.[20]

Unlike the sun's intense scorching energy, moonlight illuminates every landscape detail on a clear night without heat. Tsuki no kokoro is a metaphor for the perceptive mind without hindrance of emotion or agitation. Wilson considers tsuki no kokoro as "an acute state of nonanalytical alertness."[21] Every detail in the environment is curated by heightened senses, but without prejudice or preference. A mind like the moon is unperturbed awareness even in the face of adversity. In contrast, ramped-up emotions (fear, anger, excitability) shroud the mind, denying it of reason and hindering appropriate response.

Mizu no kokoro is the capacity for mental tranquility, as if at rest, or like undisturbed water. This state of mind strives for correct and decisive rebuke without delay, self-doubt, or excessive analysis. When a pebble enters a body of water, it is enclosed in its entirety with little agitation. Along its way, water yields as receding ripples, soon to return to its natural state as the pebble descends its natural path. Likewise, the mind of mizu no kokoro promptly returns to a calm and collected state after addressing a challenge. The responder's actions virtually merge with the attacker and then resume control in their *danse macabre*.

Renowned neuroscientist Karl Lashley has used the same analogy to describe our stream of consciousness.[22] He pictured optimal neurological processing with the fluidity of a body of water. Lashley considered resting brain activity as calm like the unperturbed surface of a lake. A low external stimulus is like a prevailing breeze that draws small ripples of activity in our attentional network. Stronger stimuli behave like gusts of wind, creating waves of action in the brain. The wave actions exaggerate the rippling effect but do not destroy the order of the process. An even stronger stimulus acts like the bow of a speed boat cutting through the water. It splits the surface momentarily, impelling responses as wakes of mental activity, yet the brain soon resumes a calm and collected state, like unchanged water in a lake.

In "The Problem of Serial Order in Behavior," Lashley theorized that thought-to-action represents the neurological process of hierarchical recall and algorithmic assembly of previously memorized tasks. Execution of sequentially aligned tasks creates the allegorical ripple effect in the form of streamlined neural signal trafficking, from the framing of speech syntax to complex motor function outputs to realize the process of walking down a street—where we have to exercise proper judgment, then take intuitive actions to avoid incoming traffic. Once committed to procedural memory, these relatively complex functions do not require effortful thinking.

Contemporary neuroscience research has validated Lashley's prescient theories on the interconnectedness of neural networks, as well as his idea on the "logical and orderly arrangement of thought and action."[23] The challenge for the martial artist is to sustain this innate sense of equanimity even at times of duress and the orderly mizu no kokoro thought process that is devoid of emotional turbulence.

Mizu no Kokoro is often considered to be synonymous with *mushin no kokoro* (無心の 心), meaning to be free from mind-attachment.[24] An unencumbered psyche is open and responsive to subtle sensations, intuition, and the capacity to respond spontaneously.[25] By concentrating on the very instant, we cast away fear, anger, anxiety, or regret that unnecessarily unhinges rational thought and burdens the spirit.

A disciplined mind brings clarity in thinking, along with heightened vigilance and enhanced perception, opening up our senses to guide our actions. This is tsuki no kokoro, which works side by side with mizu no kokoro, the trained mind-set to mount an all-encompassing response. Tsuki no kokoro is total awareness that reflects strength of the mind, whereas mizu no kokoro is the emotional attitude of total commitment, a manifestation of our spirit.

While seeking to accomplish these lofty long-term goals, the training curriculum offers more tangible ideas to shape our thoughts. A student is reminded of three coveted mind states: *mushin* (無心), the agile, "unimpeded" mind-set that sees all opportunities; *fudōshin* (不動心), the "immovable" mind-set that is steadfast under pressure; and *zan-shin* (残心), the "continuous" mind-set that sustains vigilance before, during, and after engaging an opponent. One may be disheartened to learn that these attitudes are not acquired instantly. Each is realized, hence valued, after prolonged and deliberate training. Each of these ideas is discussed further in subsequent chapters as relevant to discrete training concepts.

Heightened awareness and split-second decisiveness are two of many attributes that are acquired over the course of martial arts training. They highlight the level of our cognitive performance and the manner that we perceive ourselves and interphase with the outside world. Breakthroughs in brain imaging and animal-based studies over the past few decades have vastly improved the understanding of these attributes, which modulate awareness, thought, knowledge accumulation, judgment, and decision-making.[26] Each is the outcome of stepwise neurological events from sensory information processing, intellect output, and emotional response to motor skill control, now mapped to discrete areas of the brain (chapter 3). There is nevertheless considerable overlapped functionality among neurological networks. Interactive cross-talk appears to be the norm.

As we master a new skill, the brain undergoes adaptive processes that elevate mental performance. These include structural modifications ("hardware") that facilitate the cascade of events from cognition to motor function. A practiced skill also reflects improved neurological responsiveness ("software") that brings near-spontaneous decision-making. Evidence-based contemporary athletic training approaches also seek to condition the mind in a similar way. These ideas are examined in the next chapters as they pertain to the relevance in complementing martial arts practices.

Takeaways from "Mind Over Matter"

- Daoism and Zen (Chán) Buddhism asserted profound influence on the dō (dao) of Chinese wushu and Japanese budō arts, reflecting paths for improving the mind as well as body and spirit.

- Competition rules in traditional martial arts competitions underscore stable emotions and resolve as key attributes for successful performances.

- Budō arts adopt two mind-sets to heighten awareness and sustain composure. Tsuki no kokoro (mind like the moon) is a metaphor for the perceptive mind without hindrance of emotion or agitation. Mizu no kokoro (mind like water) strives to achieve the correct and decisive rebuke without delay, self-doubt, or excessive analysis.

3

Cut to the Quick:
How to Think Fast

The Samurai with Many Swords

Samurai in feudal Japan wore two swords as a matter of practice. The commonest pairing is known as *daisho* (big-little, 大小), comprising the *katana,* a curved blade longer than two *shaku* (about 24 inches), scabbarded with the cutting edge facing up. A shorter *wakizashi* (side-inserted sword) is carried with the katana on the same side of the body. It ranges between one and two shaku in length.[1]

While exceptions abound from personal preference or tactical demands, combat commonly begins with drawing of the katana, which is held with both hands. The wakizashi provides backup when the katana is dropped, for close-quarters conflicts, or when the samurai elects to commit ritualistic suicide *(seppuku),* the ancient practice of self-discipline to preserve honor.

I had the good fortune of training under the late Hidetaka Nishiyama sensei over the years. Nishiyama sensei, recognized as the teacher of teachers in Shōtōkan karate, often related the cautionary tale of the Samurai with Many Swords. The story began with two competing swordsmanship schools in close proximity with each other. One day, students from one school were seen milling about. Each wore three swords instead of two. They pronounced that their school was superior because it taught how to use three swords in combat.

Soon, students from the other school paraded around with four swords. "We are better because we can use four swords at the same time," they boasted. If three swords are better than two, why, having four is clearly superior. Before long, samurai were seen wearing five, even six swords.

The moral and humor of this story is as valid today as in the past. Nishiyama sensei has long held that depth of knowledge and experience is the path to mastery, as opposed to collecting a broad repertoire of unstudied techniques. Our fast-moving culture often values novelty over quality. It is easy to forget that the gathering of untested tools is a poor substitute for how one applies one's craft. This old-fashioned notion of focused immersion is now supported by contemporary athletic training concepts.

Quickness in Athletic Performance

Most competitive sports demand expertise to accelerate and decelerate (speed output and control), execute directional change on short notice (agility), then apply power against a physical load (strength and power). These skills challenge the athlete's mental constitution to quickly assimilate information and execute the appropriate actions.

Quickness refers to decisiveness to translate perception to physical action, often measured by the athlete's reaction time. To return a volley in tennis, the player has to correctly position themself before the ball arrives. They anticipate the incoming shot by predicting its direction (the opponent's forehand versus backhand shot) and the nature of the shot (lob, passing shot, dropshot, cross-court shot, or down-the-line shot). The expert player often commits to a decision even before the opponent's racquet strikes the ball.

Martial arts have long sought this level of mental acuity. Self-defense approaches are grounded on surviving against aggressors who prey on victims, perceived to be weaker in strength or number. A gang of many may pick on one victim. With little forewarning, a defender would have to muster all their mental and physical resources to overcome adversity. This mind-set is deeply ingrained during practice. A karate black belt thrives on capturing the elements of surprise and spontaneity. They promptly commit to a course of action that entails body positioning, defense, and counterattack as soon as they sense an attack.

The marital artist does so by weighing their options spontaneously after deciphering the context of the opponent's technique (intended target, invested power, estimated reach). They move into place and start their defense in the blink of an eye, well before the attack arrives. At the same instant, the karate-ka launches their counterattack, mustering maximum body power and speed to neutralize a stronger assailant. There is much at stake: delay brings about one's demise.

This is the concept of todome waza, the "finishing blow" technique (see chapter 2) in karate-dō. Todome waza threads the needle of timing to capture an opponent's split-second lapse in mental focus (*kyo*, 居). These attributes of resolve and decisiveness are also part and parcel to "athletic quickness" in competitive sports.

Technical Speed = Reaction + Decision + Technique Execution

In laboratory studies, athletic quickness is measured by requiring the participant to press a computer key when a visual image flashes on the screen. The response represents an athlete's "simple reaction time," the time interval between visual perception and completing the associated motor response.

More time is needed if the person has to choose between images that appear together in order to select images of the appropriate color (black or white), shape (circle or square), or orientation (left, right, middle). The decision-making process (including go–no go decisions) entails a longer time interval and constitutes the participant's "choice reaction time." Choice reaction time assesses the participant's mental-processing skill, also referred to as their quickness.

Finally, a competitor's technical speed measures both of their mental and physical responses. Technical speed reflects the athlete's spontaneity (and accuracy) from perception to the time they take to complete a particular task (fig. 3.1).[2]

Each of these variables is unique for any given task. When an athlete has to weigh complex variables to arrive at a go–no go decision, they will require more time than to make a simple choice (black versus white, circle versus square). Regardless of situational complexity, it behooves the athlete to decide quickly and attend to their physical action.

Research shows that athletic quickness is a learned skill set of sharpened perception and decision-making. Along with physical speed, agility, and power, quickness reflects the capacity to meld mind and body performance seamlessly.

These mental demands track remarkable parallels in competitive sports and martial arts. Quickness concepts apply equally to a karate-ka who executes a todome waza, a basketball player driving toward the basket for a layup, or an opposing lineman charging at the quarterback in American football. Each scenario involves split-second sensory input processing, consolidating information that converges to thought and decision. An unambiguous decision streamlines physical action, which precipitates as force generation to propel movement, then the application of force to complete the task (fig. 3.1).[3]

An elite athlete seeks to improve quickness through skill-specific drills that are supplemented with exercises that heighten cognitive function.[4] By comparison, mental quickness in martial arts is developed primarily through "learning by doing," the repetitive process to build acuity to wade through the complex information matrix involving self, opponent, and surroundings. "Learning by doing" is a form of attention state training (AST) that conditions the brain to stay in control under constantly evolving scenarios. AST takes longer to master than simple movement drills but offers durable cognitive improvements that extend to non–task related improvements (see chapter 5).[5] In the next section, we will examine the choice reaction performance of the martial artist as compared with elite athletes.

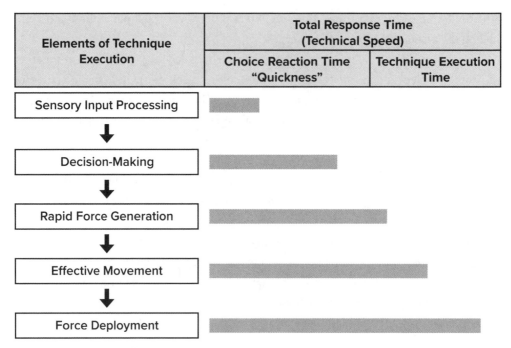

Figure 3.1 Hierarchy of technique execution.

Choice Reaction of the Baseball Batter

Baseball is a game of speed. The basic demands for hit, run, throw, and field appeal to our survival instincts. Like all things American, this leisurely pastime has been escalated to the highest level of competitiveness. Many consider Major League Baseball (MLB) batting to be among the most challenging in competitive sports,[6] therefore instructive in illustrating the reaction and anticipatory demands in athletic performance.

When a baseball player comes up to bat at home plate, they stand at a distance of approximately sixty feet from the pitcher's point of release. High school varsity fastballs arrive with an average speed of 50 miles per hour. The teenage batter has about 0.8 seconds (800 milliseconds) to decide and then hit the ball. By comparison, a seasoned MLB batter has less than half a second (400 milliseconds) against a fastball, which would arrive at up to 100 miles per hour.[7]

In *The Sports Gene,* David Epstein describes how the MLB batter, within the blink of an eye,

✓ identifies the type of pitch (fastball, curveball, slider, knuckleball, screwball, etc.);

✓ decides whether it's a strike or a ball (a go–no go decision);

✓ judges the approaching speed;

✓ plants the lead foot to extend the stride;

✓ adjusts the biomechanics to strike (or ignore) the incoming ball.[8]

To achieve consistency, they must start their swing at no closer than halfway through the ball's trajectory. This limits their decision to a scant 0.2 seconds. It is less than the amount of time to say "whoosh!"

Charlie Metro, former outfield coach for the Oakland Athletics baseball team, asserted that good hitters can predict the pitch from the pitcher's windup and release motion well before they release the ball. By then, the batter would have already committed their center of mass to the striking motion. They then decide on whether to complete the swing or abort, while the ball continues to zoom toward them.[9]

Because the projectile arrives at tremendous terminal velocity, even the best batter loses track of its final approach at around six feet from home plate. They can no longer make necessary adjustments, hence accounting for the infallibility of the breaking pitch (also called a "splitter") that dips the ball as it arrives at home plate.[10]

By comparison, the karate punch of a top-flight karate athlete achieves a top speed of close to 30 miles per hour (see chapter 12), but will come at less than six feet away. The "in your face" fist arrives in no more than 0.4 seconds,[11] challenging the defender's choice reaction response at a similar level of intensity. The defender would similarly lose track of a closing attack at about twelve inches from contact because of the diminished field of vision coverage.

The way that MLB batters rapidly assemble requisite information is similar to tennis players who intercept a return volley, the way a chess player arrives at the next move by scanning the board pieces, or how bird-watchers identify a bird from a fleeting glimpse of its color or flight pattern.[12] These attributes reflect the observer's sharpened neurological cognitive skills, which improve with highly focused repetition, also called deliberate practice or deliberate learning.[13] Through deliberate learning, elite athletes achieve superior visual search ability, confidence in judgment, and sharpened motor reflexes.[14]

Quickness in Karate Athletes

Parallel findings in karate black belts have been reported in Professor Shuji Mori's well-cited laboratory study.[15] These seasoned martial artists with four to six years of training were tested alongside novices (less than three months of training). Their simple reaction time was compared, based on participants' keystroke entry when an attack image flashed onscreen. The tallied results were segregated by training experience. The black belts' simple reaction times (from image flash to keystroke) did not differ from the novices. Both groups took about 250 milliseconds to complete the unfamiliar laboratory task.

In the second part of the study, participants responded to an image that corresponded to a punch to the face *(jodan)* or mid-body *(chudan)*. The participant's response time was recorded, along with their accuracy in distinguishing between a face versus mid-level attack.

The black belts turned out to be far more decisive (300 milliseconds) than the novices (400 milliseconds). Their aggregate choice reaction times averaged 550 milliseconds and were markedly shorter than the 650-millisecond average of the novice group. The black belts were also more accurate in distinguishing face attacks from those directed at the mid-body.[16]

Mori concluded that the experienced martial artists benefited from a cognitive advantage that was acquired from training. They have achieved near instantaneous capability to decipher the attacker's body language and motion as soon as the image flashed onscreen. This explanation is also in line with how MLB batters achieve a heightened level of athletic quickness.[17]

Yet there is always room for improvements in martial arts despite measurable gains. In Mori's study, an attacker's on-screen lead punch *(kizami-zuki)* was completed in 430 milliseconds. A front kick attack *(mae geri)* took longer, as expected, requiring 660 milliseconds.

Accordingly, the black belts' 550-millisecond choice reaction defense was adequate to fend off a front kick, but way late to address the face punch. Athletes, however, have been known to perform better under the intensity of competition. Finalists in a 100-meter sprint tend to better their times over same-day elimination races.[18] Martial artists can also be expected to be quicker during actual confrontations.

These predictions have been supported in a separate study. Reaction times of seasoned karate black belts were markedly reduced when paired against one another, according to Dr. L. Ingber at the University of California, San Diego.[19] Their best response was recorded when defending an actual attack, which triggered a preconditioned involuntary motor response originating from the body center.

Further, an effective defense is predicated on keeping proper distance away from the opponent. When the attacker is required to move into range, the defender derives the luxury of additional time to mount a proper response, which also speaks to the importance of proper distancing in sparring (see chapter 9).

Thinking Time

The biggest difference in mental performance between a professional athlete versus a novice is their thinking time.

Take the case of Derek, a college freshman sitting far back in a lecture hall. As the professor starts reviewing chemical reactions from last week's lecture, Derek's mind wanders to yesterday's epic professional football playoff between the Dallas Cowboys and the Philadelphia Eagles. His spirit is buoyed by the Cowboys' victory and his anticipated winnings in the dormitory betting pool.

Suddenly, Derek hears his name from Professor Merkel's direction. "Derek, would you provide a formula to calculate the energy released from this chemical reaction?" she asks.

Derek's shoulders tense, his blood pressure goes up, and his heart rate races. His pupils dilate as he clutches his opened laptop. Derek's mind goes into hyperdrive as he rummages through his working memory ("Where is that darned file in my computer?"). From desperation comes salvation: Aha! The formula beckons as he scrolls down the display. Seconds later, he blurts out a semi-intelligible answer, surviving yet another public humiliation in class.

Cognition is awareness of our thoughts and the events around us. When Derek struggles for a response, he calls on his cognitive systems to seek a solution for the challenge. These mental functions divert our attention to the extraordinary event, then hold our attention until the matter resolves. Our cognition control rallies to sort through prior habits in order to come to a decision. The process also interprets feelings and sensations based on past events. (It's embarrassing to be caught daydreaming.)

In athletic competition as well as martial arts, a person's technical speed depends critically on their cognitive skills. Prolonged efforts to sort through vast bodies of information delay decision and action.

Psychologists subscribe to a two-system model for cognition control, called "dual process theories of higher cognition."[20] Our mental functions operate with ease at the default level.[21] Daily thoughts arrive spontaneously with no deliberate effort and bring actions that require little or no intervention. Actions that are controlled by intuitive processing, which professors Epstein and Hammon assigned as the automatic system, or System 1,[22] operate without conscious thought. These cognition processes are tied to but not synonymous with the autonomic nervous system that controls internal organ functions subconsciously.

Intuitive processing aligns with spontaneous action and is founded on acquired knowledge and previously formed memories. Derek is at ease as he dwells fondly on earlier events that give rise to feelings of satisfaction. Through familiarity of task, we can just as easily walk to the mailbox from our front door, subconsciously avoiding minor obstacles on the way.

Research studies have localized intuitive thinking control to the caudate nucleus, one of five pairs of basal ganglia of the corpus striatum (see figs. 3.2 and 5.2). Massive sensory information is processed at the cerebral cortex, then passed along to these inner structures for learning, habitualizing routine activities (such as brewing coffee), and other automatic motor behaviors (holding a pair of chopsticks). Commands of highly specialized skills that have been mastered through deliberate practice, from musical performance to motor skills for martial arts techniques, are packaged and housed within the caudate nucleus and can be retrieved without conscious thought.[23]

In contrast, Derek's mental scramble requires heightened attention and conscious efforts in order to arrive at solutions to an unfamiliar practice. This process is assigned to our deliberative system (the effortful system, System 2), which is also central to fight-or-flight responses when we are involved in stressful situations (see chapter 5).[24]

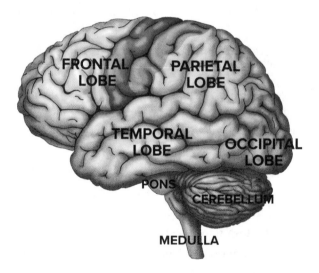

Figure 3.2 Anatomy of the brain. The top layer of the forebrain-cerebral cortex comprises the left and right frontal lobes, parietal lobes, occipital lobes, and temporal lobes. The frontal lobe governs higher-order brain functions such as reasoning and problem-solving. The parietal lobes manage sensory information for movement, orientation, recognition, and perception and house the somatosensory cortex that processes proprioceptive sensations. The occipital lobes are the visual processing center and house the primary visual cortex, V1. The temporal lobes contain the auditory cortex system and process perception and hearing, speech, and semantic memory. The midbrain is shrouded beneath the parietal and temporal lobes. It relays sensory and motor signals between the forebrain and the spinal cord, and it coordinates the assembly of sensorimotor information. The hindbrain, made up of the pons, cerebellum, and medulla oblongata, is tasked with the control of autonomous functions for balance, posture, sensory feedback processing, and other vital body functions such as heartbeat, blood pressure, and breathing.

The deliberative and analytical thinking System 2 is the final arbiter of physical action. System 1 processing gives rise to intuitions, intentions, and feelings. Before these impressions and feelings turn into responses, they are subconsciously audited by System 2 within contexts of preformed, explicit beliefs and choices.

System 2 demands singular, undivided attention to the event. It asserts self-control (impulse control) of words and actions. This multistep process involves memory recall (racking one's brain) and effortful and orderly review under the prefrontal cortex's purviews. System 2 is called into action when the mind senses "something is off." Though Derek was not faced with mortal danger, his physiology ramped up nevertheless as an instinctive response to adversity.

System 1 operates unencumbered and spontaneously, and it is entirely capable of managing the complexity of our modern daily life.[25] System 2 is slow, effortful, and requires conscious and intense focusing that consumes the bulk of our mental resources. Over time, highly practiced complex routines are relegated to System 1 control (see chapter 7). These specialized skill sets extend from touch-typing, keyboard entry, and mastery in musical instruments to the top-flight performances of athletes and martial artists.

Quickness Training Concepts

Cognitive research over the last decade has turned to sports to study the relationship of thought and skilled action.[26] Peak (flow) and impaired (choking) performances represent diametric ends of an athlete's cognitive range, hence offering an ideal paradigm to characterize mental performance.

Cognitive control is divided into perception, attention, and motor domains.[27] Heightened performance rests on an improved capacity to home in and interpret the external stimulus, prompting decision and streamlined task execution (fig. 3.1). Perception defines our capacity to gather and interpret sensory inputs into a "now" experience, based on association with past life events. Attention networks direct our mental focus to the task at hand, then consolidate, filter, and arrive at a decision based on real time information (see chapter 5). The decided task is delegated to the motor domain that translates thought into action. The athlete's trained choice reaction improves technical speed when coupled with practiced biomotor responses that mobilize the appropriate muscle groups. These attributes elevate performance regardless of whether the action is initiated at will or as an automated response to a routine sensory trigger.[28] The known contributors of athletic quickness are summarized below.

1. Enhanced Perception and Attention

Effective gathering of sensory information and prioritization can be improved through deliberate practice.[29] For example, learning to master a low block (gedan barai) to defend against a front snap-kick (mae geri) is a multistage process. After having been familiarized with the operational aspects (course of action, body biomechanics), the student is then paired with a training partner, who attacks with a front kick at modulated power and speed. This exercise renders feedback on effectiveness of the blocking application.

Apart from an understanding of timing, trajectory, and reach of the kick, the defender also gains awareness of the opponent's "body language" as they launch the attack. Partnered offense-defense drills tend to trigger a high level of awareness as soon as participants pair up. Defense against kicks is particularly sobering, as we are accustomed to only appraising another person from the waist up. Through practice, the student becomes adept at making use of central and peripheral vision at the same time, as they seek movement cues from the opponent's eyes, hands, shoulders, hips, and feet. A heightened awareness of self and opponent in turn improves effectiveness in technique applications.

The karate-ka also learns to tune into relevant sounds in the vicinity. Although light travels faster than sound, the brain is far quicker at processing auditory than visual signals. Sounds coming from the opponent at close distance, whether from breathing, technique windup, or body movement, reinforce visual information and contribute toward accuracy in judgment. Sensing an opponent's emotional state also aids in interpreting the attacker's intent and actions.

Neurological relays are streamlined over time through deliberate learning. There is enhanced connectivity between the visual cortex within the occipital lobe (V1, fig. 3.2) with pertinent, neighboring cortical regions that govern reasoning, planning, and movement. The end result (neuronal tuning) is a sharpened, goal-oriented visual search pattern that reflexively excludes extraneous information. The martial artist also gains an enhanced ability to fine-tune physical actions, having acquired heightened sensorimotor feedbacks from the limbs.[30]

2. Data Chunking

Data chunking is an internal information-processing algorithm that retrieves complex, related content as modularized "chunks."[31] Having been curated previously and committed to memory through repetition, "chunked" information deemed previously as related is retrieved as a coherent set, much like preprogrammed routines in computer software.[32]

A highly practiced elite athlete profiles their opponent's movement patterns as data chunks, such as the kinematic signature that distinguishes a basketball player's pull-up jump shot from a spin-and-drive to the basket. For example, the opponent's chunked signature could comprise the direction of their gaze, shoulder and upper body windup movements, and lead-foot placement.[33] While the weekend tennis player has to wait for their opponent's racket swing before deciding on a return, a top-flight player can rely on their opponent's "proximal kinematics" to guide their response. These are the telltale minuscule alignments of the opponent's posture before racquet movement.[34]

The black belt martial artist similarly tunes into the intent and strategy of their opponent well before the technique arrives. Their ability to collectively process and prioritize information, from the opponent's gaze to their footwork, is instrumental for formulating a spontaneous response.

Data chunking is a key contributor to confidence in judgment. Data worthwhile of consideration has been distilled from exhaustive trial and error during deliberate practice. Irrelevant information is discarded in the collective chunks, contributing to less processing time (and less informational clutter) to curate and appropriate a response. The martial arts expert now devotes their attention to tactics and strategy, thanks to an efficient mind that can quickly decipher their opponent's kinematics.

Data chunking is "synthetic" information processing, also called holistic cognition, when real time observations are instantly compared to prepackaged experience-based knowledge. By comparison, the novice engages in "analytical processing," a time-consuming response that oscillates from relevant to irrelevant sensory inputs before arriving at a decision.[35]

3. Automated Performance

Along with enhanced perception and data chunking, autonomous motor performance is the final outcome of a multistage learning process, starting out as the cognitive stage, then followed by the intermediate associative stage.[36]

In Shōtōkan karate-dō, a beginner advances through white, green, and brown ranks before attaining a black belt (table 3.1). At the cognitive stage, the novice learns "what to do" as a white belt. They are exposed to the fundamental hand and leg movements for kicking and punching techniques (kihon). The actions are placed in context by simple paired-up exercises (kihon kumite), then linked serially in basic kata practices.

This "learning by doing" process, also called implicit learning, tracks remarkable parallels to the current understanding of how a new skill is learned. New information stores in short-term memory registers, also called working memory. Working memory is the key to acquiring new skills and occurs in the cerebral frontal lobes (fig. 3.2).[37] Real-time visual-spatial perceptions are assembled into conscious thought (and coherent speech) by System 2 in the right frontal cortex. Working memory–based activities are slow and laborious and take up most of our mental resources.

At the green-belt level, students have a basic understanding of stance and body dynamics as applied to basic offense and defense techniques. They are aware of their performance requirements and can link simple skills serially into complex execution routines. They are now at the associative stage (also called the motor stage), having progressed from the "what to do" to the "how to do" learning level in about nine months.[38]

Table 3.1 Learning stages of the Shōtōkan karate curriculum.

BELT COLOR	RANK	TECHNICAL LEVEL	LEARNING STAGE
White	9th–7th kyu	beginner	cognitive
Green	6th–4th kyu	beginner–intermediate	cognitive–associative
Brown	3rd–1st kyu	intermediate	associative–autonomous
Black	shodan	intermediate–advanced	associative–autonomous
	nidan and above	advanced	autonomous

Automated motor skills, whether dribbling a basketball, shifting gears in a car, or avoiding an attack in a martial arts drill, are accomplished through procedural memory recall. Procedural memory is long-term memory that is acquired through progressive familiarization and practice.

Most students attain a brown belt with about two years of practice. They can now perform basic drills as second nature, going through technique combinations with fluidity, and muster reasonable power and speed in individual techniques. This level of competence is transitional to the default autonomous processing stage (System 1) that enables performance without conscious thought.

Procedural memory operates below the level of consciousness and is controlled by the right precerebellum and the limbic system (see chapter 5). These neuromotor relays work like the parallel networks of a computer. Sensory cues trigger instant recall of pre-established procedural memory. Memory alignment activates conditioned, simultaneous neuromotor functions to achieve complex movements. Unlike motor functions that fall under conscious deliberation, automated actions are rapid, smooth, and effortless and require minimal attentional capacity.[39]

Likewise, the elite athlete's neuromuscular control is relegated to procedural memory during competition, in order to accommodate skill execution "without thinking."[40] The working memory is suppressed in order to optimize performance.[41] The athlete's actions are hyperspecific to the learned skill as they perform at the autonomous phase.[42]

Mastery in the fundamentals is recognized by the rank of *shodan* (first degree black belt) in Japanese martial arts, embodying consistency and precision of autonomous execution. The resolute self-defense mind-set accompanies technical know-how of ranking black belts. There is no time to ponder "what to do" or "how to do" in these situations. The martial artist entrusts their techniques to autonomic control while attending to strategy and tactics to overcome the impending threat (fig. 3.3).

These benefits of implicit learning have been confirmed in a laboratory study.[43] Young women who were highly skilled karate fighters gained substantially improved visual search behavior after implicit perceptual motor training, as compared with controls who were equally skilled but received only verbal, explicit "how-to" instructions. Implicitly trained participants keyed onto fewer search locations for longer durations and achieved higher accuracy in decision-making. The findings confirmed that attentional focus is best acquired through "hands-on" training drills, as are commonly practiced in the traditional martial arts curriculum.

Enhanced perception, data chunking, and brain automation are the hallmarks of expert performance. It starts with instant and prioritized recognition of cues and patterns

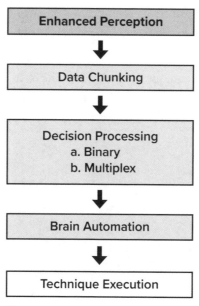

Figure 3.3 Neuromotor sequalae in technique execution.

and ends in reflexive motor response. This level of performance corresponds to structural and physiologic adaptations in the brain's primary sensory and motor cortexes.[44]

Mastery of the same skill sets moves the karate-ka's decision-making process well ahead of the opponent's incursions, allowing for the delivery of a preemptive response before the attack arrives. These concepts cannot be attained without the expertise of heightened cognition and autonomous execution.[45] More on kumite timing will be covered in chapter 6.

Takeaways from "Cut to the Quick"

- Quickness refers to decisiveness to translate perception to physical action in sports psychology. It measures the athlete's mental capacity to meld mind-body performance seamlessly.

- Quickness is improved through focused and highly repetitive practice that leads to durable benefits in performance. Quickness, along with physical speed, agility, and power, defines the athlete's overall competency.

- Deliberate practice leads to autonomous performance without conscious thought. This attribute enables intuitive task performance that is unencumbered and spontaneous under challenging conditions. Martial arts strive for a similar mental acuity to capture the opponent's split-second lapse in mental focus. Attentional state training brings enhanced perception and decision-making, heightening overall performance.

- Proper martial arts technique execution benefits from autonomous performance or "execution without thinking." Skill-set development should incorporate drills toward enhanced perception, data chunking, and brain automation.

Elevating Performance Across the Brain-Body Continuum

Mental Trigger in Sprint Launch

Athletes in short distance sprint competition are conditioned to instantly accelerate from the start. Before 1929, sprinters dug holes into the dirt[1] so that they could push off from the first step. Starting blocks are used nowadays as a matter of course. This technical innovation offers a time advantage of 0.2–0.4 seconds, which exceeds the time split between first and last place finishers in a 10-second 100-meter race.

The starting block is a valuable field-research tool for studying technical speed and neuromuscular connectivity.[2] Multivectoral force sensors mounted on the block can detect the flux in reactionary force from the sprinter's foot pressure. These devices enable time-lapse measurements from the firing of the starting pistol to when the sprinter clears the starting block.[3]

The elapsed time between the crack of the gun and the sprinter's push into the block represents their simple reaction time (see chapter 3). The sprinter's rear foot clears the block as they complete their launch, and their reactionary force drops to baseline.[4] This interval constitutes the athlete's total response time (also called movement time or technical speed; see chapter 3).

A college competitor's total response time is about half a second (508–612 milliseconds), by and large corresponding to the technical speed of elite athletes in other events.[5] However, their almost spontaneous simple response times of 143–169 milliseconds far surpass laboratory-measured values (200–250 milliseconds).

The most straightforward explanation is that our response to sound (the firing of the nearby starting pistol) is faster than of sight. Sprinters are conditioned to direct their gaze inward and focus on audio-motor response. A reduced reaction time corresponds to the much faster neural processing speed for auditory content, which beats visual analysis by about an order of magnitude (milliseconds versus tens of milliseconds).

Sprinters assume a balanced, crouched stance, priming their single-choice "go" response. Field studies found that athletes who fully engage their lower body's medium to large musculature coax a higher level of neuromuscular excitability than laboratory-based unpracticed aiming tasks.[6] The intervening "ready" and "set" commands further prime mental anticipation.

All world-class sprint finalists (though not among less-experienced competitors) can trim their reaction times further between heats and the final race.[7] This is evidence that competitors can intensify reaction performance within short notice. Brain EEG studies confirmed that the finalists generated measurably higher neural signaling spikes in the frontal premotor cortex, giving rise to a more intense motor-triggering signal. A stronger and farther-reaching neural excitation state corresponds to reduced "decision-to-act" execution time.[8]

Even after the explosive launch, the sprinter continues to focus on core compression and tensed leg muscles toward acceleration, commonly known as the coiled spring effect (see chapter 9). The runner takes off immediately after pushing off each stride, applying leverage against ground resistance, and makes use of the returned energy.[9] World-class 100-meter competitors achieve peak velocity after the first 50 to 60 meters by devoting all their resources to acceleration. Each stride demands intense mental focus and maximal physical assertion against the ground in order to sustain peak velocity to the finish.

Streamlined Mind-Body Continuum: A Peak Performer's Forte

Through deliberate practice, the sprinter's streamlined mind-body continuum achieves peak performance by promptly translating intention into motor execution. Motor skill adaptation, also called procedural learning, is the linchpin to achieve autonomous performance after having acquired heightened perception and attention (see chapter 3). Procedural memory enables the speed, accuracy, and consistency of autonomous movement.[10]

Voluntary and autonomous motor functions are both controlled by the motor cortex, which engages in planning, control, and execution of muscle responses. This domain also audits autonomous performance in the background. Areas of the motor cortex straddle the left and right hemispheres in the rear of the frontal cortex and comprise M1 (previously referred to as the primary motor cortex), the premotor cortex, and M2 (supplementary motor areas) (fig. 4.1). The left motor cortex controls movement of the right side of the body, and the right motor cortex controls left side motor function.

Sensory information from different parts of the forebrain converges at the prefrontal cortex. The occipital lobe is responsible for vision. The parietal lobe informs on the body's real-time position relative to the surroundings, temperature, taste, and touch; the

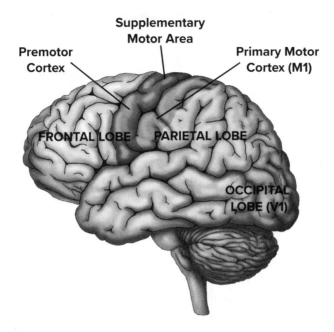

Figure 4.1 Anatomic distribution of the motor cortex. The primary motor cortex (M1) controls voluntary movement execution in all parts of our body. The premotor cortex assigns action attributes to the physical task. Supplementary motor areas support selective and sustained attention and may be involved in planning and bilateral movement coordination. M1 performance is closely linked to the adaptive visual cortex located at the occipital lobe.

prefrontal lobe informs on the intent of the action (goals, strategy), while the temporal lobe furnishes past strategies.[11] Commands are relayed to M1, which plans and executes movements in coordination with other motor areas of the brain.

Motor function execution for different parts of the body (trunk, head, neck, shoulder, arm, eye, nose, and so on) is mapped to specialized (but overlapping) areas within M1 in a layout called the motor homunculus ("little man"). The premotor cortex bridges intention and response by selecting among movement options based on external events.[12] This domain guides body movements by controlling muscles that are closest to the body's central axis, synchronizing complex actions such as orienting the body by visual cues before the hand reaches out for an object. The supplementary motor area plots complex movements and coordinates left- and right-hand movements.

M1 connects to the brainstem, then nerve fibers of the spinal cord and the peripheral nervous system. This motor network is also called the corticospinal tract. It engages over a million neurons, a neural-signaling highway that conveys activating signals from the prefrontal cortex to the spinal nerves and extremities. Feedback signals that travel in the opposite direction (ascending neural networks) return to M1 to guide fine adjustments.[13]

M1 function is highly adaptable. Repetitive practice modifies and streamlines task-specific performance by M1 and the primary visual cortex (V1) that is housed within the rear occipital lobe. These adaptations enable a conditioned and more effective visual search pattern, known as perception-action coupling.[14]

Even short-term training (hours to days) reorganizes cortical structures, promoting memory recall and informational processing.[15] The conditioning process elevates connectivity and lowers the excitation threshold for the involved neuromotor circuitries.[16] A heightened signaling strength accompanies the reduced activating threshold. Altogether, these enduring changes produce a more efficient mind-body continuum for motor performance.

Deliberate training also increases connectivity between M1 and areas of the cerebellum that subconsciously sustain balance and movement coordination.[17] Once committed to procedural memory, the premotor frontal cortex can bypass M1 and triggers autonomous skill performance through the right cerebellum and the limbic system.[18] These skill-specific motor programs account for the speed in rapid physical activation without conscious thought (see chapter 3).

Technical Speed in Karate-dō

As covered in the previous chapter, deliberate, repetitive practice in traditional karate-dō brings parallel improvements in perception, cognition, and motor skills (also see chapter 7). Kihon, or basic training exercises, calls for awareness of stance and posture (see chapter 9) to achieve balance and stability, along with other operational aspects of foundation techniques. These concepts are reinforced by kata practice as sequential technique execution that elevates attentional performance.

The student is awakened to their mind-body connectedness with kumite (sparring) exercises, which require instantaneous connectivity between perception (total awareness), cognition (mental stability and focus, decision-making), and motor functions (maximized physical output with precision). Partnered drills acclimatize response to potential threats under a controlled setting (see chapters 2 and 6). Starting with basic sparring *(kihon kumite),* familiarized punching, kicking, and blocking techniques are applied against an opponent. Each participant alternates as designated attacker or defender. As the student gains an understanding on proper timing and distance, varied formats place escalating demands on mental and physical performance, eventually progressing to the constantly evolving scenarios of free sparring *(jiyu kumite;* see chapter 6).

In the mid-1980s, I was a young *sandan* (third-degree black belt) in the Nishiyama-Shōtōkan karate-dō system. Having recently moved to Dallas and opened a dōjō, I made annual sojourns to the International Traditional Karate Federation (ITKF) summer training camp in La Jolla, California (see chapter 14). Training with hundreds of black belts for a week helped to refresh my skills.

A highlight of the camp was the opportunity to practice under the guest teaching faculty, comprising kata and kumite legends of the 1960s and 1970s. They included Keinosuke Enoeda, Hiroshi Shirai, Takeshi Oishi, and Masaaki Ueki.

Each was larger than life, yet Oishi sensei impressed with his calm and understated demeanor. Standing at 5-foot-7 and about 160 pounds, he was slight in build by Western standards. Yet this JKA standout was ranked as *Black Belt* magazine's top karate competitor in 1970 and 1971.[19]

Professor Oishi has been a four-time All-Japan Champion in kumite (1968, 1969, 1971, and 1973). The Shōtōkan stylist at age twenty-nine put on a dazzling performance at the First World Karate-dō Championships and led his kumite team to victory.[20]

While a young shodan at Komazawa University, Oishi trained with Hidetaka Nishiyama sensei, then the head instructor of the JKA University Club. A decade later, Oishi still remembered Nishiyama's personal advice, that "karate requires a combination of speed and an instantaneous driving force."[21]

Oishi usually appeared reticent at the beginning of his matches. As he moved in and out of reach with fluidity and ease, Oishi seemed hesitant and indecisive, in reality gauging his opponent's tactics and reactions. Then, abruptly, Oishi would surprise and pierce his opponent's defense without forewarning. He had the speed of a cheetah, his quick and largely unanticipated foray aided by his opponent's panic.

Oishi's favorite technique was the *choku-zuki,* a straight punch to the face or mid-level. He could cover three or four paces in the blink of an eye from a relatively high kumite stance. Oishi's impeccable timing, ability to close distance, and near perfect techniques epitomized the dynamics of technical speed in karate-dō.

Sensei Oishi continued to display this transitional fluidity decades later. At the ITKF training camp, his prescient anticipation of an attack opening seemed infallible. Oishi's spontaneous movements exemplified the seamless mind-to-body continuum. It is inconceivable, but true, that Oishi's kata was deemed to be hopeless when he was a brown belt, to the extent that none of his *senpai* (senior colleagues) knew how to correct him.[22]

In traditional martial arts, the ability to control and rapidly accelerate is critical for effective technique execution. Much like a sprinter, the martial artist focuses on core and lower body joint compression to achieve the coiled-spring effect for balanced acceleration (fig. 4.2). Even without the benefit of a starting block, the martial artist is conditioned to push off from the ground to accelerate the center of mass. Oishi's mastery enabled him to stop his opponents in their tracks. Yet his efforts would have come to naught without the high level of vigilance, unshakeable decision-making (also called fudōshin, 不動心), and his lightning-fast mind-body continuum.

Figure 4.2 Technique execution utilizing back-leg reactionary force. Body power originates with reactionary force that is derived by applying back-leg pressure against the floor. Release of stored coiled-spring energy from lower body muscles enables hip rotation, followed by punching-arm extension, opposing-arm pullback, and *kime* at impact.

Corticospinal Facilitation: The Lightning-Fast Reflexes of Karate-ka

In a recent Italian study, karate black belts were recruited to examine the effects of neuroadaptation on movement, technique speed and power, and body rhythm. The participants were veterans of international championships for more than ten years, and thus represented an elite crop of martial arts competitors. Electromagnetic coils were attached to their M1 motor cortex. They were then pulsed with an electric current of increasing strength. Their muscular response (motor evoked potential, or MEP) was measured by a surface sensor placed over the right-hand muscle (the first dorsal interosseus muscle) between the thumb and the index finger.

Similar to world-class sprinters, the black belt subjects' response threshold was markedly lower than untrained controls.[23] The highly trained black belts also generated a higher MEP spike that corresponded to a conditioned neurological trigger for hand-muscle activation. Their execution times to the pulsed stimulus were also shorter.

The phenomena of increased M1 sensitivity and higher neuromuscular activation are shared by elite soccer and basketball competitors. This increased latitude in performance, known as corticospinal facilitation,[24] is indicative of an improved mind-body continuum from repetitive and intensely focused practice.

Motor activating signals that are generated within regions of the brain's white matter travel through axons in the spinal cord, terminating in nerve endings that control muscle cells in the limbs. The relaying axons are wrapped in myelin sheaths, made up of fatty tissues that insulate electric impulse and minimize signal degradation.

Animal studies show that repetitive physical practice thickens the myelin layers of participating neural axons. Increased insulation promotes faster informational transfer, giving quickened neuromotor responses.[25] According to neural mapping studies, karate black belts displayed prominent brain white-matter and neural-fiber reorganizations, as did athletes who are required to meet high coordinative demands.[26] These neurological "hardware" modifications are common among elite athletes, musicians, and dancers who spend upward of fifty or sixty hours a week to refine their skill set and to improve areas of weakness.[27] Increased excitability, connectivity, and structural adaptions all work in concert to enhance mastery in movement, speed, and accuracy in performance.[28]

Hidari Ashi Kamae (Left Foot Forward)

Humans and some primates are the only known species with a handedness preference (directional asymmetry), a trait that is tied to lateralized performances in our brain.[29] Our brain's left hemisphere differs from the right in both configuration and function. Scientists continue to debate about selective forces that shape this evolutionary feat, although most concur that brain specialization broadens our cognitive capability for attention, reasoning, decision-making, and learning.

Lateralized functioning accommodates parallel processing of sensory signals and allows for separate and more intricate formulation of downstream motor responses.[30] The right hemisphere controls motor skill execution for the left side of the body. The right frontal cortex also coordinates motor functions to achieve positional accuracy, timing, and duration of action (spatiotemporal and attentional processing).

The left hemisphere controls muscle functions for the right side of the body. It is responsible for analytic thought, logic, reasoning, and numbers skills, and specializes in motor execution of speech and fine adjustments of motor functions through feedback.[31] The left hemisphere (and the right side of the body) is more adept for modulating complex motor tasks,[32] most likely tied to its dominant role in deciphering rapidly evolving events.

The lateralized behaviors in sprinters add intrigue to these concepts.[33] Both right- and left-handed sprinters react faster to the start signal by planting their left foot on the rear starting block. Yet both groups produce a faster launch (execution time) when they push off with the right foot to the rear.[34]

Remarkably, these findings serve to confirm functionality of the respective hemispheres. By focusing on push-off with the right leg (placed in the rear block), motor control is delegated to the left cerebral hemisphere that governs motor feedback and fine adjustments, leading to a reduction of about 80 milliseconds in execution time.[35]

This lateralized advantage diminishes as highly skilled athletes become ambidextrous.[36] Studies on judō athletes are particularly informative. Expert judō-ka tend to have asymmetric upper bodies. Right-standers (left-foot-forward stance) and left-standers (right-foot-forward stance) are evenly divided among highly qualified competitors.[37] A majority have overdeveloped right-side musculature, indicative of preference for right-handed throws for both groups.[38] Interestingly, these elite competitors all exhibited right-hemispheric-biased activation.[39] Likely explanations include the key role of the right hemisphere functions for positioning and timing, the competitor's dexterity to engage the opponent with both the right and left upper limb (see fig. 13.3), and the simultaneous demand on left-legged pivots even for right arm throws.

Beginners in most martial arts as well as boxing are taught to assume a left-foot-forward opening stance (*hidari ashi kamae*, 左足構え). Right leg deployment in the rear is consistent with left-hemisphere activation for quicker execution. With the dominance of right-handedness in our society, push-off with the right leg also works to promote physical power and speed during forward transitions and kicks. Punching and striking from the right arm while in place can be executed through rotation with the trailing right hip (fig. 4.2).

Kendō is an exception to the rule. Traditionally, the *kendō-ka* carries the long sword (katana) over the left hip. The sword is drawn with the dominant (right) hand with the cutting edge (*ha*, 刃) facing upward. The kendō-ka's left foot is placed to the rear to accommodate the length of the weapon, with the heel slightly raised (*okuri ashi*, 送足)

in the opening posture of the low crouching (*iai goshi,* 居合一腰) or standing (*tachi ai,* 立合い) stance.

The right foot in forward position (*migi ashi kamae,* 右足構え) supports the center of mass and helps to offset angular momentum as the drawn katana arcs across the body, transitioning smoothly into the first cut. The relative positions of both feet are maintained over the course of subsequent movements.

Kinesthesia: The Sixth Sense of Martial Artists

In 1827, Scottish scientist Charles Bell discovered that separate nerve fibers were involved for transmitting motor and sensory signals bidirectionally from the brain to muscles in the extremities and back.[40] Bell and Magendie coined the phrase "muscle sense" to describe a "sixth sense" of movement awareness and position, which is distinct from sight, hearing, smell, touch, and taste.[41]

Quickness and precision in movement are predicated on an exquisite awareness of our limbs' positions and their motion. This kinesthetic "sixth sense," now called proprioception, consciously and unconsciously tracks the directionality, speed, and load of limb movement as well as calibrates force produced by the limb (limb limit).[42]

Proprioception, vision, and the inner ear vestibular system provide essential sensory inputs to sustain balance. Peripheral mechanosensory receptors that populate muscles, tendons, and joints orchestrate proprioceptive responses. They connect to nerve fibers emanating from the spinal column to sustain neuromuscular sensation and control. The mechanoreceptors work in concert with exteroceptors, a second group of proprioceptors that sense external stimuli (touch, pressure, vibration, temperature). Seismic sensors in our hands and feet are exteroceptors that tune into the strength of contact with solid objects and the resulting frequency of vibration.[43]

A sum total of these somatosensory inputs addresses the necessary feedback requirements to modulate voluntary movements and reflexes.[44] As movements trigger proprioceptive sensations, "bottom-up" signals feed through the spinal cord and onto the basal ganglia, a deeply seated structure inside the forebrain that governs movement initiation and continuously modulates movements in real time.

Proprioceptive intervention is part of the cognitive process, triggered when our balance is affected by load or movement, such as when carrying a heavy box, tripped by an unexpected barrier in the street, wielding a staff or bat, or when our punch is blocked by the opponent.

The proprioceptive network connects to the thalamus within the limbic system and brainstem (see fig. 5.2). Its inputs are assimilated into procedural learning, habit learning, eye movement, and cognition.[45] As an athlete becomes familiarized with a new task, proprioceptive connectivity allows for constant audit, planning, and movement control by the motor cortex, and it is eventually integrated into task-specific procedural memory.

Our understanding of kinesthesia underscores the importance of broadening awareness beyond audiovisual inputs in everyday life. An overarching goal in martial arts training is to achieve spontaneous response in rapidly changing scenarios. This level of performance requires consistency in interpreting our body's orientation in reference to the immovable ground (see chapter 9), underscoring the importance of proprioceptive awareness in addition to vision and inner-ear equilibrium. Kinesthetic feedbacks provide rapid, real-time updates on the status of structure and balance, enabling the motor cortex to modulate and augment automated motor responses.

Takeaways from "Elevating Performance Across the Brain-Body Continuum"

- Procedural memory is acquired by deliberate practice and elevates neuromuscular responsiveness.

- Effective technique execution shares a common objective with the sprinter's launch start. Both demand controlled, rapid acceleration of the center mass.

- The left hemisphere of the brain controls the right side of the body and is more adept in adjusting complex motor tasks in rapidly evolving situations. Right-foot-dominant postures activate the left hemisphere. Accordingly, sprinters commonly launch by pushing off the right foot in the rear. Traditional martial arts similarly adopt hidari ashi kamae (left foot forward). Both pay homage to laterization of brain functions.

- The heightened responsiveness of karate black belts and elite athletes has been attributed to neuroadaptations in the brain. Awareness of our limbs' positions and their movement course can further improve performance quickness and precision. This "sixth sense" was previously called "muscle memory" and is now known as proprioception.

The Jekyll and Hyde Elixir: How Adrenaline Affects Performance Outcomes

Sports experts have known for a long time that athletes perform better with intensified emotions such as excitement, fear, or anger.[1] A perceived threat heightens emotions by triggering catecholamine release, a family of hormones and neurotransmitters that include noradrenaline (norepinephrine), dopamine (the "happy hormone"), and in particular, adrenaline (epinephrine). This "fight or flight" response, preserved over millennia in humans and other animals alike, is housed in the brain's primordial limbic system. It triggers catecholamine release by the adrenal glands that sit above the kidneys and by a small number of neurons in the brainstem (medulla oblongata).

An adrenaline surge is a welcomed elixir for competing athletes. Adrenaline brings euphoria to moderate and high intensity exercise. The "rush" revs up body physiology through the sympathetic nervous system, elevating autonomic functions that include pupil dilation that heightens cognition, sweating, and an increased cardiac and respiratory output to accommodate heightened physical demands. Elevated blood flow to muscles and arousal of the central nervous system increase strength and intensify sensory processing. Endorphins also circulate throughout the body. This chemical, released from the hypothalamus and pituitary gland, binds opiate receptors and reduces pain sensations. It enhances muscle tone and strength, elevates awareness, and further stimulates the release of adrenaline. Within seconds, these events boost preparedness and mental clarity and bring a buoyant mood.

Fight-or-flight responses (also called acute stress responses) can be triggered by both real and imaginary threats. The catecholamine cascade primes the body to either stand firm to defend against the threat or run to safety. However, when stress overwhelms the mind, a high adrenaline load shunts rational thought into uncontrollable panic that is accompanied by loss of physical coordination.

An important aspect of martial arts training is in recognizing the polar-opposite psychological impacts of acute stress responses. Training exercises simulate potential threats. Competition constitutes confrontations under a controlled setting. Both scenarios bring exposure to the positive and negative effects of adrenaline surge. This chapter will discuss the facts and fiction of adrenalized performance, how intensive training serves to trigger augmented performance, and approaches that broaden the positive bandwidth in high-stress situations.

A Potion for Mr. Hyde

In the novel *The Strange Case of Dr. Jekyll and Mr. Hyde,* Robert Louis Stevenson wrote about the fictional Dr. Henry Jekyll and Mr. Hyde, alter egos of the same person. Mr. Hyde's propensity for violence drew public outcry, yet upon his death he was found to be one and the same as the very private Dr. Jekyll.

The introverted Dr. Jekyll had discovered a potion capable of separating the individual's good and evil natures.[2] By taking the potion, Jekyll transformed to Hyde, his dominant evil twin. Eventually, Mr. Hyde was hunted down and met his demise when he could no longer revert back to Dr. Jekyll. The duality of man's character has long intrigued clinical psychologists, well before the novel was published in 1886. To date, modern humans' vulnerability to inward rage remains a popular theme in literature, stage, screen, and politics.

Stevenson's story gave rise to speculations that the Jekyll-Hyde character was based on psychotic patients who suffered from dissociative identity disorder. Psychotic episodes of disordered thinking and paranoid delusions may also be caused by a great excess of adrenaline release.[3]

Adrenaline and Task Performance

In the same year that Stevenson's novel was published, the American physician William Bates discovered a crystalline substance that was produced by the adrenal gland, later known to be adrenaline. Decades passed before adrenaline was fully characterized as a hormone and neurotransmitter.[4] The polar-opposite psychological effects of adrenaline are now well established, although its association with antisocial behavior is tenuous at best.[5]

In 1908, professors Yerkes and Dodson published their seminal findings on the influence of adrenaline on animals. When rats were given a low electric shock that led to mild adrenaline release, they became far more capable in navigating through a maze. Yet the same animals panicked and scurried about aimlessly after having received a higher voltage that vastly increased adrenaline output. The researchers summarized this phenomenon in a biphasic dose-dependent bell-shaped curve (fig. 5.1), now called the Yerkes-Dodson law.[6]

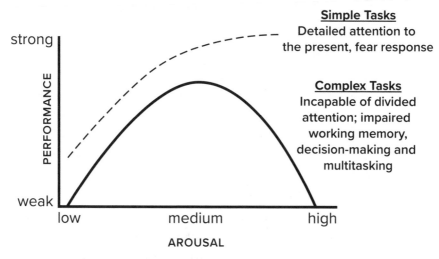

Figure 5.1 Yerkes-Dodson law on arousal and task performance. The performance of simple and well-practice tasks improves with increasing stimulus by adrenaline (dashed line). Overstimulation, however, is decremental for the performance of complex, unfamiliar tasks that involve decision-making (solid line).

Attention and motivation improve initially when stress induces adrenaline release. One feels energized and focused at the peak of arousal and is capable of performing at a higher level. However, massive adrenaline that is brought by sustained stress induces despair from a sense of crisis. Interestingly, simple task performance improves at the same time while awareness and analytical skills plummet (fig. 5.1).

The negative effects of adrenaline overload were evident even among well-trained fighter pilots. Many World War II aviators committed fatal errors when faced with prolonged high-stress combat. Their cognitive functions deteriorated rapidly under extreme duress, according to interviews with aviators who survived these ordeals. The pilots could not translate real time perception into appropriate response in the state of panic. They experienced tunnel vision, reduced situational awareness, and were completely incapable of rational thought. Increased reaction time and hampered motor functions ultimately led to loss of their planes.[7]

Mental performance even in everyday life is acutely sensitive to the negative effects of adrenaline overload, such as when taking a difficult mathematics examination or memorizing complex information. One feels jittery and restless right before public speaking or a job interview. We are overcome with helplessness as sundry physiological symptoms set in, including a pounding heart, uncontrollable shaking, dry mouth, shortness of breath, lightheadedness, profuse sweating, sensations of a fever, and nausea.

Sustained anxiety continues to ramp up adrenaline secretion, driven by the cascade of released hormones, including CRH (corticotropin releasing hormone, from the hypothalamus), ACTH (adrenocorticotropic hormone, from the pituitary gland), and cortisol from the adrenal glands. Their abundance precipitates in a panic attack.[8] A person is overwhelmed by external light and sound in this hypervigilant state. Persistent aggravation impedes clear thinking. Chronic adrenaline assaults from stress are deleterious to health and lead to weight gain, cardiovascular ailments, and a suppressed immune system.

Tilting the Balance to Dr. Jekyll

A delicate balance of circulating catecholamines sustains our mental composure from day to day. Thoughts and intent originate from the prefrontal lobes, while the parietal cortex interprets sensory information from all parts of the body and manages our body positioning (fig. 4.1). Noradrenaline receptors inundate these areas of the brain. The prefrontal cortex regulates its own performance by prompting the hindbrain (pons cells) to dispense small doses of noradrenaline. The low level bolsters homeostasis by binding receptors in the prefrontal cortex, completing a "delicious cycle" through positive feedback.[9] Dopamine that is released from the hypothalamus and midbrain reinforces a sense of well-being.

Meanwhile, our basic emotions (fear, pleasure, anger) and drives (hunger, empathy, sex) are governed by a complex array of neural networks beneath the depths of the temporal lobe (fig. 5.2). Known as the limbic system, the circuitries attend to subconscious bodily functions (heart rate, digestion, respiration, arousal) through the autonomic nervous system and the endocrine system, in effect transforming emotional experience into physiological response.

The amygdala is the seat of our emotions, particularly fear, and is an important component of the limbic system. When the amygdala perceives psychological stress, it triggers system-wide fight-or-flight responses through the neuroendocrine axis (hypothalamic-pituitary activation), in effect placing our physiology on high alert.

Neurotransmitters produced by the amygdala, including noradrenaline, instantly reshuffle brain activities. A high level of noradrenaline tamps down higher functions at the prefrontal cortex (perception, memory, thought, and voluntary muscle control), rerouting mental resources to autonomic functions. This bottom-up, emotion-driven event usurps top-down, "delicious" feedback that was previously in place.[10] Increased noradrenaline gives rise to a "vicious cycle" that hampers rational thought and escalates fear and paranoia.

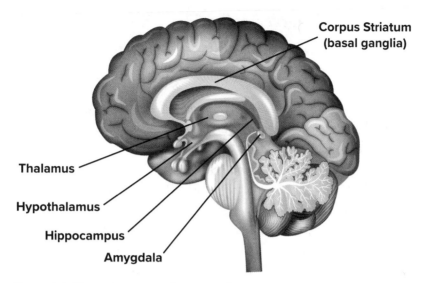

Figure 5.2 The limbic system. Part of the forebrain, it functions as our emotional center, comprising the thalamus, hypothalamus, corpus striatum, hippocampus, and amygdala. These areas are embedded beneath the cerebral cortex and govern the fight-or-flight response on perception of danger or high stress.

Sports scientists are fully aware of the negative impact of adrenaline overload. Stress reduces the effectiveness of critical thinking. An athlete's chained, automated task execution slows as their prefrontal cortex tries persistently to override procedural memory recall. As the competitor senses a loss of control, they begin to second-guess their automated skill set, which further delays and creates errors in their actions.[11]

However, high stress also elevates the capacity for simple or highly practiced tasks.[12] By learning to recognize the detriments of adrenaline overload, the athlete can direct their attention to practiced, automated tasks (see chapter 3) to sustain performance. When placed in a highly challenging situation, the martial artist would gravitate to well-practiced techniques that can be executed with high precision and consistency, in effect calling on their autonomic repertoire to sustain their performance while under high adrenaline stress.

Standout athletes condition themselves to harness this tenuous state as others succumb to it. Repetitive training and feedback give rise to an inner strength to calibrate one's emotions, as well as heightened awareness to address the negative effects. Regular exercise also reduces resting level of adrenaline production, which contributes to an even mood and stable metabolic output.[13] As well, stimulated catecholamine levels can be lowered through physical training, which allows the sympathetic nervous system to better handle the initial adrenaline spike.

Apollo Astronauts' Nerves of Steel

A heart rate of 115 to 140 beats per minute (bpm) for the average person represents an optimal, heightened response to sustain coherent thinking and quick reactions. At over 140 bpm, the sympathetic nervous system begins to divert body resources to essential functions for survival. We start to lose complex motor control at around 150 bpm, which accounts for the panic-induced disorientation and loss of balance when people jostle to escape from a burning building.

On July 20, 1969, the Apollo 11 spaceflight mission landed the first two humans on the moon, then actualized their safe return to earth. This feat ranks among the highest technological accomplishments of the twentieth century.

Mission Control at the Houston Space Center maintained continuous communication with the Apollo astronauts, except when the Lunar Module, named *Eagle,* started its descent to the designated lunar landing site. At halfway through the approach, the flight crew realized that the landing site was within a football field-sized crater, its surface laid with large uneven rocks as big as automobiles.

Neil Armstrong, as mission commander, made the decision to manually fly the Lunar Module to an alternative landing that was four miles away. By touchdown, only thirty seconds' worth of fuel (about 1 percent) remained.[14] One could fathom that the last-minute improvisation was a matter of life and death.

Throughout the Apollo 11 mission, astronauts' heart rates were monitored continuously from launch to splashdown. Armstrong's heart rate went from 75 to 150 bpm while transiting to the alternative lunar landing.[15] The heart rate of Edwin "Buzz" Aldrin, lunar module pilot, rose to 110 bpm. The excitement of actually setting foot on the moon brought Armstrong's heart rate to 125 bpm. By all accounts, both astronauts functioned at exemplary mental and physical capacity, reflecting their capacity to handle adrenaline-pulsed execution from past training and piloting experience.

Adrenalized Learning

Noted scientist Hans Selye coined the term *good stress* (eustress) to describe a person's capacity to convert the heightened state into higher energy, better mental focus, and increased confidence.[16] Our pulse quickens to the hormone surge, and we feel completely in control. The key is to learn how to manage "bad stress" (distress) when fear and anxiety set in. Stressors should be recognized for their transitional nature, which can be properly managed through training and conditioning of our neurological responses.

Aviation is recognized as the world leader in safety among today's industries. Fatal accidents in 2019 numbered 0.51 per million flights globally, a reduction by twelvefold since 1970. At 40 per trillion passenger-kilometers flown,[17] the low level of risk is the envy of all industries and has been attributed to identification and appropriate actions that mitigate mechanical failures and human error.

Seventy percent of all aviation accidents came from human error, according to error management records.[18] Pilot in command error was the main contributor.[19] Compromised procedural execution (followed procedures but with wrong execution) and improper decision-making (wrong decisions that increase risk) figure prominently among the five most common causes.[20] The contributing factors that affect performance under duress are grouped into three psychological elements: a situational demand interpreted by the person to be far from ordinary; the person's own perceived ability to cope; and self-determination on the importance of coping.[21]

A person is more prone to stress-induced impairment when they feel out of control. This is a common predicament when struggling to arrive at an unfamiliar solution, as mentally intensive deliberative thinking is severely compromised in the fight-or-flight state.[22] By undergoing training that is based on scenario-based familiarization, the participant acquires intuitive processing skills (see chapter 3) that enable them to handle the negative impacts of adrenaline over a broadened range of arousal (fig. 5.1). The perception of control, even if misplaced, still helps to reduce anxiety.

Present-day aviation training programs incorporate adrenalized learning. Computer-simulated upset prevention and recovery training (UPRT) is designed to condition pilots to circumstances of mishaps, with the expectation of elevating response skills, knowledge retention, and overall performance.[23] The simulations seek to foster high situational awareness and stress management by improving visual cues and flight control execution under escalating conditions.

Peace officers and other professionals also train to elevate cognitive control while under high stress situations, with the capacity for planned execution in spite of fear and anxiety. They participate in reality-based training (RBT) that simulates adrenalized performance in order to broaden their perceived ability to cope, also known as the individual's "zone of optimal functioning."[24]

Adrenaline-induced high anxiety leads to rapid and shallow breathing. One can nevertheless sustain conscious control of the breathing cycle. Emergency first-responders are taught a deep breathing technique known as combat-tactical breathing to reduce adrenaline shakes.[25]

Combat tactical breathing involves deep inhalation, holding one's breath while counting to four, exhaling, then counting to four before breathing in again. This approach is applicable in our fast-moving society to maintain equanimity, such as when we are provoked unnecessarily or are in a road-rage situation.

Purposeful, controlled breathing is part of Zen meditation (see chapter 11). Diaphragmatic breathing (chapter 11) and mindfulness practices (chapter 15) divert autonomous neural control from the adrenaline-prone sympathetic system to the parasympathetic nervous system, which slows the heart rate and increases intestinal and endocrine control. The brain associates slowed breathing with a sense of well-being, hence serving to reduce the adverse effects of the fight-or-flight response.

Alert, Orient, Execute

Studies have found that adrenaline-induced loss in mental focus and reasoning originates within our attentional neural networks. Attentional networks are tasked with ensuring promptness and accuracy in our decisions. Like field officials who monitor the real-time progress of a football game, attentional control slots between perception (sensory organization and interpretation) and motor function oversight (see chapter 4) in order to sustain proper response to an external stimulus (fig. 5.3).[26]

The vigilance network (alert and arousal) maintains awareness of our surroundings that present as a complex informational matrix. This network picks up extraordinary events (loud noises, bright flashes, sudden movement) among the sea of incoming sensory signals and alerts the mind to initiate an appropriate response.

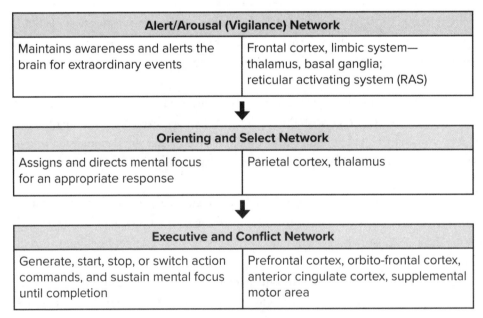

Figure 5.3 Attentional networks.

The orienting and selection network acts as gatekeeper of cognitive awareness. It directs mental focus to the target (stimulus-guided search) by suppressing unattended information, thereby heightening our attention to the cued event. Focal attention reinforces the relevance of sensory inputs within the context of memory recall.

The executive and conflict networks wean out contradicting information, then issue top-down commands to initiate action.[27] Speed and accuracy of decision-making rest largely on this reconciling process.[28] Two key functions of the executive network are responsible for real-time control: to generate, start, stop, or switch action commands; and to sustain mental focus until the task is completed.

Impending task requirements are aligned with response parameters, as the executive network correlates real-time evolving sensory signals (events) with memory recall (expectations). The network assimilates reasoning, action, and emotion into conscious thought while ignoring unattended information.[29]

For example, when a martial artist takes part in a sparring exercise, their vigilance network continuously audits the movement pattern (kinematic profile) of the opponent. When the network identifies actionable changes in the opponent's mental attitude or telltale physical indicators, orientation and selection circuitries focus attention to the impending attack, at the same time filtering out irrelevant distractions, such as cheers and jeers from the audience. The executive network prioritizes defensive actions based on past parallel experience to select among all the possible engagement options versus evasion.

Our mind can calibrate action parameters between self and attacker in the blink of an eye, thereby anticipating when and where the attack would arrive (its temporal-spatial context). These networks are conditioned to sustain our attention until one realizes a sense of safety—this is the martial arts concept of zanshin (residual mind, 残心), the practice of maintaining total awareness well after completion of physical action (see chapter 7).

The focused engagement of an opponent in kumite subconsciously triggers a fight-or-flight response, conditioning the student to the effects of adrenaline surge. Like emergency first responders, controlled breathing practice before, during, and after sparring is effective for reducing anxiety. The martial artist learns to rely on a repertoire of straightforward well-tested techniques that can be applied with high consistency under autonomous control, thereby bypassing the jitteriness of conscious intervention.

Attentional State Training

The martial artist's skill set is complex, engendering a broad repertoire of techniques and strategic options. It calls on psychological (confidence in decision-making, commitment to technique) and physiological (shortened reaction time, synchronization of mind and body) attributes in order to achieve mental clarity and physical dexterity under duress. In particular, mastery of the mind takes precedence over that of the body.

When tested for attentional network performance in a laboratory setting, young martial artists with two or more years of training (karate, taekwondo, kickboxing, jujitsu, tai chi, judō, Thai boxing, or kung fu) demonstrated markedly higher scores in vigilance network performance. This overall heightened alertness contributes to quickened defense decisions, echoing findings in the Mori study (see chapter 3).[30] Their orienting and executive scores did not differ from controls, probably because the martial artists were just as unfamiliar with the laboratory-based drills as control subjects.

Heightened vigilance does not lead to increased hostility. In a separate study, the assaultive and verbal hostility scores of male undergraduate students trained in karate or jujitsu were lower than untrained controls. Longer periods of martial arts training corresponded to lower overt expression of hostility.[31] School-age children similarly exhibited better compulsive behavior control with martial arts training. Their working memory, attention, and concentration improved as well.[32]

Training approaches that simultaneously condition mind and body are ascribed as attention state training (AST). AST relies on "hands-on" learning and differs from the attention training (AT) that is commonly associated with task-specific "brain training games," such as computer-based adrenalized aviation learning approaches.[33]

Eastern cultures' meditation, being one with nature, yoga, and martial arts, are representative of AST practices.[34] AST but not AT produces crossover benefits in mindfulness that are not limited to the specific task.[35] AST improves cognition, emotion, and social behaviors but requires longer immersion periods of more than six months.[36] Attentional benefits come from adaptations in the autonomic nervous system, which regulates breathing, heart rate, and circulation. There is increased connectivity between the limbic system's anterior cingulate cortex and the striatum. The anterior cingulate cortex engages in complex cognitive functions such as impulse control and decision-making, whereas the striatum serves as the seat of executive network function for motor learning, sometimes referred to as the "mind-body connection."[37] Their enhanced performance reflects heightened capacity to translate novel events into complex cognitive functions, such as conflating judgment and emotion into decision and physical response.

To maintain focus and intensity while controlling the side effects of excessive excitability, martial arts training acclimates and channels the practitioner's adrenalized state into heightened awareness by the attentional networks. Mental dexterity is constantly tested in budō arts through paired exercises, when the martial artist confronts the unpredictability of incoming attacks. I fondly recall younger competition days when taking part in daily intensive training. I felt the conditioned adrenaline kick as soon as I stepped onto the dōjō floor. The "rush" could be sustained at will but receded naturally as the workout ended. In subsequent chapters, we will examine underlying elements that promote vigilance and cognitive control in martial arts practice.

Takeaways from "The Jekyll and Hyde Elixir"

- The welcomed adrenaline surge that heightens performance also accounts for sharp declines under sustained stress. Adrenaline overload leads to subpar competition performance.

- Repetitive training and feedback enable control of one's emotional state and a heightened awareness for the negative effects of adrenaline surge.

- Martial arts training is a form of attention state training (AST) that conditions both mind and body. The traditional curriculum incorporates hands-on challenges to simulate stress under controlled conditions as well as nonconfrontational mindfulness practice (such as kata and meditation). These approaches combine to condition and stabilize the practitioner's emotional state.

Our Time-Bending, Attention-Shifting Mind

Our Concept of Time

Our body's internal clock is separate and distinct from our biological or circadian clock.[1] This internal clock is more accurate in gauging short time intervals, matters that transpire in seconds to minutes. Several internal clocks may run concurrently, independent of one another, so that the brain can arrive at a best estimate.[2]

Time perception is also subjective, hence relative and not absolute. The internal clock operates through the brain's parietal lobe at its lateral intraparietal area (LIP), guided by eye movement control within that same brain area.[3] The brain tracks time by pulsed neurotransmitter secretion that paces neighboring neurons within the frontal cortex.

This process, known as psychological time, is swayed by mood, emotion, and intensity of concurrent cognitive events such as attention. Elapsed time estimation, from waiting in line to running to catch a bus, requires attention and recall. More time is deemed to have transpired when processing unfamiliar new events, as the mind has to devote more attention than to familiar tasks.[4]

The first half of the long journey feels agonizingly slow. After acclimation, the second half "takes no time at all." According to psychologists, a youngster thinks that a new task takes forever because they have to absorb, assemble, and interpret large amounts of information. Our distressed brain operates in hyperdrive in the midst of a traffic accident. Accordingly, time slows because of intensified brain activities. The massive informational dump over the course of danger gives the impression that time moves in a crawl. The knack to slow the perception of time is known as "subjective time dilation."

Conversely, athletes lose track of time's passage when they are completely immersed in the moment (see chapter 4). Psychologist Mihaly Csikszentmihalyi coined this timeless state as "flow,"[5] an optimal experience when the athlete achieves complete awareness and performs at the highest level of absorption. They assert control over the minutiae of the process without conscious effort or awareness of time. Flow performance is the counterpoint of timing strategies in martial arts. This state of effortless, totally immersed performance allows one to exploit the opponent's loss of mental (and physical) dexterity from over-fixation to their own techniques. These concepts are discussed further later in this chapter.

To achieve a unified (spatiotemporal) impression of space and time, our brain has to integrate multisensory data (visual, auditory, touch, and so on) within the same instant, a process called "temporal binding" or "neural integration."

Temporal binding takes about 80 milliseconds (0.08 seconds) to assemble all sensory inputs into the "now" experience.[6] While sound from afar arrives at a much slower rate than light, auditory processing takes a far shorter time than the informationally intense visual signal. Transduction of audio signals takes milliseconds, versus tens of milliseconds for optical inputs. The association of sight with sound is also a learned process that is founded on past experience.

Our analog-processing brain is nothing short of amazing, capable of extracting visual data that requires over a hundred times more bits of high-definition video files to reproduce. So long as sight and sound are synchronized to within 0.1 seconds, such as in a movie, the signals are comingled without any sense of miscue. This holds true even though one is aware of close-range auditory cues before the brain "fills in" visual content.

From force of habit in everyday life, novice martial artists tend to over-rely on visual input, which does not necessarily work in their favor. Deliberate practice broadens overall awareness and attention by consolidating information from all the senses. Multisensorial focus encompasses inputs from the visual cortex (occipital lobes), auditory cortex (temporal lobes), and somatosensory cortex (parietal lobes) (fig. 3.2). One accepts that awareness to sounds in the surroundings does not detract or delay proper response. A heightened awareness to touch, vibration, or even temperature enhances the accuracy in interpreting the attacker's timing and distance (spatiotemporal elements). Overall vigilance promotes clarity in thinking, adds confidence to decisions, and improves precision in our actions.

For example, hacking into an opponent's seismic signature (breathing, footfalls) informs on their body rhythm. Augmented awareness waives the need to divert one's visual field to hunt for these telltale signs. This comprehensive assessment is completed in milliseconds, allowing the martial artist to sharpen their strategy and tactics.

Even having mastered the dynamics of technical speed (see chapter 4), it remains counterproductive to trade blow for blow, particularly when outnumbered or outmatched in strength. A more effective approach is to blunt aggression before it arrives.

This is the strategy of timing. Timing is like surfing on an ocean wave, or the counterpoint in a musical composition. It is based on catching the opponent's attack rhythm, then disrupting that rhythm with a highly unanticipated counterattack.

At the Karate-dō World Championships, team competition events are the international audience's perennial favorites, in particular the Women's Team Kata (form, 団体形) and Men's Team Kumite (sparring, 団体自由組手). National kata teams comprise three individuals who perform the same form in synchrony. Scores are assigned according to collective focus, overall technical strength, and synchronization of movements. The best teams often impress with grace and precision, as all three spirits merge as one over the course of performance.

Unlike Team Kata, Team Kumite is known for explosiveness and brutality. The event comprises three separate one-on-one sparring matches in succession, pitting the best fighters of one nation against another. In the event of a tie, an additional match is carried out between a representative member from each team.

As soon as the first pair of competitors bows in, the stadium erupts with cheers, jeers, and air-horn bleats. Stomping feet reverberate like thunder. Just as quickly the crowd hushes in eager anticipation for the match to begin.

Team Kumite is generally slotted as the final event of the day. Earlier, athletes in this event would have gone through many challenging rounds in the Individual Kumite event. They are worse for wear, many having earned badges of honor in the form of incidental bruises and minor nicks and cuts.

Frustrations set in as attacks don't pay dividends, or when defenses turn inadequate. Victory ultimately goes to athletes who sustain mobility while patiently seeking out the opportunity to strike with the highest precision.

Team Kumite is as much a showcase of athletic talent as tenacity from depth in training and resilience in spirit. Oftentimes one side presents the carelessness of an opening. As their opponent enthralls at the unexpected opportunity, the athlete charges in, turning the tables in their own favor.

This thrilling display of audacity seems precognitive, yet it falls within the toolbox of timing strategy. As an adversary is trapped in the mental bubble of fixation, the athlete's explosive charge freezes them at mid-action, in effect neutralizing the opponent's power and reach.

This preemptive strategy is known as *sen* (先), the ability to take the lead in defining position and timing of engagement, striking before the opponent can complete their technique.

Lost in the Shuffle: The Invisible Gorilla

The handicap of mental fixation is exasperating even in everyday life, as illustrated by Christopher Chabris and Daniel Simons's study, fondly known as the Invisible Gorilla.[7]

Participants were required to watch a ninety-second video that featured two teams of players, one in white shirts and the other in black. The players passed a basketball back and forth among their own team. The scenes were intentionally chaotic, as everyone shuffled positions endlessly in a tight circle, while two basketballs changed hands in opposite directions, at the same time. Participating viewers were instructed to keep count of the number of passes made by the white team.

Halfway through the drill, someone in a black gorilla suit wandered in, casually weaving through players from one end of the field to the other. Rehearsed players continued to pass the balls as the gorilla casually thumped their chest, then exited after having been on the screen for nine seconds.

Of thousands who watched this video, a majority congratulated themselves for having arrived at the correct number of passes by the white team. Remarkably, half of the respondents were unaware of anything unusual. They insisted that there had been no gorilla in their first viewing, even after they watched the video for the second time. You can search online for a video of the "Selective Attention Test" to see the study in action.[8]

This study confirmed that people lose sight of perspectives when they are intensively focused on a singular task. Fixation on the white team members has taken up the entire attention span of many viewers.[9] Two take-home lessons stand out from the study. First, that we tend to miss the forest for the trees, and second, that we are even oblivious to this oversight (undersight?).

Stuck in Time

Sen timing goes hand in hand with the Japanese concept of *suki* (付), denoting a brief, vulnerable lapse in mental attention.[10] The literal translation of *suki* is "affixed," in the sense that one is preoccupied or "stuck." When a person falls off a building while taking a selfie, they have committed an extreme form of *itsuku* (the verb form of *suki*).

Itsuku (付居) in kendō refers to fixation on a singular thought, hence being suspended in a state of flux due to loss of mental dexterity. The attacker is so bent on striking the intended target that they ignore all other aspects of space and time.

Tunnel vision and reduced situational awareness are commonly associated with this hypervigilant state. We have all experienced this subjective time dilation, when we felt helpless in addressing an impending danger because of a preoccupied mind (and adrenaline overload, see chapter 5). Our focus is lost to incoming traffic while stooping to pick up an errant dollar bill. An opponent places themself in jeopardy when totally invested in launching their attack, while being incapable of responding to an unexpected counterattack.[11]

Kyo (居) is a similar concept in traditional karate training, referring to a breach in mental or physical awareness, or both.[12] The aggressor's kyo is an exquisite opportunity to intercede in their foray, a moment that enables the defender to harness the opponent's mental lapse and their wasted physical momentum. It is at that moment in time that a counterattack is unstoppable, in part due to the speed of execution, but more so because the aggressor's attention is blinded by their zeal.

Timing Strategies: *Sen* and *Go*

Self-defense experts consider victims of an assault to have a narrow safety window of ten seconds or less to survive unscathed. The brevity in opportunity incentivizes the mind to decisive, proactive technique execution.

An optimal vigilant state precludes hyperexcitability that slows the reflexes, which bends time to one's own demise (see chapter 5). The martial artist strives for the mind state of tsuki no kokoro (mind like the moon, see chapter 2) that sustains heightened awareness of the surroundings and their opponents' actions, yet without being hindered by emotion or agitation that delay decision or physical response. They mitigate overexcitement by coming to terms with the symptoms of adrenaline overload and are adept at channeling its detriments (see chapter 5). Rather, they embrace the attitude of mushin (無心), the agile "unencumbered" mind-set that sees all opportunities. Competition further instills mizu no kokoro (mind like water), which entertains the pursuit of any and all solutions of offense and defense.

Recklessly charging into an attack is foolhardy and counterproductive and incurs a high likelihood of injury. Instead, the seasoned competitor keeps high vigilance while maintaining a safe distance, their body coiled and ready to strike.

As both competitors maneuver for position, they gauge and challenge one another's mental and physical preparedness. Both sides jockey for an opening for a finishing blow technique (todome waza, 留め技). Once sensing kyo, there is but a split second to close in and take charge of the situation.

Todome waza constitutes the perfect solution in kumite as in self-defense, much like boxing's knockout punch. Most traditional karate tournaments specify that the competitor must withhold full application of force, yet require a demonstration of effectiveness in dismantling the opponent's mental composure and incapacitating their offense. This razor-thin criterion involves the utmost mental focus and physical control.

In the Japanese language, current time is referred to as *gen* (the present, 現). *Sen* (before, 先) refers to events that transpired before the present, while *go* (after, 後) will occur after gen.

During training, paired kihon kumite (basic sparring, 基本組手) exercises introduce beginning students to the application of learned techniques against an opponent. The attacker announces the intended technique and target (such as *oi-zuki,* jodan, stepping punch, face) before proceeding, allowing the defender to process the verbal information before the start of the attack.

Students begin kihon kumite practices with *go-no-sen* timing (after, then before, 後の 先), carried out operationally as "block, then counterattack." Go-no-sen trains for technical precision by the attacker and defender, while maintaining a measure of safety. The defender executes a blocking technique (or an evasive body movement) *after* (go) the first attack arrives, but seeks situational control with a counterattack *before* (sen, 先) the opponent's follow-up actions.

Students stay with go-no-sen timing until they have a strong sense of self-control, mentally and physically. This rule of thumb applies to green- and brown-belt students as well. When in the role of an attacker, the student presses the defender with intensity, power, and speed and the confidence of fudōshin (an unshakeable mind-set) and zanshin (total awareness even after completion of physical action; see chapter 5). As defender, the student learns to achieve spontaneity in executing blocking and counterattack sequences at increasingly challenging scenarios, and the mental attitude of mushin (an agile, unimpeded mind-set) as well as zanshin.

Go-no-sen forsakes physical but not mental initiative, as the defender locks into the psyche of the attacker and seeks the opportunity to gain an upper hand. Go-no-sen timing is highly effective against a faster opponent with less than powerful techniques, and it continues to be indispensable even among seasoned competitors.

Kihon kumite drills are also useful to instill proper sen timing—the defender now seeks mental and physical initiative before an attack arrives. However, it behooves one to master go-no-sen timing before proceeding to sen exercises. Akin to two bulls charging at each other, the smaller margin of safety imposes added risks to unskilled practitioners.

For advanced practitioners, the coveted imperative of sen timing is parsed into even finer options. At the fundamental level, *tai-no-sen* (body, then before, 体の先) refers to completion of a preemptive strike (sen) as soon as the opponent initiates physical (*tai,* body) movement (fig. 6.1).

Figure 6.1 Sen timing. The defender (Robert Fusaro sensei, right) executes a preemptive side thrust kick *(yoko kekomi geri)* to halt the aggressor's intended face attack *(jodan gyaku-zuki;* Michael Fusaro sensei, left).

Kake-no-sen (mentally transfixed, then early, 掛の先) is intervention at an even higher level. The expert hacks into their opponent's emotional contents, landing a preemptive strike within the window of the opponent's mental commitment *(kake)* but well ahead of their physical action. Kake-no-sen is analogous to *sen-no-sen* (before, then before, 先の 先) that alludes to intervention and completion of execution at the earliest opportunity (table 6.1).

By all measures, go-no-sen and sen timings fall within a continuum, given the reality that both parties seek the initiative of engagement. In the next chapter, we will examine training concepts to improve mental preparedness in martial arts, in order to correctly execute these timing strategies.

Table 6.1. Budō timing strategies.

STRATEGY	TIMING	ACTION
Ōji waza (counterattacking techniques, 応じ技)	sen (先)	kake-no-sen (sen-no-sen): before opponent's physical action
		tai-no-sen: concurrent with opponent's physical action
	go-no-sen (後の先)	*amashi waza:* through repositioning
		uke waza: with blocking techniques
Kake waza (charging-in techniques, 掛技) or *Shikake waza* (offensive techniques, 仕掛/け技)		self-initiated attack with (shikake) or without (kake) disrupting opponent's mental equilibrium

Takeaways from "Our Time-Bending, Attention-Shifting Mind"

- Perception of time is swayed by mood or concurrent attentional processes. As martial artists, we should be mindful that intense, singular focus can lead to a loss in perspective.

- Martial arts timing strategies subscribe to the concept of suki (attached), an attentional vulnerability in the self or the opponent who is preoccupied or "stuck in time."

- Advanced practitioners covet the imperative of sen timing (before, ahead) that produces a preemptive strike to an opponent.

Key Elements for Response Spontaneity

Performance experts consider "a regimen of effortful activities for a minimum of ten years" as key for mastery of a skill set at the elite level, also known as deliberate practice. Without discounting natural talent, this level of devotion is commonly found among high achievers in sports, music, and chess. Dr. Anders Ericsson named this attribute the "ten-thousand-hour rule" in his highly influential study in 1993.[1]

In the book *Outliers,* journalist Malcolm Gladwell extends this concept to experts of other mentally intensive practices, such as business and informational technology.[2] He considered deliberate practice as critical to any endeavor that requires a sharpened decision-making mind-set.

The *Outliers* best-seller was preceded by an earlier work, *Blink: The Power of Thinking Without Thinking.*[3] Giants of industry who have mastered their field of expertise, asserts Gladwell, can arrive at brilliant decisions within the blink of an eye. The mastery of "thinking without thinking" enables experts to accurately appraise a situation, then commit to action by merely weighing selected key variables, instead of having to devote lengthy deliberations on a vast body of information.

Gladwell called this "thin slicing," the end result of deliberate learning after exhaustive processes of trials and errors over the course of a career.

Martial arts demand the skill of extreme thin slicing as a matter of course. The intent is to prepare for the unexpected when under intense threat, and to be able to arrive at solutions in the blink of an eye. Unsurprisingly, psychologists Leslie Zebrowitz and Nalini Ambady consider thin slicing to have evolved from our innate need to quickly interpret and respond to danger. This level of decisiveness commits the defender to act promptly and without hesitation.[4]

Thin slicing can now be explained by the close ties between creativity and memory. The rendering of appropriate judgment without conscious thinking is based on previously acquired knowledge and practical experience. Our brain arrives at a current solution by reviewing previously curated information that is stored in long-term memory.[5] The process is also called "borrow and build." Deliberate practice archives an abundance of past parallel experiences, which helps to tackle real-time challenges in a big way.

The acquired cognitive skill of thin slicing bypasses conscious analyses and performs at the autonomous level (see chapter 3).[6] This acumen is the prerequisite for mastering sen strategies (see chapter 6) that demand confidence in judgment and spontaneity in action.

Warming Up the Brain

Athletes warm up religiously before competition. Major League Baseball players practice batting for at least three hours before the first pitch.[7] Pro-Am golfers undergo structured warm-ups for more than an hour before tee-off. The pugilist shadow-boxes incessantly before stepping into the ring.

The compulsiveness for warm-ups is mind-boggling. The body does not need to be warmer—it is already at its natural operating temperature. Explanations range from a desire to increase muscle blood flow or to jump-start the adrenaline rush to attain a higher performance level.

A professional athlete's quaint rituals are similarly confounding. When in a tie-breaker, the tennis champion bounces the ball incessantly before launching their serve. Professional basketball players subconsciously touch their face or uniform before releasing the free throw. Surely these actions do not offer a performance edge—or do they?

Studies show that warm-ups and rituals direct our attentional networks to the associated motor function at hand. Complex, automated motor routines apparently are modularized and scattered across multiple long-term (procedural) memory registers. While the brain takes hardly any time to put routines in order, warm-up drills preassemble automated execution commands, giving that extra edge by priming the cognitive process to attend to essential cues.

Repetitive warm-up routines pair up specific mental and physical skill sets. The brief respite allows the athlete to run through action plans in their head. However, these benefits gradually recede within thirty minutes after the athlete stops the routine, a phenomenon called "warm-up decrement."[8] This explains the urge of athletes to be constantly moving around, mentally focusing on the action even though they are not in the game.

Preperformance rituals intensify the athlete's perceptions as they scroll through options in their head. These trivial yet comforting processes calm the nerves, consolidating their attentional neural networks to execute "without thinking."[9]

Martial artists practice the meditative process of mokusō (table 7.1) before competition to sooth the psyche and offer an interlude for strategy review. At the start of the match, they assume their preferred sparring stance (*kamae,* 構え) that frames their attention to the current event.

Ichi Gan, Ni Soku, San Tan, Shi Riki

The budō axiom *ichi gan, ni soku, san tan, shi riki* (first, eye; second, feet; third, courage and determination; fourth, strength and effort, 一眼二足三胆四力) speaks of the hierarchy of committing mind and body for technique execution.

Ichi gan (first, eye, 一眼) calls for vigilance to an impending threat. The martial artist adopts the attitude of total awareness, primed for spontaneity of decision and action. They embrace the tsuki no kokoro (mind like the moon; see chapter 2) mind-set of unperturbed awareness without prejudice. Reference to *gan* (the eye) underlines the pivotal importance of not merely looking, but seeing, to notice and be aware. A well-trained visual search pattern appraises the opponents and the surroundings, instantly capturing information that is pertinent for prioritizing decisions. Techniques that rely on speed and precision are appropriate against a bigger, yet slower assailant, whereas an opponent of weak stature or spirit is susceptible to powerful maneuvers that upset their emotional or physical balance.

Within a choice reaction time of half a second, the martial artist has to arrive at solutions (see chapter 3) that embody proper timing, positioning, and technique in order to take control of the situation (see chapter 6).

Ni soku (second, foot, 二足) refers to the physical initiative to secure proper positioning,[10] whether to stay within striking range, move out of harm's way, or shift to the opponent's side to attack their exposed flank. Soku embodies proper balance, center of mass placement, and stance functionality (see chapter 9). Harking back to the concept of mushin no kokoro (being free from mind attachment), or mizu no kokoro (mind like water, see chapter 2), one responds to an attack intuitively from the ground up, without delay or self-doubt.

The ancient Chinese word for bile (*tan,* 胆) is a metaphor for courage and aligns with the Western idea of "gall." Represented as the third nexus, san tan (third, courage, 三胆) calls for the display of confidence and resolve, but predicated on the assurances of gan and soku.

The final element of shi riki (fourth, strength and effort, 四力) commits all physical resources to maximize speed and force in technique delivery.

Wits and Feet: Vigilance and Spontaneity

Modern sedentary lifestyles dwell on intensive audiovisual experiences, arm and hand dexterity, and emphasis for upper body aesthetics. As a result, the average person places little attention on "leg health" and lower body strength.[11]

When first introduced to paired sparring exercises, beginning students tend to stay in the same spot, flailing their arms in futile attempts to ward off an attack. This approach is ineffective for getting rid of insects, and even less so against a face punch.

To realize effective technique execution, ichi gan, ni soku specifies proper positioning that immediately follows the primacy of heightened awareness and a focused mind. The application of courage (tan) and physical action (riki) is otherwise ill-advised.

Emphasis on balance and foot placement reinforces the decision to engage the opponent. Instead of scrolling through all possible options, the defender considers situational outcome to be hinged on arriving in the right place at the right time, whether to confront or to evade.

We instinctively become uncomfortable when someone "leans in" and encroaches our personal space. One would do well in martial arts to keep others farther away than at arm's length. When an opponent rushes in, they crowd our visual field and adversely affect our capacity to arrive at proper judgment. The safety margin drops precipitously when an adversary comes within breathing space. Closed-in conflicts favor strength over finesse. Grappling and takedowns destroy composure in quick succession unless one is well versed in groundwork.

Beginner students tend to take up positions that are too close for comfort in paired kihon kumite exercises (see chapter 5). Tight separation improves their reach for counterattacks, but the novice forgets that this advantage applies to both parties, leaving them vulnerable to quick jabs and front kicks. In particular, I recommend starting positions to be at least two paces apart, particularly as exercises escalate to free sparring. This rule of thumb offers reasonable safety and sustained mushin. An adversary's kinesthetic signature is easier to interpret at four or five feet away. Both attacker and defender are expected to cover the intervening distance quickly, an essential ingredient for kumite success.

Anticipatory foot placement is the foundation for a well-timed countermeasure. Correct placement and balance of the body center is sustained through stance (see chapter 9). Like the track sprinter, one launches into sharp and balanced footwork from forethought, through the coiled spring actions of the lower body musculature (see chapter 9).

The same ideas are also common practice in other competitive sports. In tennis, basketball, or soccer, proper anticipation neutralizes an opponent's intended foray and is hence a key determinant of strategic success. After having grasped the concept of proper distancing, a second lesson is to attend to one's "wits and feet," which couples mental preparedness with physical action during sparring.

Integrative Deliberate Practice

Proper distance coverage is an important concept among budō arts, including karate-dō, judō, kendō, and aikidō. This remarkable eye-brain-body coordination is evident among second- (nidan) and third-degree (sandan) black belts. In the Shōtōkan Karate-dō system, a rank of sandan represents eight to ten years of immersive training, an interval that falls comfortably within Ericsson's ten-thousand-hour rule.[12]

In a more recent publication, Ericsson reminds readers that the ten-thousand-hour rule is more of a "provocative generalization." He considers quality of practice to be more important than total duration.[13] Having a growth mind-set, meaning a willingness to learn from mistakes and integrate them as lessons for improvement, can bring expertise in a far shorter time, such that athletes can achieve elite performance status with just three thousand to four thousand hours of deliberate learning.[14] This accelerated process, however, requires an intense desire for improvement and the integration of sports science and neuropsychology concepts into practice.[15]

Regimented training has been shown to produce neural adaptation among elite athletes.[16] However, four key training elements have to be incorporated in mastering complex motor performance in order to promote procedural memory development and recall.[17]

1. Establish specific and attainable goals in areas that need improvement.

2. Maintain a high level of focus—which should be intense, sustained, and repetitive toward achieving set goals.

3. Accept immediate feedback on practice performance.

4. Aspire for higher performance beyond one's comfort zone.

Table 7.1. Deliberate practice in traditional karate-dō training.

| | DELIBERATE PRACTICE CURRICULUM* | | | |
	MOKUSŌ 黙想	KATA 形	KIHON 基本	KUMITE 組手
Training Goals	achieve a state of mindfulness to elevate spirit	improve integration of psychology and proper serial technique execution in a nonconfrontational setting	continuous improvement in the operational aspects of individual techniques	excel in technique application when under challenge, a synthesis of mind, body, and spirit
Intensity and Sustained Focus	calm introspection involving sustained focus	moderate to high level of mental focus	moderate to high level of mental focus	promotes highest level of awareness and spontaneous execution
Immediate Feedback	self-correcting practice for the "wandering mind"	challenges mental equanimity and the balance of speed, power, and precision	improper execution leads to loss of balance or uncoordinated movements	mistakes lead to immediate loss of situational control
Reaching Beyond Comfort Zone	striving to extend mental reach and inner awareness	coherence of mind-body continuum over the course of the performance, then integrating spiritual aspects of the kata	maximize individual technique, power, speed, precision, and energy transfer in coordination with mental focus	paired exercise elevates emotional stability when challenged, seeking to understand one's strengths and weakness while under duress

*Mokusō: the practice of meditation by directing one's "thought of no thought" inward; kata: performing prearranged movement sets that embody multidirectional defense and counterattacks against imaginary opponents; kihon: the repetitive study and practice of fundamental punching, kicking, striking, blocking, and throwing techniques; kumite: sparring exercises for application of defense and offense scenarios.

The traditional karate-dō curriculum comprises basic training (kihon, 基本), the study of form (kata, 形), partnered sparring (kumite, 組手) drills, and meditation (mokusō, 黙想) at the beginning and end of each session. These unique but complementary practices happen to align closely with contemporary, deliberate practice goals toward mind-body synergy (table 7.1). Students can expect measurable gains in motor function recall, visual search strategy, and decision-making in six to nine months.[18] Specific capabilities as applied to martial arts performance (table 7.2) will continue to improve with immersion. Over time, students will be accorded their own preference insofar as area of emphasis. While mokusō and kihon practices remain as constants, one may elect to focus on kata or kumite based on personal temperament.

Table 7.2. Cognitive conditioning through traditional karate-dō practice.

MENTAL CONDITIONING	CURRICULUM			
	MOKUSŌ 黙想	KATA 形	KIHON 基本	KUMITE 組手
Perception	✓	✓✓	✓	✓✓✓
Data Chunking	✓	✓✓	✓	✓✓✓
Brain Automation	✓	✓✓✓	✓✓	✓✓
Choice Reaction	✓	✓✓	✓✓	✓✓✓
Mind-Body Continuum	✓✓✓	✓✓✓	✓✓	✓✓
Stressor Level	low	low to moderate	low to moderate	moderate to high

Discipline of the Mind

Mental discipline brings confidence and consistency in competitive performance. It engenders focus, the ability to perform under pressure, and a capacity to manage physical and emotional pain.[19] These attributes enable the athlete to perform without second thoughts. Nicole Forrester, the Canadian high-jump Olympian, considers mental discipline as an unshakeable self-belief, resiliency, and motivation. Jackson Yee, expert fitness trainer and self-proclaimed "mental toughness guy," defines self-discipline as the ability to do a task even when one does not want to.[20]

Studies show that most Olympic and professional athletes look forward to meeting the challenges of training. They derive more satisfaction from working hard and meeting a higher standard of excellence than from money or fame.[21] Based on personal experience, the author considers that the majority of traditional martial arts experts are driven by these same intrinsic motivations. They practice their craft with gravitas, fully realizing that carelessness could lead to dire outcomes beyond the bruising of one's ego. At worst, negligence leads to loss of life or limb, or that of the opponent.

The intensity of conflict in kumite by nature demands strong mental discipline, purging the mind of extraneous thoughts. Competitors must focus on the very moment, and it calls on one's fortitude to sustain performance. An unwavering attitude is also essential to sustain consistency in kata performance. Abstract thinking is engaged herein to meld mental and physical agency for individual movements, then tie them in succession with resolve and precision.

The sensei and senpai are expected to lead by example. The sensei by virtue of knowledge and experience is tasked with formulating incremental goals to spur the student's progress (table 7.1) and to assign supplementary physical conditioning or mental focusing exercises as needed. The student gains technical mastery in a matter of years, along with an understanding of perseverance even when mentally worn or physically exhausted. By reaching for one's limits, the student acquires a stronger sense of self-reliance and purpose. Improved mental clarity and focus give rise to a strong spirit (see chapter 14).

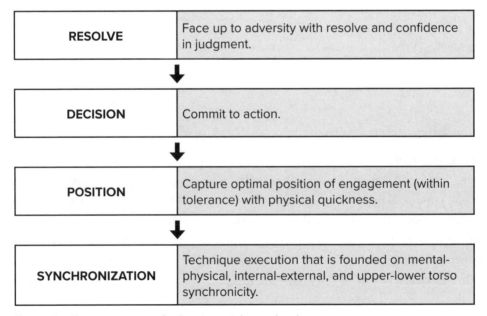

Figure 7.1. Key components for kumite quickness development.

Quickness Conditioning in Martial Arts

The key elements of a successful military campaign include soundness of strategy (planning and execution of tactics), situational control, and effective deployment of force. Considering self-defense as a campaign on a personal scale, a student is urged to attend to their decision, position, and synchronization (the three N's) to ensure consistency in outcome (fig. 7.1). Attention to these three elements ensures quickness and agility of the mind and precision in action. The training points to accomplish these aims are summarized below.

1. Forethought and Preparedness

Situational awareness, resolve, and accepting all odds contribute to decisiveness. The tolerance to confrontational stress is strengthened by partnered training and competition. These experiences broaden our perspective and condition us to overcome uncertainties (table 7.2). Like elite athletes, martial arts development is best driven by one's own desire to excel, or to aspire for higher values such as dignity, duty, and honor.[22]

2. Perception in Action

The expert martial artist constantly calibrates their actions over the course of an unraveling situation. Sparring drills fortify the performance of the orienting and selection attentional network (see fig. 5.3) by promoting mental focus to the cued event.

A drill to elevate this skill set is to place the well-trained individual against two opponents at the same time, who launch controlled attacks from separate directions in no particular order. The defender learns to control their anxiety with heightening overall awareness. They rely on mental dexterity and stable emotions to capture, process, and prioritize responses to the two attackers within the blink of an eye.[23]

Unlike learning by verbal instructions (explicit learning), learning by doing (implicit learning) enables the student to realize how techniques and movements come together dynamically. These experiences form an essential part of the subconscious procedural memory. There is firsthand feedback when a punch feels weak because of improper biomechanics. As a student stumbles when they kick, they learn to adjust their balance or lower their center of mass. When an over-the-shoulder throw *(seoi nage)* is held up, the judō-ka reorients their body center reflexively to improve leverage. Repetitive practice leads to mastery of anticipatory skills and performance while under pressure.[24] Acclimations to incidental physical contact during practice tamper negative emotions in confrontations. Repeated practice also conditions the midbrain to release dopamine, which curtails fear and anxiety.[25]

Each of us is wired differently in our innate neural makeup and our neural-adaptive development, which is founded on personal experience. Implicit learning engages the individual's learning matrix, aligning cognition control with motor functions that are distinctively "us." Textbook-perfect techniques rarely apply in real life. Feedback from practical learning shapes our perceptive skills through successes and failures and helps to sustain consistency by forming the following habits.

a. Calibrate response by an opponent's skill and kinematics, such as movement speed, technique, and their tactical mind-set (aggressor versus counterpuncher): Quick assessments of their strengths and weaknesses can elevate the chances of success in an engagement.

b. Broaden sensory inputs: Learn to entrust a totality of sensory inputs through practice and trial and error, including sight, sound, touch, vibration, proprioceptive feedback, and the impression of the opponent's kinesthetics. This all-encompassing perceptive skill expands our grasp on time and space, generally recognized as vastly superior to relying merely on visual input. A broadened informational base fully utilizes the diverse processing functions of the brain, thereby heightening neuromotor performance.

c. Sustain confidence in judgment: Place unwavering confidence in one's decision-making process that has been honed by exhaustive trial and error during training.

3. Binary Trigger

Over the course of a match, the martial artist streamlines their actions with serial yes-no triggers, scrolling down a preestablished hierarchical algorithm. For example:

a. To engage or not engage the opponent.

b. Once deciding to engage, to either take sen initiative or execute a block and counter sequence (go-no-sen).

c. Singularize a go–no go decision according to the likelihood of meeting a game-changing objective, such as proper positioning (see the following section), or exploiting the opponent's psychological or physical limitations.

d. Streamline tactics by favoring well-tested techniques with a high likelihood of success. Commit mind and body to ensure proper biomechanics and body rhythm. Exercise due diligence to hone these combinations to the level of autonomous execution during practice, so that each set can be unleashed spontaneously.

4. *Optimized Position of Engagement*

The optimized position of engagement is realized by a predeveloped intuitive sense of reach in order to properly execute defense and offense techniques. The knack to arrive in the right place at the right time depends on confidence in judgment, allowing the competitor to take charge and steer conflict to a favorable end.

A common error in competitive sports is to be out of position, broadly known as "failure to anticipate movement in time and space."[26] To arrive in position before the opponent, one has to be conscious of balance, neuromuscular control, and resolve to adhere to the practice of "wits and feet" discussed earlier.

5. *Synchronicity*

Effective martial arts techniques are the end results of well-coordinated biomechanical outcomes (see chapter 10). The three synchronicity levels are:

a. Mental-physical: The martial artist is mindful of "qi follows yi" (yi yi yin qi, 以意引氣) and realizes that physical action (qi) is guided by mental intent (yi) (see chapter 15). Therefore, mental resolve is key to effective performance.

Stress-related failure in performance is often attributed to overactivity of the left cerebral cortex, which governs speed, feedback, and fine adjustment of movements (see chapter 4). An "automation squeezing" drill allows competing athletes to overcome this detriment. By repetitively squeezing a rubber ball with the left hand, the athlete redistributes their awareness toward right cerebral hemisphere control. The exercise reduces demands on the left cerebral cortex that otherwise audits and slows autonomous performance.[27]

b. Internal-external: Preparatory routines that focus on breathing, such as by mokusō or combat-tactical breathing (see chapter 5), are effective for relieving stress and performance anxiety.[28] Diaphragmatic breathing also enables conscious control of abdominal core muscle contraction and expansion during actual task performance. Controlled exhalation (often referred to as internal tension practice) is a critical aspect of traditional martial arts practice that strengthens and coordinates upper and lower body movements (external tension) (see chapter 11).

c. Upper-lower: Our hips and core muscles are critical links for our body's kinematic performance. A practiced awareness of their stabilizing performance is essential for efficient energy transfer onto the target. A detailed discussion on this subject is provided in chapter 10.

6. Zanshin

Zanshin (residual mind, 残心) is a valued practice in karate-dō, judō, aikidō, and kendō as well as other budō arts. Mushin and zanshin form the mental bookends for technique execution.

Often referred to as "relaxed alertness," zanshin sustains total awareness even after completion of physical action by remaining vigilant until the situation is resolved. The practice ensures spontaneity in response in case the opponent springs another surprising attack.

Zanshin roughly corresponds to "follow-through" practices in baseball, tennis, or billiards. This mental state ensures commitment to technical execution and primes the martial artist's attention for further action. Over the course of defensive strategies, zanshin is coupled with moving to a position of safety, away from the point of conflict.

7. Ichi-Go Ichi-E

Ichi-go ichi-e (one lifetime, one meeting, 一期一会) signifies that each encounter is unique, as the same circumstance will not materialize ever again.[29] Though mostly designated to the Japanese tea ceremony (chadō, 茶道), this Zen-inspired concept applies equally to budō arts when facing an opponent.

Chaji (tea event; 茶事) entails a series of formal and elaborate rituals, each with a specified protocol. The ritual begins when guests are admitted through an ornate anteroom with *tatami* floors and a hanging scroll that marks the season and occasion. As the guests are invited into the tearoom, they are required to sit in *seiza* (proper sitting, 正座). Each receives a multicourse light meal served with sake and undergoes ritual purifications of hand-washings and rinsing of the mouth before and after. The guests are then served a round of thick tea, then another round of thin tea.

Formal etiquettes are observed by host and guests to ensure total immersion. The guests are expected to scrutinize the furniture, wall displays, and particularly the ornate antique tea-serving wares and offer praises to the host. The host takes pride in conducting chaji with focus and economy of motion, for which they are accorded respect and a series of bows throughout the ceremony.

As one's senses are inundated by ritual and process, concerns of past regrets and anxieties for the future are cast aside. Participants exercise studied precision to treasure the moment and assign utmost attention to the ritual. Each embraces their connectedness to the surroundings, the host, and other participants. This attitude aspires for an unperturbed spirit (see chapter 17).

This attitude of finality is also emphasized in kendō training. The kendō-ka devotes complete attention to performing "the cut" and without second thought, as if it were the last action in their life. They will to "place themself at the opponent's disposal." By investing all mental and physical resources, the kendō-ka is at peace after the fact, having given their utmost.

An absence of self-doubt is required in quickness development for all budō arts. As the karate-ka fully immerses themself in an at-risk situation (kumite, self-defense), their dignity is preserved only through total commitment and determination that leads to decisiveness.

Takeaways from "Key Elements for Response Spontaneity"

- Martial arts training reinforces the ten-thousand-hour rule, the requirement for effortful activities for ten years to achieve expertise. Enduring, deliberate practice enables the practitioner to commit to a decisive course of action with no hesitation. This level of expertise can now be achieved in three thousand to four thousand hours of intensive training by integrating sports science and neuropsychology practices.

- Elite athletes carry out preperformance routines that intensify their attentional focus. A martial artist assumes their favored kamae (fighting stance) to achieve a similar purpose.

- Vigilance and proper positioning (anticipatory foot placement) is critical in achieving a successful outcome in competitive sports. This concept is equally important in martial arts.

- Contemporary cognitive science supports the three key tenets in martial arts techniques—namely, decision, position, and synchronization.

Shēn

Endurance is one of the most difficult disciplines, but it is to the one who endures that the final victory comes.

—GAUTAMA BUDDHA

The heights by great men reached and kept

Were not attained in sudden flight,

But they, while their companions slept,

Were toiling upward in the night.

—HENRY WADSWORTH LONGFELLOW

To know and to act are one and the same.

—WANG YANMING, FIFTEENTH-CENTURY NEO-CONFUCIAN SCHOLAR

PART III

The Body

Bipedalism: Our Template for Speed and Power

Much of martial arts training is about self-discovery, most fundamentally that of the physical body—from understanding its native strengths and weaknesses to reaching for full potential through disciplined training.

The attitude of finality (ichi-go ichi-e, see chapter 7) is central to martial arts practice, based on the premise that a person's mental and physical limits are brought to bear in self-defense. Accordingly, the martial artist seeks the utmost in mental quickness, physical power, and technical precision. Different practice styles have sought unique and specialized approaches to this end. Nevertheless, human performance cannot stray far from the bounds of biomechanics without risking injury. Our capacity to fully balance and travel on two feet has come to define our physical aptitude and behavior. Traits that are tied to the behavior of full bipedalism have evolved over millennia by choice and selection, superseding the development of a larger and more complex brain.[1]

Bipedal and Four-Dimensional

The modern human's identity is ascribed to how we traverse time and space. Our imagination, for the most part, is tethered to our physical horizon.

According to fossils discovered in the 1930s and 1940s, tree-dwelling *Australopithecus* inhabited the earth some 4.5 million years ago.[2] These partial bipeds had short legs, long curved fingers and toes, long forearms, and "stooped" shoulders with ample muscles that connect to the head and neck.[3]

Our current physical likeness as full bipeds was not evident for another 2.5 million years. Theories abound regarding survival pressures that favored full bipedalism.[4] In particular, the endurance-running theory considered bipedal distance travel as survival advantage for pursuit of prey, and by outlasting other scavenging predators.[5]

In the 2004 article "Endurance Running and the Evolution of *Homo*," Bramble and Lieberman theorized that many of our anatomic and physiological features were coevolved to improve survival from predators or to mitigate injury when combating other males.[6] These traits, highlighted below, undoubtedly influenced martial arts development.

INCREASED VIGILANCE.   The bipedal anatomy enables the human head to be held upright. Our raised head provides a far superior field of vision and enables multidirectional sentience over distance. The human's flatter head and face, together with smaller teeth and shorter snout, contribute to balance for an upright gait.

MODIFIED POSTURE.   Unlike quadrupeds, the human head is positioned farther back, with its center of mass aligned vertically with the spinal cord. A straightened spine better sustains the upright posture (fig. 9.1). An enhanced range of motion of arms and legs functions as first lines of defense, protecting command nerve centers (the brain and spinal cord) that are now recessed toward the back. By comparison, a primate's skull is jutted forward, with its backside aligned to the rear of the spine, giving a stooped natural posture with its head hanging toward the ground.

ROBUST FACIAL STRUCTURE.   Our evolved facial features include a shortened jaw of enhanced musculoskeletal robustness and strength. The strong human cheekbones are more substantial than those of primates. Prominent periorbital bones better protect the eyes.[7] These rugged features are vastly improved from those of early humans, which were more prone to impact fracture.

EXPANDED UPPER-BODY MOTION.   Our four-jointed shoulder complex articulates hand and arm movements in a 360-degree range. The configuration decouples shoulder movements from the head and neck, enabling torso rotation without having to divert the head. One can therefore maintain forward attention while turning the hips to muster power or to avoid an attack.

IMPROVED ARM AND HAND DEXTERITY.   The upper limbs are no longer needed for balance or weight support. An increased reach and manual dexterity enable building and wielding tools, holding the young for nurturing, and fighting. Our shortened forearms reduce the sway of turning motions and are more adept for counterbalancing the lower body while running or kicking. Accordingly, modern humans can better control our lobbing hand movements with a circular trajectory such as a roundhouse or an open hand strike (haito-uchi, see chapter 13).

The highly evolved human hand has shorter palms and fingers compared to other primates, but proportionately longer thumbs (fig. 8.1). A reproportioned hand with enhanced dexterity and the strength of opposing thumbs enable humans to grip with strength and precision. These attributes are crucial for wielding tools and for performing skills that require intricate manual dexterity.[8]

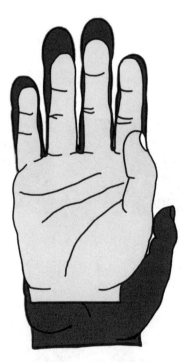

Figure 8.1 Anatomic proportions of the human and chimpanzee hand. The human palm (foreground) is shorter, with proportionately longer fingers as compared with the primate hand (background). The human hand's shorter and more muscular thumb enables a broader range of motion with added dexterity. Shorter finger digits wrapped against the palm improve buttressing in closed-fist punching.

The human hand can deliver the same magnitude of peak force with an open hand or a closed fist. Shorter fingers allow for tight wrapping of the distal and proximal phalanges against the palm, thereby elevating impact energy transfer from the increased stiffness of a closed fist. Effective force penetration doubles when force is delivered on the protruding knuckles (the proximal phalanges) of the index and middle fingers instead of the entire surface of the fist.[9]

A balanced and upright posture can generate more impact force than four-legged animals.[10] A downward strike is aided by gravitational acceleration, thereby garnering more power than quadrupeds' upward or sideways forelimb swipes.

DECOUPLED UPPER AND LOWER BODY MOVEMENTS.   A narrower torso and waist enable independent movement of the upper and lower body. The upper body is able to counteract the twisting reactionary force of the hips when the legs are in motion. Separate movement control enables complex three-dimensional task execution, such as moving forward while simultaneously launching a rotational forearm strike (uchi waza, fig. 8.2) or grappling coupled with a leg throw (ashi waza).

Figure 8.2 Independent control for upper and lower body movements. Dallas karate practitioners Campo Caceres (right) and Vo Dinh (left) demonstrate a forearm strike concurrent with a leg throw.

INCREASED MANEUVERABILITY.   As hominins, we are endowed with proportionately longer legs that extend our stride. Our upright posture also improves maneuverability. The body's turning radius equals the width of our pelvis, whereas quadrupeds turn with entire lengths of their torso. Our pelvis is broader and flatter, which offers lateral stability during movement transition (and possibly to facilitate birth).

Anterior and posterior ligaments fortify the front and back sides of the spinal column. These ligaments run longitudinally from the back of the skull and neck to the thoracic vertebrae, acting like shock absorbers during rapid weight shift. They stabilize the vertebral column during twisting, turning, and lunging motions.

Even for the simple process of walking, our center of mass has to be balanced with one foot on the ground while the opposite leg advances, pulls back, or move sideways. Our inwardly canted thigh bones help to center movements when pushing off the ground.

Intricate foot and ankle anatomies accommodate the necessary biomechanics to achieve traction during movements, balance, and leverage for acceleration.[11] A flexible ankle allows full heel contact to the ground, which gives better support. An enlarged heel bone provides better shock absorption when jumping. Moreover, bones of the human foot form a stiff arch to sustain rigidity. Shorter toes and inwardly angled big toes combine to grip the ground to promote forward momentum transfer.

Human buttocks are overdeveloped compared with other primates.[12] The back-hip extensor musculature comprises the hamstring and the gluteal complex. Apart from connecting the femurs—our body's largest bones—to the pelvis, the gluteal muscles (maximus, medius, minimus) work to sustain upright posture. These muscles on the backside counteract forward pitching tendencies of the upper body during sprinting or jumping. The same musculature sustains our balance during rapid directional change, push-offs during acceleration, and sharp hip-rotating movements (see chapter 10).

As full-time bipeds, our sense of balance is distilled from the finely tuned trifecta of inner ear vestibulum, vision, and proprioception (see chapter 4). These sophisticated sensory inputs enable the martial artist to sustain balance on one leg while the opposite leg executes powerful kicks or sweeps (see chapter 13).

LOWER BODY BUILT FOR POWER.   Quadriceps and hamstring muscles in our thighs, both crucial to bipedal movements, are far better developed than upper-arm biceps and triceps. The thigh muscles promote torque actions from the twisting of the hips and also support lunging movements that are initiated from the back leg.[13]

Meanwhile, the highly developed hip musculature maintains stability of the pelvis during movements. A properly anchored hip architecture promotes biomechanical energy transfer from the legs to the upper body in order to achieve power, speed, and agility (see chapters 10 and 13).

ACHILLES TENDON FACILITATES CENTER-OF-MASS TRANSITION.   Elite athletes achieve remarkable center of mass acceleration by pushing off the ground in each step while sustaining exquisite balance and stability from the hips and core (see chapter 10).[14] The martial artist produces a comparable level of ground-reactionary force by applying sharp heel pressure onto the floor (see chapters 9 and 12). By pushing off with the heel, they achieve the capability to launch their technique explosively, then restore the energy as the foot and toes land ahead of their center of mass. This alternating cycle of plantar flexion and dorsiflexion is facilitated by the Achilles tendon at the heel.

As our largest and strongest tendon, the Achilles work to transmit the explosive power output from the calf muscle to the heel and the foot. The tendon begins near mid-calf, anchoring the calf muscles (gastrocnemius, soleus, and plantaris) to the heel bone (calcaneus). It performs like a spring as the leg flexes and extends during hopping, walking, and running. The tendon compresses as the heel strikes the ground, temporarily storing elastic energy. As the center of mass shifts to the front of the foot, stored energy is released to enable heel lift.[15] The unique energy-conserving feature improves peak running speed by over 80 percent, at the same time reducing energy expenditure by more than 75 percent as compared with nonhuman primates.[16]

INDEPENDENT BREATHING CYCLE.   Unlike four-legged mammals, the human breathing cycle is uniquely decoupled from the stride. A sprinter can consciously hold their breath during a burst of speed in order to consolidate control with their center core. This effort can be sustained until autonomous reflex overrides the anaerobic state.

They can also alternate between an aerobic and anaerobic state by taking a breath for every other stride. We intuitively hold our breath to sustain core muscle tension when lifting a heavy load or while launching a flurry of punches. The practice of conscious breath control is also the hallmark of mindfulness meditation and for maximized physical performance (see chapter 11).

While bipeds cannot match the speed of the fastest quadrupeds with a flexible backbone, we can sustain farther travel distance with better energy efficiency. Our highly regulated homeostasis contributes to physical endurance (endurance-running theory)[17] and sustains peak performance for a longer duration while under duress.

Traversing the Fourth Dimension: Time and Timing

Humans are perhaps the only evolved beings to have mastered the abstractness of present, past, and future. This appreciation goes beyond built-in instincts and conditioned responses. We realize the passage of time objectively from the movements of planets and clocks[18] and subjectively by our body's internal clock (see chapter 6).

With apologies to Einstein and aficionados of the superstring theory, we will consider time as a separate entity from space.[19] The concept of time provides an appreciation of timing, which is akin to being able to arrive at the right place at the right time.

Timing determines the interactivity of self with the surroundings at any given moment (see chapter 5). As applied to self-defense, proper timing is the capability to close in and take control of a distracted (or slower) opponent. This anticipatory skill involves mental fluidity to steer the present and the immediate future (see chapter 7). Mastery of our biomechanical capabilities is also essential for spontaneity in action.

Our body configuration is embedded with capabilities to cover a hundred yards in less than ten seconds, clear a height of over eight feet, or strike with fifteen hundred pounds of force. Contemporary training approaches have enabled modern humans to maximize mental and physical potentials. Yet these insights track remarkable parallels with prescient martial arts practices that elevate mind, body, and spirit. Discussion in subsequent chapters will examine the convergence of these ideas, with the aim to maximize martial arts performance.

Takeaways from "Bipedalism"

- Our bipedal traits give rise to coevolved vigilance and an exquisite sense of balance.

- Bipedal posture increases upper-body range of motion and arm and hand dexterity.

- The decoupling of upper and lower body movements and independent breathing control elevates performance capability for complex motor functions. Our stronger lower-body musculature serves to enhance rapid center-of-mass transition.

- Martial artists should be conscious of our biomechanical strengths and weaknesses in skill development.

Posture and Stance: The Foundations of Martial Arts

As the magician mesmerizes with sleight of hand, they achieve the illusion from the other hand and body maneuvers unawares. Likewise, the martial artist's impressive kicks and strikes are their mastery of biomechanics in action.

Yoko Okamoto sensei, Founder of Aikidō Kyoto in Japan, holds the rank of seventh dan and is one of the best-known instructors in the budō art of aikidō. Throughout international teaching seminars, Okamoto sensei's calm demeanor belies her mental focus, precision, grace, and speed. This diminutive *aikidō-ka* never ceases to amaze as she weaves in and out of attacks by far bigger opponents, tossing them about despite a handicap in size and strength.

For proper technique execution, Okamoto counsels focus on the *tanden* (energy center; see chapter 16), posture, lower body and hip control, and correct positioning.[1] Okamoto's advice mirrors the teachings of the late Morihei Ueshiba, founder of aikidō, that "good stance and posture reflect a proper state of mind."[2] These crucial elements underline the proper attitude in budō arts, wushu as well as combat sports.

Issun no Ma'ai (One-Inch Interval)

Traveling back in time, we peer into the final duel of Musashi Miyamoto, Japan's best-known swordsman. The historic event took place at Ganryū Island in April 1612, having since transcended fact into near-myth.[3]

The challenger, Kojirō Sasaki (also known as Ganryū Sasaki), was regarded as Musashi's most formidable opponent. Sasaki had an undefeated record against the very best. He could rapidly change direction and strike unpredictably. His weapon of choice was the *nodachi,* a long two-handed field sword with a blade of around three shaku (about thirty-six inches).

The nodachi blade has a significant reach advantage to the katana, the two-shaku (about twenty-four inches) blade favored by most samurai, including Miyamoto. Mastery of the nodachi requires incredible strength and dexterity. Sasaki's signature move, the *subame gaeshi* (turning swallow cut), had been lethal to his opponents. The move is preceded by a feint—a missed cut that brings the sword toward the ground. As the opponent advances, the swordsman instantly sweeps his blade upward, slicing his opponent from groin to chin.

Legend has it that Miyamoto arrived with a *bokken* (wooden sword, 木剣) that he carved from an oar during the long boat ride to Ganryū Island. The bokken was intentionally longer than the nodachi.

Miyamoto was three hours late for the duel. Many considered this to be a ploy to irritate Sasaki. The two squared off on the beach in an intense battle. Sasaki came close to victory many times. Then Miyamoto positioned himself with the sun in Sasaki's eyes. He goaded his opponent again, triggering Sasaki's pent-up frustrations for his lack of success. Provoked, Sasaki charged in angrily for another attack. And the duel was over in a single pass.

Sasaki's last thought was his deadly vertical *men* (face) strike. Yet Miyamoto countered and killed him instantly, breaking Sasaki's ribs on the left side and puncturing his lungs.

As Miyamoto walked off, witnesses noticed a trickle of blood running down his face. His *hachimaki* (headband) had been sliced in half, his forehead nicked by the tip of Sasaki's sword.[4] Yet Miyamoto prevailed by superior swordsmanship, wile, and exquisite judgment of relative distance.

The story has been retold countless times with increasing embellishments. It never ceases to enthrall, at the same time informing on the importance of psychology, precision, and Miyamoto's mastery of *ma'ai* (distance, 間合).

Miyamoto is worshipped to this day as Japan's *kensei* (sword saint, 剣聖). Having perfected his skill and strategy with the katana, he developed the two-handed fighting style known as *niten ichi* (two heavens as one, 二天一), the opening presentation of two swords held above the head. This stance became the hallmark of the niten ichi school, which he founded as first described in *The Book of Five Rings*.[5] To date, Niten Ichi-ryū continues as a kobudō (old martial art, 古武道) swordsmanship art in the Kansai region of Japan.

Correct judgment of distance (ma'ai), alongside timing and technique, form the key elements that define budō arts execution. During combat, the measure of distance is not absolute. As both parties jostle for position, a difference in approaching speed at any given time figures into the separating space.

In the unarmed art of karate-dō, practitioners recognize that success or failure in conflict rests on a reach differential of one inch or less (issun no ma'ai, 一寸の間合). Armed budō arts such as kendō consider that life or death rests within a hair's breadth (*kami shi to e,* 死髪と会). It is therefore important to have a strong grasp of relative distance when engaging an opponent. Research studies show that correct distance estimation depends critically on proper posture.

Posture and Preparedness

Our estimation of distance differs by the way we hold our head (head orientation), which in turn is affected by posture.[6] Through correct posture, the practiced precision in gauging separation serves to enhance confidence when intercepting an attack.

In the simplest terms, posture refers to the way we keep our body upright. Awareness of posture gives a proper frame of reference of the physical self with the surroundings, a must for accurate perception and precise action. Visual cues contribute critically to posture reflex, alongside sensory feedbacks from the ears, nose, skin, muscles, and joints, all the way down to our feet.[7] This "top-down" informational hierarchy provides the basis for orienting the center of mass to balance our head and trunk when standing still (static posture) or when the body is in motion (dynamic posture). Proper balance reduces body misalignment that places unnecessary load on the spine. In turn, posture reinforces the visual field, streamlining gaze input and interpretation.[8]

Proper posture is defined by the sustained natural alignment of the spinal column as we stand, sit, or move. In its unstressed state, the spine is shaped like an *S* from top to bottom. Three natural curves are positioned at the neck (cervical), mid-back (thoracic), and lower back (lumbar) (fig. 9.1). The curvatures accommodate fine motor control to sustain dynamic equilibrium of the body frame.

Extraneous joint pressure and muscle tension are minimized in all dimensions when the spinal column maintains its natural curvature. Correct orientation of the head and spine gives a proper reference of the vertical axis. Alignment with gravity brings a strong spatial invariant to sustain balance.[9]

Our information-intensive society increasingly gravitates toward sensory overload and physical inactivity. The inordinate amount of sitting, chronic stress, and oblivion to physical surroundings contribute to bad posture.[10] The common lean-in, head-forward posture, also known as "scholar's neck" or "text neck," chronically strains neck and shoulder muscles as well as the cervical vertebral joints.

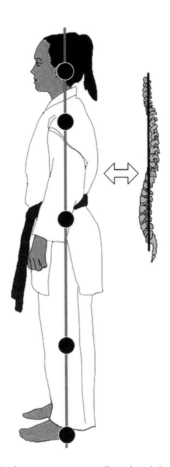

Figure 9.1 A proper posture aligns key joints and maintains natural curvature of the spine. The properly aligned head is held perpendicular to the shoulders. The neck and shoulders are in line vertically with the hips, knees, and ankles.

Joseph H. Pilates, exercise expert and inventor of the Pilates methods in the mid-1900s, firmly believed that his contemporaries' poor health was due to bad posture and inadequate breathing.[11] "Never slouch," preached Pilates, "As doing so compresses the lungs, overcrowds other vital organs, rounds the back, and throws you off balance."[12] These claims were not supported by scientific facts. Nevertheless, there is strong evidence that bad posture places mechanical stress on affected muscles and joints. The pressure on the cervical (neck) vertebrae is increased by 50 percent for every 15 degrees that the head is off alignment.[13]

Posture also affects attentiveness. We tend to be more engaged with our surroundings when standing or sitting straight. Whether lounging on the sofa or lying in bed, we reflexively bolt upright when surprised by an errant sound or when a cat or dog vaults into our field of vision. The act of "springing into action" is ingrained into our psyche as a precondition for vigilance and physical readiness.

Posture Awareness

We normally do not need to attend to our posture, which is sustained by back muscles and hamstrings. However, day-to-day issues such as excessive biomechanical demands, chronic stress, and overexcitement during a fight-or-flight response contribute to postural imbalance, as the body compensates for additional breathing requirements and muscle hyperactivation.

Posture awareness and our state of attention are managed by an intricate system of neural controls and feedbacks. A mesh-like neural network called the reticular activating system (RAS) connects the information-processing forebrain (cerebrum) with the hindbrain (cerebellum) for procedural memory recall, coordinating posture and balance with voluntary skeletomuscular movements. RAS centralizes in the brainstem (fig. 5.2) and ties together our three attention networks. The RAS network also relays activating motor signals downstream by way of the spinal cord (see chapter 4). This neural-coordinative infrastructure sustains readiness and binds postural control to awareness, problem-solving, and motivation.[14]

As our senses are flooded with stimuli from the surroundings, RAS acts the gatekeeper of cognitive awareness, directing our attentional networks to focus on information deemed worthy of response.[15] Conscious thought is based on one-one-millionth of the sensory data that has been filtered through RAS, which is then given to further scrutiny and subsequent action.

Meanwhile, proprioceptors in key muscles and joints continuously inform on positions and movements of our limbs. Sensory signals converge from all parts of the body (skin, muscle, organs; the somatosensory system) at any given instant (see chapter 4) through ascending nerves to the higher-functioning forebrain. Peripheral sensory feedback is reinforced by visual and vestibular (inner ear) information to sustain body alignment through fine motor adjustments. Collateral control minimizes extraneous corrective actions, thereby streamlining motor responses.

Back in 2004, Professor Robert Skinner determined that the RAS is the key driver of fight-or-flight responses (see chapter 5).[16] Its intricacy likely evolved from primordial selective pressures to survive from predators by linking posture to elevated awareness, perception, and problem-solving.[17] When an alerting stimulus rallies our attention and voluntary actions, it also rouses postural and locomotor controls. Conversely, fight-or-flight malfunction has been tied to posture and gait abnormalities.[18] Martial arts training is grounded on the ability to arrive at a viable solution instantaneously. This practice elevates RAS's capacity to properly screen our informational inputs (see chapter 7). By attending to our posture, we alert the RAS system to home in on matters of immediate importance.

Postural Demands in Martial Arts

A martial artist prepares for the situational demands of continuously evolving adversities. This skill set calls on mental and physical resources, both of which involve attention to posture, in particular an awareness of the following considerations.

VIGILANCE.   An awareness of our stationary posture is necessary to sustain vigilance. Proper posture reflexively defaults to a steady gaze, promoting sentience rather than fixation on events in propinquity. Our peripheral vision is engaged, which alleviates tendencies of tunnel vision.[19] As our mind welcomes comprehensive sensory inputs, it accommodates posture-dependent proprioceptive feedback to fine-tune movements (see chapter 4).

DYNAMIC EQUILIBRIUM.   The martial artist requires a heightened awareness of upper and lower body placement to achieve complex motor functions, such as moving in and out of position while parrying or striking with arms and legs. Correct posture offers consistency in positional reference during rapid defense and counterattack measures.

OPTIMIZED BIOMECHANICS.   Proper posture serves to optimize power and speed. Force transfer that is mediated by hip rotation (angular momentum), or forward- or back-shifting of the body center (linear momentum), is most efficient when the hips are aligned in the same vertical plane as the shoulder cuffs (see fig. 10.5). This configuration stabilizes the spine and enables balanced core performance by the front abdominals and side and back muscles.

Unhindered by the need for posture correction, our body's core components can now focus on transmission of power and speed and on sustaining movement precision (see chapter 10). A properly aligned spine and pelvis streamline the airway, maximizing air intake and exhalation to sustain an aerobic state.

SHOCK ABSORPTION.   The natural spinal curvature helps to distribute shearing pressure on individual vertebral discs. Anatomically, each lower disc is only required to partially support the weight of its upper vertebra. Correct posture ensures proper performance of this shock-absorbing springlike function during movement transitions such as torso twisting or jumping, serving to reduce vibration and shock, and minimizes wear and tear.

This attribute takes on increased relevance when having to redistribute the impact shock and vibration from an external load, from wielding a weapon like the katana (samurai long sword) or *bō* (fighting stick), to leveraging one's center mass to upset the balance of an opponent.

The Barefooted Martial Artist

Training in bare feet is a long-held tradition in Japanese and Okinawan martial arts. Qigong practitioners also exercise in bare feet or wear thin and flexible footwear when outdoors. Likely explanations vary from the profound to the mundane, starting with the cultivation of energy flow.

For centuries, southern Chinese gōngfu practices have been vastly influenced by Daoist thought. This philosophy subscribes to awareness of inner energy channeling, seeking to nourish human qi *(ren qi)* with those from earth *(di qi),* and cosmos *(tian qi)* (see chapter 16). Bare soles of the feet were believed to enhance the channeling of di qi into the body.

The practice also makes scientific sense. Even over the mundane task of walking, the brain requires an accurate assessment of ground position (and its horizonal slant) in relationship to the body. Our brain relies on continuous positional feedback to sustain posture during movement, in part from proprioceptors that inform on joint position, joint angle, and the change thereof. In particular, significant input originates from exteroceptors on the soles of our feet that are sensitive to touch, pressure, and vibration (see chapter 4). Accordingly, bare feet heighten awareness by informing on the subtle changes in balance as well as body position.[20]

Exteroceptor signals alone can trigger adjustments for posture and body orientation well before visual confirmation.[21] Proprioceptive feedback from the front-foot heel strike to the ground sets off a practiced chain of events (kinematic chain; see chapter 10) for streamlined motion. As we wander in bare feet in the house, these heightened sensations elevate awareness of our equilibrium when we are in motion.

Barefoot training also brings attention to foot anatomies (heel, ball of the foot, toes, inside and outside edges) and their contributions in kicks, sweeps, and throws (see chapter 13). With repetition, the martial artist builds strength in skill-specific foot muscles in coordination with the leg muscles. We learn to appreciate the subtle but important balancing process of toe grip and release during movement transition. Push-off with the toes aids forward propulsion and promotes traction in hip rotation. One becomes more conscious of foot placement to leverage lower body muscle actions in order to maximize speed and the generation of force.

The Concept of Stance

Stance is defined as the geometry of the hips, knees, and feet in reference to the body center. Stance practice dates back to the genesis of fighting arts, evident in ancient artifacts since the Greek Hellenic era, circa 440 BCE.[22] The general public associates martial arts as fighting techniques in deep stances, some of which are seemingly too low for mobility.

This misperception has to do with the training expectations in Chinese wushu and Japanese budō arts. Both emphasize sinking of the body's center of mass to better understand stability and balance control (particular when wielding a weapon; see chapter 10). The practice also conditions lower-body muscles and joints for generating maximized reactionary force from the floor.

Most styles incorporate full-length as well as shorter stances to address differing operational requirements (figs. 9.2 and 9.3). After a thorough understanding of stance functionality, a martial artist can realize the same biomechanical advantage while foregoing the low, external configurations.

Naihanchi-dachi (内畔戦立) is one of most recognizable stances associated with the weaponless art of karate.[23] Its name reflects application for close-range (*nai,* 内) sideways (*han,* 畔) combat (*chi,* 戦) techniques. Movements are performed with straddled feet, summoning total body power in an immovable posture (fig. 9.2).

The Naihanchi kata were popularized by karate master Sōkon Matsumura and his student Yatsutune (Ankō) Itosu in late-nineteenth-century Okinawa. Though small in stature, Itosu had incredible speed, was as strong as a bull, and possessed an uncanny ability to withstand blows. The three Naihanchi were among Itosu's favorites, all performed in the stance of their namesake, the Naihanchi-dachi.

When Ginchin Funakoshi brought these kata to Japan, he renamed them Tekki (iron ride, 鉄騎)[24] to call upon the imagery of an immoveable armored samurai on horseback. Naihanchi-dachi was revised to Kiba-dachi (Calvary Horse Stance, 騎馬立) for correspondence. The lower, wider version (fig. 9.2) was intended to strengthen leg and core muscle control through practice.

Figure 9.2 Naihanchi-dachi and Kiba-dachi. Left: Historic rendition of Naihanchi-dachi by Choki Motobu sensei. Right: Demonstration of Kiba-dachi by the author. Traditionally, the width of Naihanchi-dachi is about one to one and a half times shoulder width; the Kiba-dachi about twice shoulder width.

Tekki kata inculcate enormous body power through explosive hip actions without repositioning the center of mass. Mastery of Naihanchi or Tekki instills an unyielding spirit and powerful technique execution harnessed from successive joint compression from the hips and through the knees and ankles to generate stance pressure (see "Stance Pressure and Biomechanics" on page 115). The resulting ground reactionary force is transmitted into the shoulders, elbows, and wrists. There is strength in spirit from holding one's ground, giving no quarter to the opponent.

Stance Classifications

Different wushu and budō styles classify stances according to their own preference. Nomenclature in budō arts was based on the configuration of the legs and feet or a simile. For example, the literal translation of Forward Stance (Zenkutsu-dachi, 前屈立) is "front leg bent-posture"; Back Stance (Kōkutsu-dachi, 後屈立), "back leg bent-posture"; High Back

Stance (Nekoashi-dachi, 猫足立), "cat foot–like posture"; and so forth. Fine details such as foot angle and toe positions are elaborated during training to ensure balance and stability.

Stances in companion to the twenty-six Okinawa-te kata introduced to Japan by Gichin Funakoshi were among the most curated in budō arts. They can be divided into close-range (the opponent is within arm's reach) or full-range (the opponent is more than a step away) stances according to intended applications (table 9.1). Close-range stances tend to be higher and narrower. Each category incorporates outside-tension stances that seek to optimize power and stability and inside-tension stances that favor mobility and quick body transition.

An outside-tension stance asserts floor pressure by consciously canting the knee and ankle joints away from body center. The structure is fortified by quadriceps, hamstring, and calf muscle tension. Inside-tension stances draw the knees and ankles inward toward the body center. This geometry is supported by squeezed inside thigh muscles and hip flexors (see chapter 10). Both types are reinforced by core muscle activation to assert a downward pressure toward the ground (chapter 10).

Table 9.1 Commonly practiced stances in budō karate-dō.

STANCE		TENSION*		CONFIGURATION**
		OUTSIDE	INSIDE	
CLOSE-RANGE STANCES				
Hachiji-dachi	Natural Stance (八字立)			
Kiba-dachi	Calvary Horse Stance (騎馬立)	✓		
Nekoashi-dachi	Cat Foot Stance (猫足立)		✓	
Sanchin-dachi	Three Battle Stance (三戦立)		✓	
Shiko-dachi	Square Horse Stance (四股立)	✓		

*Outside-tension stances involve the application of outward pressure and outward canting of knee and ankle joints that is stabilized by the quadriceps; inside-tension stances involve inward pressure and canting of knee and ankle joints that is stabilized by the inside thigh (abductor longus, psoas major) muscles and hip flexor tension.**The circle denotes the position of the center of mass; the lighter footprint denotes less stance pressure compared with the opposite foot.

FULL-RANGE STANCES				
Hangetsu-dachi	Half-Moon Stance (半月立)		✓	
Kōkutsu-dachi	Back Stance (後屈立)	✓		
Sōchin-dachi or Fudō-dachi	Sōchin Stance (壮鎮立) or Immovable Stance (不動立)	✓		
Zenkutsu-dachi	Front Stance (前屈立)	✓		

Stance: The Root of Posture and Power

The key measures of an athlete's power output are their muscular force and the speed of transferring that force onto the target.[25] In any action, biomechanical force and speed outputs bear an inverse relationship to one another (see appendix B). A baseball player can generate a faster swing with a lighter bat. The martial artist's punch is faster than their kick, which is slower but delivers more force because of the mass of the kicking leg.

Punching and kicking derives power from acceleration of the center of mass, accomplished either through body rotation, forward and sideways shifting, or dropping and rising of the body center (see chapter 10). However, these processes are woefully inefficient without leveraging stance functionality, as further elaborated below.

1. Stance Versus Stance Pressure

Movies portray the inept martial artist as someone who assumes a stilted pose, then bounces about without purpose. The audience instantly recognizes their limitations from the lack of stability, dynamics, or control, which are the key features of a proper stance.

Stance without dynamic pressure is like pitching a tent without guy wires. While configuration is framed by the tent poles, staked multidirectional guy wires add dynamic tension that sustains the structure and counteracts gravitational force.

Proper stance starts with our skeletal framework, which provides weight support by the geometry of the front and back legs, and proper placement of the center of mass within the perimeter. Differential activation of core and lower body muscles (and gravity) creates dynamic tension to reinforce this structure.

The Shōtōkan front stance (Zenkutsu-dachi; see table 9.1) favors the front leg with a 90-degree bent knee. The back leg is extended, its knee slightly bent to accommodate the rotational biomechanics of the hips and to promote back-leg thrust during movement transition (fig. 9.3). Front and back legs apply differential downward pressure at a ratio of 70:30 front to back. The activation of knee and ankle-joint compression is biased toward the front, stabilized by isometric activation of core muscles. The body center is positioned at approximately one-third of the distance toward the front.

Dynamic stability resulting from this differential leg muscle tension offers a division of labor for technique execution. The front leg caters to balance support and anchors hip movements. The back leg is freed to drive forward, propel directional change, or execute kicking techniques.

Figure 9.3 Shōtōkan stances by master instructors of the Japan Karate Association, circa 1974. From right: inside-tension stances: Hangetsu-dachi (Masatoshi Nakayama) and Nekoashi-dachi (Hidetaka Nishiyama); outside-tension stances: Sōchin-dachi (Teruyuki Okazaki), Kōkutsu-dachi (Rajiro Mori), and Zenkutsu-dachi (Kenichi Haramoto).

Conversely, the back stance (Kōkutsu-dachi) favors the rear leg for balance support. The back knee is turned away from the body center and bent at close to 90 degrees (see table 9.1). This configuration promotes pressure application from the back leg (a 40:60 ratio front to back) to achieve the compression of hip and knee joints. The center core is recessed toward the back, placed at approximately two-thirds the distance from the front foot (fig. 9.3). Kōkutsu-dachi accommodates backward transition during defense (see "Center of Mass Transition," below), also freeing the front foot for kicking, sweeping, or body-pivoting actions.

Both Zenkutsu-dachi and Kōkutsu-dachi are outside-tension stances that emphasize stability, and they are commonly engaged for delivering powerful techniques. By comparison, inside-tension stances such as Hangetsu-dachi or Nekoashi-dachi (see table 9.1) promote quick body-shifting and positional changes.

The various stances activate different mid-foot muscles and tendons. Inside-tension stances cant both legs inward, emphasizing dynamic control with the foot's inside edge. Outside-tension stances exert pressure with mid-foot muscles and tendons on the outside edge and correspondingly aligned knees to the outer perimeter. Stability is reinforced by floor-gripping toes in both instances. Through deliberate practice in bare feet, the martial artist entrusts subtle proprioceptive feedback to trigger spontaneous body movements.

2. Stance Pressure and Biomechanics

In a blog post titled "Throwing a Cross Punch—Body Movement Analysis," health and fitness expert Antonio Srado divides the complex movement of a boxing counterpunch into five phases.[26]

In phase I, the boxer starts with an orthodox fighter's stance (left foot forward, right foot back, knees bent) to prepare for a right-hand cross punch. Their posture is stabilized by compression of the core muscles and joints.

The fighter intensifies the isometric contraction of core and back-leg calf muscles as they go into the preparatory phase (phase II). Unbending the back knee, the boxer drives their weight into the ground sharply, pivoting their hips to transmit waves of kinetic energy onto the shoulders (phase III, movement). Impact energy penetrates the target through extension of the punching arm (phase IV, impact and follow-through), as the boxer quickly recovers back to the fighting stance (phase V, recovery).

Up to 80 percent of a boxer's punching power comes from the legs and torso, according to a study of Russian Olympic athletes. Close to one-half of this power is derived from sharp floor pressure application that overcomes the moment of inertia. Angular momentum through hip rotation contributes another 40 percent.[27] Accordingly, thrusting momentum from the punching arm accounts for only 20 percent of the total punching power.

Striking and kicking actions in traditional martial arts differ in trajectory and execution (see chapters 10 and 13). Remarkably, torso and lower body musculature contributes the same level of biomechanical outcome.

Sports physicist Randall G. Hassell divided the karate counterpunch (gyaku-zuki) into three distinct kinematic phases.[28] Starting with the driving motion phase (phase 1), the body recruits the necessary power source to set the technique in motion, including the application of muscle tension from the legs and hips to generate a ground reactionary force (see chapter 12). Differential core muscle activation brings directional velocity through the hips and toward the target (phase 2, the speed phase). Integrative mental and physical efforts work to deliver force and momentum at the impact phase (phase 3).[29]

Astute readers may notice analytical parallels with regard to the critical roles of lower body muscles and joints to power upper body movements. Stance dynamics accounted for up to half of the gyaku-zuki's technique power and speed, according to Hassell, which is in line with the findings on Russian Olympic boxers.[30] These practices are as valid today as centuries-old precepts of tai chi and aikidō. Both disciplines preach that "power is rooted in the feet, released by the legs, controlled by the waist, and expressed by the hands."[31]

3. Foot Placement and Technique Delivery

Proper stance and foot placement play key roles for meeting specific demands in technique applications. Without breaking posture, stance determines orientation of the upper body, while front-foot placement establishes the reach of hand and foot techniques. Differential stance pressure governs the apex of peak power delivery in reference to the frontal plane (fig. 9.4).

For example, the 70:30 pressure bias of Front Stance (Zenkutsu-dachi) favors momentum transfer from back to front leg; hence, it is optimized for forward force delivery. The side-to-side configuration of Kiba-dachi limits the reach to the front but offers the capability of maximal force delivery to either side.

Support-leg placement determines the effective distance of kicking techniques. Kicking force delivery hinges on core muscle control at the time of impact. Balance and stability of the hips and spine are essential for momentum transfer (see chapter 13), resting entirely on the sustained balance of the One-Legged Stance (Sagiashi-dachi). Otherwise, the kicker would bounce off the target on impact due to a lack of body integrity. Blazingly fast transitions are managed by skeletomuscular compressions at the hip,

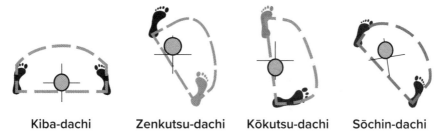

Kiba-dachi   Zenkutsu-dachi   Kōkutsu-dachi   Sōchin-dachi

Figure 9.4 Stance determines the apex of optimal force delivery. Stance configuration is a key determinant of upper body orientation. Front-foot placement defines the reach of hand techniques. Differential stance pressure governs the apex of power delivery (the apex of the dotted curved line). Circles denote the position of the center of mass. A lighter footprint denotes lower stance pressure as compared with the opposite foot.

knee, and ankle joints, and reinforced by full sole-of-the-foot contact and toe grip with the ground.

4. Deashi

A sprinter's launch exemplifies the well-practiced sharp-foot pressure to harness ground reactionary force (fig. 9.5). As the athlete receives the "ready" and "set" commands in the starting position, they compress their major joints like coiled springs. Their weight is mostly supported by the front leg, while the muscles, ligaments, and joints of the rear leg are tasked with producing directional force vectors.[32]

At the crack of the starting pistol, the intensely focused sprinter explodes off the mark, launching off the starting block with a sharp push of the back leg (F_y, fig. 9.5). They maintain core muscle compression to consolidate center-of-mass acceleration. These efforts counteract gravitational pull (F_g). The horizonal force vector (F_x) overrides the moment of inertia to achieve forward acceleration.

As the athlete continues to accelerate down the track, each step makes contact with the ground for less than one-tenth of a second. This split-second change in momentum creates a reactionary force that is more than two and a half times their body mass.[33]

For centuries, the martial artist has utilized the same mechanical advantage to achieve forward impetus for a straight punch (oi-zuki, fig. 9.6). Lacking the benefit of a starting block, the martial artist makes use of their understanding of stance pressure to achieve speed of movement. The harnessed ground reactionary force enables a start to technique impact in less than a second and can reach a peak punching velocity of more than 10 meters per second.[34]

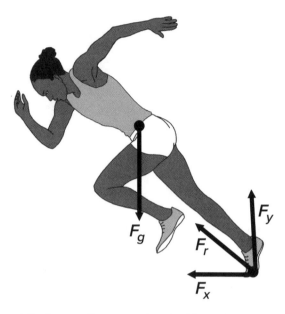

Figure 9.5 Center of mass acceleration. The sprinter uses ground reactionary force (F_r) to power their explosive launch and acceleration. F_y, the vertical component of F_r, counteracts gravitational pull (F_g), while the horizonal component (F_x) produces forward impulse.

Figure 9.6 Center-of-mass acceleration in karate technique.

Back in the 1970s, U.S. audiences were introduced to Karate World Championships, hosted by the All-American Karate Federation (AAKF), now known as American Amateur Karate Federation. They were instantly mesmerized by the remarkable speed and precision of the JKA competitors from Japan.

Time and time again, these athletes charged into their opponents' personal space. They could complete their techniques while their stronger and bigger opponents were stuck in windup.

"What I particularly noticed about the contestants was their slowness both in the start of their techniques and in switching from one technique to the next," commented Master Masatoshi Nakayama after the First World Friendship Tournament in 1971. "Contestants from Great Britain and the United States all had powerful and variable techniques. But ... they were often beaten to the draw or hit between the techniques by the Japanese contestants."[35]

Deashi (start, 出足) is the tactic of an explosive launch initiative when engaging the opponent. Mastery of deashi was a key determinant for the Japanese athletes' success. As the opponent presented an opening for sen timing (see chapter 6), the Japanese athletes' quick starts through stance pressure and release bestowed an early arrival, catching the opponent in mid-thought. Apart from a high level of vigilance (see chapter 7) and technical command, their stance-enabled acceleration gave the defining, competitive edge to the smaller Japanese athletes.

5. Technical Precision

Disciplined practice instills the hierarchical order of technique execution, from leveraged ground reactionary force to its transmission through the hips and into the striking arm or leg. Stance is the base that stabilizes the biomechanical cascade for kicking and punching. Proper posture and connectedness to the floor work together to sustain body-center control, which minimizes the impulsivity to reach with the limbs.

Athletes tend to rely on upper body strength as they lose composure from intense competition. This physiologically "turtling" by shrugging the shoulders amounts to a loss of body integrity as upper-body actions disconnect from stance-based stability. Wild swings of the arm or foot commit one's body momentum prematurely, disrupting posture and balance. These practices hasten the point of no return and are detrimental to technical precision.

6. Center of Mass Transition

Tai sabaki (defense via body movement, 体捌き) is a common maneuver in karate sparring. It refers to repositioning of the body away from an opponent's reach. Tai sabaki can be accomplished by retracting from a full-distance stance to a close-range stance. One example is to withdraw the front foot from front stance (Zenkutsu-dachi) to a cat foot stance (Nekoashi-dachi, fig. 9.7) without moving the back foot.

The defender pulls back the front leg sharply by squeezing the inside-thigh muscles, leveraging off their back-leg floor pressure. Their body center is now drawn backward near the back leg and away from the opponent.

This foot movement is facilitated by contracting the back hip and inner thigh muscles (see chapter 10). The defender in effect switches from an outside-tension full-distance stance into an inside-tension close-distance stance (see table 9.1) through the controlled compression of the back-leg joints.

Tai sabaki is commonly coupled with a well-timed blocking technique (top frame in fig. 9.7). The defender soon returns to a full-distance stance, reclaiming their territory before another attack arrives. Their balance, now reinforced by front-leg stance pressure, enables a stabilized counterattack (bottom frame in fig. 9.7). These events unfold in a matter of seconds, requiring the fluidity for back, then forward, weight redistribution.

7. Heel-Toe Pressure Transfer

Our natural walking motion adopts a rolling heel-to-toe stride. Primary support moves from the heel to the ball of the stationary foot in each advancing step. The same process is repeated by the leading foot as it lands on its heel and then pushes off about eight to ten inches ahead from the ball of the foot.

Researchers proposed that this cyclic motion replicates the swing of a compound (jointed) inverted pendulum.[36] Like the weight at the top of a metronome, our center of mass near the pelvis functions as the pendulum's weight (bob). Heel-toe motion effectuates the pendulum arc, moving the center of mass forward from an equilibrium position that is aligned at mid-foot. This rocking motion, when it extrapolates from a virtual pivot point several inches below ground, presents the geometry of a longer "virtual leg," in effect enhancing the biomechanical efficiency of our physical form.[37]

The reflexive heel-to-toe gait has been integrated in martial arts practice in our favor. The quick bending (flexion) of ankle and knee redistributes weight support from the heel to the ball of the foot, enabling accelerated forward motion without a change in posture. The reverse process facilitates back motion from sequentially extending the same joints.

A low and rooted stance offers stability but tends to reduce mobility, thereby hindering rapid positional changes. Over the course of continuous maneuvering, the martial artist overcomes this limitation with a subtle pressure-and-release rhythm against the floor. Maximum stance pressure is applied at the beginning and end of a technique.

Figure 9.7 Stance transition in defense and counterattack. The defender (right) withdraws their center of mass by pulling back the front foot into Nekoashi-dachi. Sustained back-leg floor pressure gives impetus for an open back-hand block *(jodan haishu-uke)*. Using the recoil energy from engaging the face punch, the defender recovers into a Front Stance (Zenkutsu-dachi) to execute a counterattack (jodan gyaku-zuki).

Lower-body joint compression signals the start of a kinematic chain of events (see chapter 10). This process is repeated at the time of impact, together with assertion of total body muscle tension (*kime*, "focus" or "cut," 決め). Kime bolsters biomechanical efficiency through sharp momentum transfer onto the target (see chapter 12).

A seasoned practitioner can apply the same stance pressure from a high fighting stance. Like the elite sprinter, they peak and then sharply release floor pressure in no more than one-tenth of a second.[38] The uncoiling of their lower-body muscle tension channels the explosive reactionary force to power their techniques.

8. Shock Waves of Impact

Impact feedback is integral to traditional karate training so that the practitioner becomes aware of the sensation of force delivery and recoil energy that returns from a struck target back to the body. This is often carried out with a makiwara (training post, 巻藁) (fig. 9.8), a rudimentary contraption that is made of a long slender wooden board anchored into the ground. Impact strikes are directed to the top of the board, wrapped traditionally in canvas over natural rough hemp cords to blunt the impact and minimize slippage of the fist.

Depending on construction, anchored makiwara vary from high to moderate resiliency. A pliable version is recommended for novices, as flexing of the board absorbs most of the transferred momentum. Over time, the karate-ka trains with increasingly stiff devices that are braced to give immediate feedback.

Cadenced, repetitive punching with a closed fist is the basic makiwara drill, as the student practices gyaku-zuki, the quintessential karate counterpunch that utilizes hip rotation, launched in stationary Zenkutsu-dachi with the opposite leg forward (fig. 9.8).

Kime is critical to the exercise. It is the conscious effort of skeletomuscular activation throughout the body at the split second of impact. The karate-ka stabilizes their hips through sharp downward stance pressure, followed by deliberate compression of the major joints from the legs, through the hips and shoulders, and through the punching arm. They focus mentally on the tanden, modulating their center-core muscle tension (intra-abdominal pressure) to send energy into the target (see chapter 11).

As the back of the makiwara stiffens at the limit of mechanical compression, an equal and opposite recoil force reverts along the same skeletomuscular path. Connectedness to the floor, when coupled with proper kinetic chain performance (see chapter 10), channels this recoil energy back into the ground, mitigating joint injuries or muscle sprains.

This uncanny phenomenon has been recorded by high-speed photography, when Shōtōkan karate masters were invited to take part in a 1970 study at the University of Oregon.[39] As these experts struck a mounted wooden board with gyaku-zuki, cameras captured their kinematic flow at over 250 frames per second.

Figure 9.8 Impact training with the makiwara. Historic rendition of Gichin Funakoshi sensei in makiwara practice, circa 1924. Correct posture and body alignment are essential in impact training to ensure proper force transmission through the hips and to the arm. Contemporary practices adopt the Zenkutsu-dachi with the front knee bent.

Slow-motion analyses revealed sequential skeletal compression that originated from the legs, converging through the torso and into the punching arm. A reversing shock wave emitted from the resilient mounted board immediately followed. Involuntary joint compressions now came in reverse order, coursing back through the arm and the body, the hips, and then the knee and ankle joints.

A second wave of sequential involuntary muscle contractions instantly returned toward the target. Oscillating forward and reverse cycles of shock waves were observed until the impact energy was dissipated completely.

The struck boards flexed on initial impact, then broke into two only after having received the second and third impact waves of energy. The long-held belief of stance-based biomechanics for impact was henceforth validated through high-speed photography.[40]

Shōtōkan karate practitioners consider back-heel contact to the ground as a necessary component to achieve this mechanical phenomenon. The stable base facilitates the transfer of biomechanical recoil back into the ground, then toward the target again. Dorsiflexion of the back foot and ankle (which brings the foot toward to the shin) increases balance and stability. Downward heel pressure has been shown to promote sharpness in hip rotation by coordinately activating calf muscles and hamstrings.

9. Rituals to Promote Cognitive Recall

When placed in a high-stress situation, professional athletes engage in repetitive personalized rituals as a way to calm themselves. These practices "warm up the brain" (see chapter 7) by priming our attentional networks toward the impending task.[41]

For the same reason, a martial artist focuses their attention by adopting a personalized kamae (opening position) when facing an opponent. Then they redistribute their stance pressure continuously between front and back feet. This self-imposed rhythm reinforces kinesthetic feedback, heightening awareness of the body's center of mass, hips, and other major joints. Proper posture ensures precise distancing between self and opponent. The mind is unburdened from anxiety as it continuously monitors connectedness to the floor while seeking the optimal position for engagement. The frontal cortex now attends to the state of heightened vigilance, rallying for appropriate mental solutions.

Our body is a highly sophisticated and versatile machine, adept at performing a broad range of physical tasks. Without the infrastructure of posture and stance, it would be physically untenable to meet the overarching power demands of martial arts techniques.

For centuries now, Daoist and Zen Buddhist thought have subscribed to the concept of the tanden (lower *dantian* in Chinese), a hypothetical energy center within our body. Positioned below the navel within the abdomen, the tanden is considered the seat of our physical consciousness. The tanden corresponds to the contemporary interpretation of our static center of mass, situated at the front of the second sacral vertebra, cradled by the pelvis and braced by core muscles. Center-core integrity is universally recognized as key for achieving peak athletic performance, as the robust musculature of the hips, thighs, abdomen, and back orchestrate upper-body movements in coordination with the lower extremities. Discussions in the next chapter will examine the role of these muscles in transmitting speed and force.

Takeaways from "Posture and Stance"

- Awareness of stance and posture gives rise to a proper state of mind. These crucial elements underscore the attitude in budō arts and wushu as well as combat sports.

- Correct posture enhances confidence through precise distancing against an attack. Proper posture reflexively defaults to a steady gaze, promoting sentience rather than fixation on events in propinquity. Properly aligned spine and pelvis optimize air intake and exhalation. The transfer of force from the floor to the target is most effective when the hips are aligned in the same vertical plane with the shoulder cuffs.

- Chinese wushu and Japanese budō stress the sinking of the body's center of mass for balance and to maximize reactionary force generation from the floor. Having mastered the fundamentals, the martial artist can realize these mechanical advantages without resorting to the external form of a low stance.

- By focusing on the dynamics of stance pressure, we can achieve sharp acceleration of our center of mass. Heel pressure to the floor and the coil-spring effects of lower-body joints and musculature create waves of penetrating force into the target.

Hips, Core, and the Kinematic Chain: The Dynamos of Performance

Kinematic Chain of Performance

The coaching staff of professional sports monitor the performance of their million-dollar players religiously. Each team jealously guards their analytical algorithms to preserve a competitive edge. Yet there is little doubt that kinematic chain assessments are high on the checklist.

The kinematic chain concept (also called kinetic chain) was introduced in 1875 by Franz Reuleaux. This engineering concept later expanded to study movement efficiency in athletic performance, sports medicine, physical therapy, and prosthetic design.[1]

The model considers the human body as a rigid, segmented structure (bones of the skeleton) linked by serial joints and attached with soft tissues (muscles, fat, organs, and so on). Segmental movement at the joint sets off motion for the contiguous parts. Complex motor functions represent the outcome of sequential joint movements. An orderly activation is required to sustain biomechanical efficiency.[2]

Segmental movement is founded on the contractile functions of attached, striated muscles. Muscles of the torso, chest, and particularly those of the legs can sustain a semi-contracted state for extended periods, enabling them to store elastic energy like compressed springs. This "coiled spring effect" primes muscles to contract or expand explosively, such as when the martial artist pushes off the ground to launch forward their center of mass.[3]

Each segmental movement generates force. Strength and directionality of the force are transmitted to the adjoining segment.[4] Accordingly, movements of the upper extremities benefit from sequential skeletomuscular activation from the legs, hips, and torso (fig. 10.1).[5]

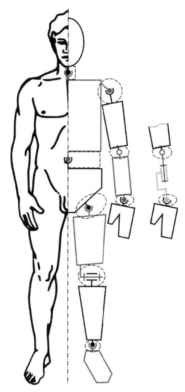

Figure 10.1 Kinematic chain model in athletic performance. A biomechanical model of the human body as serially arranged, rigid, and overlapping segments connected by joints. Effectiveness of complex motor functions depends on the sequential and orderly activation of intervening segments.

The force imparted on a target represents the final outcome from all contributing joints along the kinematic path. Final velocity is the end result of accelerated momentum transfer from major joints (hips, shoulders) to smaller joints (elbows, wrists) of the extremities.[6] Exercise physiologists consider feet and ankles, knees, hips and pelvis, shoulders, and head and neck as the five pivotal junctions in the body's kinematic chain.[7] Their coordinated performance critically defines biomechanical output.

The tension of attached muscle groups also maintains stability and ensures structural integrity. Interceding joints are susceptible to mechanical stress if kinematic transfer is disrupted. A lack of postural control or upstream segmental overcompensation (shoulders, arms) decouples dynamic performance and increases the likelihood of injury.[8]

Our intrinsic kinematic chain starts at the naturally stronger lower-body musculature. Leveraging resistance to the floor is highly effective for generating force and speed (see chapter 9). The process represents complex (involving more than two muscle segments) kinetic chain events that are "closed" when the initiating segment is anchored to an immoveable area (the "distal aspect"). Most traditional martial arts techniques engage complex, closed kinetic chain execution, starting with the heel and ankle pushing against the ground.[9]

The well-studied motion of the overhand baseball pitch illustrates this process (fig. 10.2).[10] The pitcher starts the windup phase while perched on their dominant leg, facing sideways away from the batter. Their explosive release culminates from sequential muscle-unit activations of the whole body.[11] The pitcher initiates the sequence by extending the leading foot (the stride phase). They propel their center of mass forward as soon as the leading foot lands, then rotate their hips to channel ground reactionary force up the kinematic chain (the arm cocking phase). Pushing off the mound with both feet, the pitcher uses hip rotational torque to turn their throwing shoulder (the acceleration phase; fig. 10.2).

The pitcher brings their throwing elbow away from the torso (external shoulder rotation) to complete the release trajectory. Hip and shoulder rotation powers the elbow forward, launching the ball as they halt hip movement abruptly (the deceleration phase). The pitcher completes the release arc by sharply drawing in their elbow (internal shoulder rotation), at the same time recovering their balance (the follow-through phase).[12]

The serial mobilization of engaged body segments is completed at lightning speed. The entire process from pitch windup to follow-through takes about three seconds. Less than two-tenths of a second elapses between landing the leading foot to ball release.[13] The accelerated arm releases the ball within hundredths of a second. The throwing shoulder completes a 60-degree inward arc at an angular velocity of 7,000 degrees per second, which represents one of the fastest human motions in sports.

| Windup | Early cocking | Late cocking and acceleration | Deceleration and follow-through |

Figure 10.2 Motion of a baseball pitch.

Closed kinematic chain movements permeate human athletic endeavors, from the aforementioned baseball pitch and batting to tennis, golf, and track and field events such as javelin and shot put. Highly sophisticated models have been designed in each instance to optimize the contribution of individual joints.[14] While individual segments do not contribute equally, each adds successively to the dynamic final outcome from ankle to wrist, a phenomenon known as "summation of speed."[15] These concepts are entirely consistent with the modus operandi in traditional martial arts techniques (see chapters 12 and 13).

An athlete's intention governs the trajectory of technique execution. The baseball pitcher's objective is to hurl the five-ounce baseball into the distance with speed and accuracy. Martial arts techniques are directed at delivering force and momentum on contact, whether it's a karate punch or a judō throw. Nonetheless, the effectiveness of both is founded on closed kinematic chain activations.

In particular, the segmental skeletomuscular recruitment of a karate counterpunch (gyaku-zuki) has been scrutinized by high-speed motion capture photography.[16] As the karate-ka drives their back foot into the ground, energy waves travel up the body, sequentially releasing skeletal joint compressions at the ankle, then the knee, and into the hips (see fig. 4.2). The power of hand techniques is derived from the hip-twisting angular momentum that accelerates the upper torso and shoulders, propelling forward extension of the upper arm through the elbow joint to launch the lower arm and fist.[17]

This masterful event produced up to 1,500 pounds of punching force, according to Masatoshi Nakayama, then chief instructor of JKA. The peak speed of a well-executed gyaku-zuki measured at 13 meters per second.[18] Throughout this process, core muscle strength is pivotal to ensure segmental connectivity between upper and lower body joints, a key determinant for consistency in performance.

Center Core: The Lumbopelvic Hip Complex (LPHC)

Athletic trainers have touted core stability as the holy grail of performance in the last decades, to the extent that core strength development is central to most conditioning regimens. Core integrity is considered essential for track and field, court sports, and martial arts alike, as men and women proudly display their abdominal six-packs and sculpted backsides.

Our core comprises a multilayered cylinder of musculature that extends from the rib cage down to the hips. The core houses our body's largest muscles, although there is no universal consensus for its precise composition (fig. 10.3). It is bounded by the diaphragm as its roof, abdominal muscles at the front, the paraspinals in back, and the pelvic floor and hip musculature at the bottom (see table 10.1).[19] The core's technical name is the lumbopelvic hip complex (LPHC), embracing the lumbar spine, pelvis, and related musculoskeletal frameworks.[20]

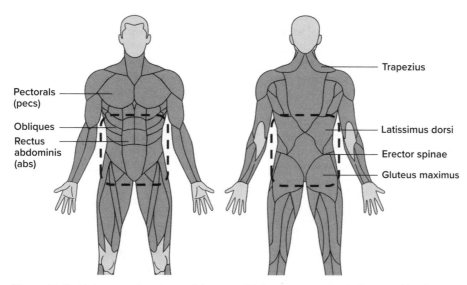

Figure 10.3 Major muscle groups of the core. Major core muscles in front and back are delineated within the dotted lines.

Dr. Ken Kibler defines core stability as "the ability to control the position and motion of the trunk, to allow optimum production, transfer, and control of force and motion in integrated athletic activities."[21] Core control comes into play when we have to modify posture to generate speed and force. Core muscles sustain balance as we twist or bend the upper torso. They work to relieve excessive load to the spine. Core strength and endurance stabilize our balance during rapid movements and frame the directionality of force transmission. Core functionality sustains the orientation of distal joints (wrists, ankles, fingers, toes) by maintaining torso control of arm and leg movements.

Core Functionality

In the late 1970s and early 1980s, I competed in traditional karate tournaments as a young black belt. Controlled body blows and minimal face contact were acceptable in sparring contests in that era. Being in the Pacific Northwest, I was often paired against competitors who were built like a mountain and kicked like a mule. At that time, I weighed no more than 130 pounds in a 5-foot-7 frame—my face lined up nicely to the six-footer's straight punch, and my torso was prime for kicking attacks.

Redemption came from my speed and a strong grasp of timing, which together with the youthful delusion of invincibility allowed me to advance over bigger opponents. Nonetheless, one collected "badges of honor" in the heat of battle in the form of black eyes, cut lips, and bruised ribs. When I returned to my research laboratory on the following Monday morning, moving around gingerly, colleagues would stare at my battered face and give the usual sanguine remarks, like, "So, been at a karate tournament last weekend?"

As training progressed over the years, I came to realize the critical nature of core integrity toward performance. Core control translates to sustained awareness, spontaneous and unitized body movement, augmented punching and kicking power, and last but not least, critical protection against impact force.

The semi-rigid framework of core muscles wraps and protects the spine and fragile abdominal organs. Frontal abdominal muscles work in concert with muscles in the back and side to sustain posture and balance. Dynamic core muscle actions stabilize spine and hip architecture, enabling sharp multidirectional body movements to achieve functional tasks.

Most identify core strength by well-developed "six-pack" abs. These paired rectus abdominis muscles are but part of an outer layer also made up of the lats (latissimus dorsi), spinal erectors, gluteus complex, quadratus lumborum, and hip flexors (table 10.1). External and internal obliques form the abdominals' mid-layer that affects upper torso rotation and side-bending (lateral flexion).

Table 10.1 Core configuration.

POSITION	MUSCLE GROUP	
Top	diaphragm	dome-shaped sheet of muscle and tendon, key muscle of respiration
Front	rectus abdominis	paired abdominal muscles commonly known as six- (or eight- or ten-) packs; responsible for flexing (bending forward) of the lumbar column (lower back); plays important role in forceful exhalation duration exertion by modulating intra-abdominal pressure
Front and side	external and internal obliques	top and outer flat muscles that cover the front and sides. The paired and layered oblique muscles ensure body stability for bilateral (on both sides) movements to flex the trunk, or unilaterally (one or the other side) to flex and rotate the torso.
	transverse abdominis	deep-seated muscle layer beneath the internal obliques that sustains stability of the spine, fortifies the spine and pelvis during lifting motions, and is engaged during child-birth labor and delivery

POSITION	MUSCLE GROUP	
	iliopsoas	the iliacus and psoas major flex the hip joint, rotate the hip joint externally, and sustain posture during transitions of the body center
Back	erector spinae (sacrospinalis)	includes the longissimus thoracis and the central semispinalis muscles. The muscles straighten the back, pull the head back, critically sustaining forward and back upright posture and during lifting. Unilateral contraction turns the head.
	multifidus	short, deep-seated back muscle fibers on either side of vertebral spinous processes that stabilize the spinal column, extending the vertebral column through bilateral action, or unilaterally rotating the vertebral column
	quadratus lumborum	deepest abdominal muscle that extends from the lower rib to upper pelvis (iliac crest). Controls lower back forward (flexion), side (ipsilateral) motion, and back (extension) motion of the lower back (lumbar column) and assists the diaphragm in inhalation
Bottom	pelvic floor/pelvic diaphragm	levator ani, coccygeus muscles, and associated connective tissues that support lower abdominal organs and sustain their functions
Other core muscles	The latissimus dorsi (upper back), gluteus maximus, and trapezius (lower back) stabilize our posture and balance.	

Deep-seated transverse abdominis and the obliques work together to sustain dynamic balance when repositioning the body center. The entire abdominal complex fortifies the spine as the upper body bends forward (flexion), thereby averting the tendency to topple over.

The abdominals also augment the performance of sundry daily tasks, in particular support for breathing during forceful exhalation (see chapter 11). Alongside with pelvic floor muscles, the activation of abdominal muscles effectuates reflexes such as coughing, vomiting, urination, voiding of waste, and childbirth.

An estimated 50 to 80 percent of force produced in an upper body action (throwing, punching, blocking) originates from the lower body musculature.[22] Apart from postural control, the LPHC contributes to half of this energy through hip movements, which also serve as conduit to transfer biomechanical energy from lower body to the upper body.[23]

Training to achieve speed and force in martial arts techniques requires maximized as well as coordinated efforts of the front, side, and back core muscles. They constantly adjust in order to sustain proper orientation of the thoracic cage and shoulders on top and the pelvis at the bottom, and to maintain pelvic joint compression to transmit energy. These functions optimize balance and redistribute joint pressure from foot to lumbar spine.[24] Hip destabilization due to a lack of core muscle control decouples force transfer, disrupting the body's equilibrium and increasing risk of injury from muscle strain or shearing forces on the joints.

The Yin and the Yang of Tai Chi

Susan Radtke and I were young black belts in the late 1970s, training under Art "Butch" Cherry sensei at the Oregon Martial Arts Center in Portland. Cherry sensei has been a pillar of the Portland martial arts community for the past several decades and headed the largest Shōtōkan karate network in the area. At the time, Susan and I also served as instructors of the Portland State University Karate Club.

An affable Englishman came to visit one day when Susan was teaching and I was out of town. He politely inquired about joining the club. As was proper etiquette in that era, he put on a white belt, since his rank was not earned at our dōjō. It didn't take long, however, for Susan to realize that the Englishman was as much a white belt as he was a native of the city.

We got to know Raymond Kerridge sensei quite well over time as senpai and a long-time friend. Before he and his wife, Sarah, moved to Portland, Kerridge had completed the prestigious JKA Kenshusei Instructor Program at the Tokyo Honbu Dōjō (head dōjō, 本部道場). To the present day, the Kenshusei Program is known for the extreme mental and physical demands placed on interns.

Prior to Japan, Kerridge trained in the famed London dōjō of Keinosuke Enoeda. Enoeda sensei was the unofficial tiger on the Shōtōkan emblem in light of his indomitable spirit and powerful techniques. Kerridge was captain of Enoeda sensei's National Kumite Team, known worldwide for its members' ferocity and speed.

During the time when Kerridge undertook the Kenshusei course, non-Japanese nationals had to maintain constant vigilance as soon as they entered the Honbu Dōjō. Japanese senpai were inclined to test the foreigners' mettle by singling them out for kumite practice.[25] Kerridge was not a big man, perhaps 5-foot-9 in height, but built like granite. He had the ability to break anyone in half, politely. He obviously made a good account of himself, having graduated from the course and earned the rank of sandan (third-degree black belt) before departing Japan.

Kerridge often regaled us with stories about training with Masatoshi Nakayama, then Head Instructor of JKA and senior student of Gichin Funakoshi. He visited Nakayama sensei's personal dōjō regularly, which was located in the basement of the sensei's apartment building, within walking distance from the Honbu Dōjō.

He considered Nakayama (and Hirokazu Kanazawa sensei) as aficionados of taijiquan (tàiji quán or t'ai chi ch'üan, 太極拳), which he practiced in their company. When we trained together at the Portland State University Karate Club, Kerridge passed along the simplified tai chi Twenty-Four Forms to Susan and me.

I already had more than three years of traditional Shōtōkan training by then. Still, I was confounded by the lack of urgency in the movements. My experience in karate hardly mattered at the beginning. Apart from an attenuated tempo *(andante!)*, movement dynamics can be perplexing to the uninitiated. As I revisited the form decades later, uncanny parallels began to emerge.

Two of today's most practiced tai chi forms date back to the sixteenth century. The Eighty-Three Forms (first road, 一道形式) were favored by the original Chen style, while the overlapping Eighty-Eight Forms were practiced by the Yang style in the nineteenth century.[26] Both take about thirty to forty minutes to complete, but decades to master.

The contemporary entry-level Twenty-Four Forms (also called Beijing Form) are now more popular. This shortened version was introduced to the public by the Chinese Sports Commission in 1956. Four of that era's most renowned experts assembled the Twenty-Four Forms, intended as physical exercise for the masses.[27]

Though abbreviated, the Twenty-Four Forms embody key principles of taijiquan. Movements start by limbering the upper and lower body, then progress to poses that promote segmental awareness and center-of-mass control. The third and fourth segments, performed at the same leisurely pace, escalate to challenge balance and stability. Movements that require low stances transition to one-legged poses (Golden Rooster), kicks above the waist (right and left Heel Kicks), and swooping hand motions at ground level (Snake Creeps Down, Needle at the Sea Bottom). The forms close in even slower tempo for warm-down.[28] With balanced poses repeated on the left and right sides, the forty-plus movements take about six minutes to complete.

The current emphasis of tai chi on mental-physical conditioning and health improvement belies its Wudang quan (武當拳; see chapter 1) roots. Wudang quan aligns with the Daoist belief that our physical constitution is a microcosm of the universe. Tai chi movements emulate qi flow under the influence of nature's Eight Energies (heaven, earth, lake, mountain, wind, thunder, water, and fire) according to the *I Ching* (3000 BCE), the Chinese book of oracles (fig. 10.4). Soft and graceful arm movements contrast with the demands on core and leg strength.

Yang and yin qi represent nature's complementing entities in masculine versus feminine, hard versus soft, and creative versus receptive aspects of nature. Tai chi movements simulate yin and yang flux within each of the eight energetic portals (fig. 10.4). Rooted stances and proper posture draw external qi from earth (yin qi) and the heavens (yang qi), while channeling internal human qi (ren qi) through hip movement and controlled breathing (see chapters 11 and 16).

Like mathematical calculus, each movement integrates a totality of minutiae of the very moment. It consolidates visualization, breathing, and body control. Slowed actions direct the practitioner's mental focus to unify mind and body before transitioning to the next pose.

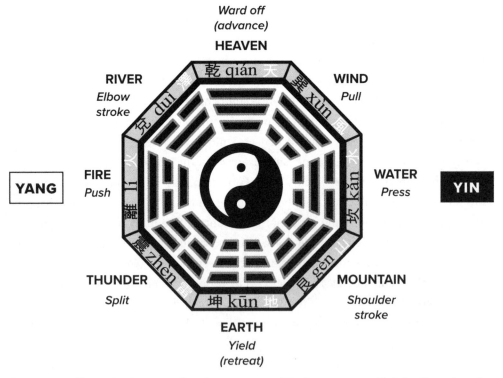

Figure 10.4 The eight elements of tai chi movements. Tai chi movements (italicized) emulate the yin and yang balance for heaven, earth, fire, water, thunder, wind, lake, and mountain. Each element is governed differentially by yang (solid bar) and yin (broken bar) qi.

A lack of speed limits momentum from the upper torso and arms. Rather, one relies on the strength of the legs and core to sustain rhythmic flow, drawing qi from the immovable yet receptive elements (earth, mountain). Muscles guide the unlocked fluidity of shoulder, arm, and wrist joints that work in unison. Deliberate, slow, and perpetual motion directs the ebb and flow of internal qi (the pliable, everchanging and penetrating nature of wind and water).

There is exquisite demand for core control and hip stability. Circular hand and arm motions are actuated by movements of the hips. These actions simulate parrying to deflect rather than blunt incoming momentum. Redirection of incoming force serves to upset an attacker's balance to their own demise.

Transition from one pose to the next is coupled with rooted and unhurried lower-body transitions. Over the course of redistributing center support, there is great reliance on balancing on one leg, then on the other. The flow of circular arm movements synchronizes with rhythmic breathing, preceded by the "sinking of (internal) qi" (*chen qi,* 沉氣) to the lower dantian (energy center; see chapter 16). Operationally, this is analogous to a deep stance that lowers the center of mass while focusing on abdominal breathing (see chapter 11). The goal is to master balance and control through core awareness and energy flow.

Tai chi appeals to both novices and experts. The expert unquestionably performs at a far higher mental and physical level, much like elite athletes in table tennis, golf, badminton, or softball as compared to "weekend warriors." With decades of practice, the tai chi master continues to seek nuances of the art. The studied sequence of energy transfer from the feet, through hip movement and to the arms, is the body's kinematic chain in action, practiced long before the concept's formal introduction in performance athletics.

Core Integrity and Budō Arts

Budō techniques are known for the focused application of explosive power and speed, at first glance diametrically opposite to tai chi concepts. Ultimately, both endeavors seek a unified mind and body, such that techniques can be executed with precision at any speed. It is therefore instructive that both quick and slow martial arts styles emphasize dynamic movements of the hip.

Masatoshi Nakayama devoted an entire opening chapter on this subject in his book, *Best Karate, Volume 2: Fundamentals.* He considered the LPHC as our body's primary source of power, and that smooth and rapid hip rotation was essential for impelling punching and blocking techniques. Discussions extended to biomechanical parallels in baseball, golf swing, and the shotput. Stance was the subject of the second chapter, which Nakayama attributed as key for maintaining hip stability and balance.[29]

These concepts have now been updated as proper kinematic chain activation from feet to fist (fig. 10.5), whereby sharp hip rotation transmits and augments ground reactionary force to the extremities. As tremendous force and speed course through the body, back core muscles also work to anchor the spine from shearing forces, while front abdominals and pelvic muscles sustain hip mobility.

Similar biomechanical concepts apply when wielding a weapon, such as a three-foot bokken (wooden sword), a katana, or a six-foot bō (staff). The samurai katana approximates the weight and dimensions of a heavy baseball bat. It weighs 2.7 pounds on average, with an overall length of about thirty-nine inches and a twenty-four-inch blade. Its "sweet spot" (the point of optimal energy transfer, also called the center of percussion) is at two-thirds the distance from the grip (the pivot of the lever), hence requiring strength and dexterity to control. The LPHC is now required to perform at a higher level: to balance additional mechanical demands and manage the implement's angular momentum.

Figure 10.5 Dynamic hip movements consolidate technique power and speed. Leveraged ground reactionary force from stance pressure is used to propel hip rotation. Angular momentum from the hips, coupled with contractional forces of the core muscles, serves to power shoulder movement and arm extension.

Kendō masters counsel sinking the hips and bending the knees. Precision rests on firmly anchoring the spine from the core so that key joint movements (hips, shoulders, and elbow) are under control while transmitting speed and force.[30]

Biomechanical Aspects of the Pelvis and Hip Muscles

As the largest joints of the body, our hips serve as a central pivot for body movements. The LPHC comprises muscles that are attached to the hip joints in various orientations. Their selective activation enables multidirectional upper-body movements relative to the upper thigh (femur).

The LPHC complex houses our body's strongest muscles. However, none of the muscles act along the axis of leg movements. Accordingly, they create movement with torque, a force that turns an object around an axis. Performance depends on the muscle's strength (magnitude of force), its angle of insertion relative to the femur (directionality of force), and the perpendicular distance between muscle action and the hip joint (moment arm) (fig. 10.6).[31]

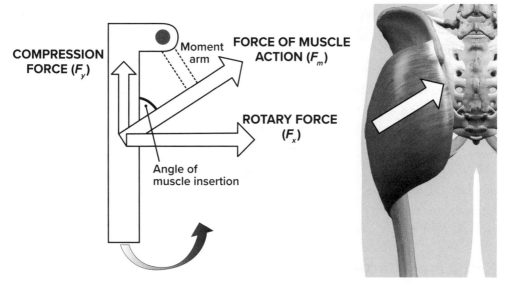

COMPRESSION FORCE (F_y)

Moment arm

FORCE OF MUSCLE ACTION (F_m)

ROTARY FORCE (F_x)

Angle of muscle insertion

Figure 10.6 Biomechanical aspects of hip muscles. Muscle torque force is dependent on orientation of the joint and the muscle's insertion points. Left: The force from muscle activation (F_m) can be resolved into horizontal (F_x) and vertical components (F_y). F_y compresses and sustains joint stability, while F_x produces rotary force. Torque is calculated as force (F_m) × moment arm, the perpendicular distance between directional force and the joint (axis of rotation). Right: The gluteus maximus is one of the strongest muscles, stretching across the pelvis, and it attaches to the femoral joint, the pelvic bones, and the tailbone. The glutei maximi work in concert with the hamstrings to enable thigh extension and lateral rotation.

Hip and thigh muscles work together to change direction, or propel our center of mass forward, backward, or sideways. When stationary, directional torque from the concerted efforts of hip muscles enables sharp hip rotation, bringing angular momentum to upper-body movements. Quick body shifting or stepping generates linear momentum. The rate of change of momentum (acceleration or deceleration of the center of mass) is a key source of speed and power in martial arts techniques (see appendix A).

Kinesiology of Forward Motion

A stabilized LPHC ensures proper balance control during movement transition, as illustrated by oi-zuki (step-in straight punch) execution, one of the fundamental techniques in budō karate (see fig. 9.6).

The biomechanics of oi-zuki resemble a sprinter's standing start.[32] Both rely heavily on sharp, linear acceleration that originates from the back-leg reactionary force. Starting in left-foot-forward opening stance (hidari ashi kamae; see chapter 4), applied pressure is biased toward the leading foot with the front knee bent, hence favoring mobility of the rear leg. As the back leg drives off the floor, properly activated core muscles sustain upper and lower body integrity, keeping the body upright to consolidate forward center-of-mass transition (see chapter 9). When the right back leg moves past the body, balance is maintained entirely by the left leg and hips (chapter 9). This is the single support phase, while the bent left ankle and knee stores elastic energy in maximum flexion.

As the body continues to move past the support leg, its extension adds to acceleration of the body center. Within tenths of a second, the right foot achieves full stride and lands in front stance (*migi* Zenkutsu-dachi). Forward motion is halted at that point by the anchored front foot. Meanwhile, extension of the punching fist transforms momentum into penetrating force through compression of the center core muscles (see fig. 9.6).

Enablers of Hip Movements

Mother Nature designed the hip joint for weight-bearing and for anchoring the spine. The sacroiliac joint connects the spine to the pelvis at the top. Our thigh bones (femurs) are fitted to the two triplanar ball-and-socket joints at the pelvic acetabulum.[33] Twenty-one sets of muscles traverse the pelvis, orchestrating body movements in all three dimensions. In addition, a thick band of connective tissues (fascia) braces the hip and knee joints, extending vertically from the outer surface of the pelvis to below the knee. The fascia aids in consolidating muscle torque. An awareness of their coordinative functions informs proper technique execution and helps to minimize injury.

Directional movements of the hips are divided into forward versus backward positioning of the upper thigh relative to the hip (flexion and extension), side movements of the upper thigh away versus toward the midline of the body (abduction versus adduction), and turning of the femur bone outward versus inward (lateral-external versus medial-internal rotation).

1. Abduction and Adduction

These processes reposition the hips as we side-step. When moving to the right, one moves the right leg away from the body (abduction), then pulling the left leg toward the body's midline (adduction). The same muscle complexes assert control for kicks to either side (yoko geri; see chapter 13).[34]

Table 10.2 Mediators of hip and torso movements.

MUSCLE FUNCTION	HIP MOVEMENTS		
	FLEXION	**EXTENSION**	**ROTATION**
Agonist*	iliopsoas, tensor fasciae latae, gluteus medius	gluteus maximus, gluteus medius	gluteus medius; gluteus minimus; tensor fasciae latae; adductor longus, brevis, and magnus
Antagonist*	gluteus maximus, gluteus medius	iliopsoas, tensor fasciae latae, gluteus medius	opposing gluteus medius and minimus
Stabilizers*	core stabilizers: transverse abdominis; intrinsic muscles of the spine, diaphragm, and pelvic floor		
	TORSO MOVEMENTS		
Agonist*	rectus abdominis	erector spinae (iliocostalis, longissimus, spinalis), latissimus dorsi	contralateral external oblique and ipsilateral internal oblique
Antagonist*	erector spinae, latissimus dorsi	rectus abdominis, external obliques, internal obliques, psoas	opposing external and internal oblique
Stabilizers*	core stabilizers: transverse abdominis, intrinsic muscles of the spine, diaphragm, and pelvic floor		

*Agonist muscles contract while antagonist muscles relax to achieve a movement. Stabilizer muscles stabilize the joint so that the desired movement can be performed.

Kicking places a high biomechanical demand on the supporting leg, tasked with providing sole weight support as the kicking knee is raised above the waist. When executing a side-thrusting kick (yoko kekomi geri), balance is sustained by the support leg. Dynamic tension of the inside thigh muscles enables the kicking leg to extend toward the target.[35] Real-time calibration by the hip adductors and pelvic-floor muscles fortifies this one-legged support (Ippon-ashi-dachi) (see table 10.2). Meanwhile, abdominal core muscles at the front (transverse abdominis, multifidus), back (quadratus lumborum), and side (obliques, iliopsoas) work in concert to stabilize the spine.

With the knee tucked up front in the preparatory position, abductors of the kicking hip and thigh contract to enable kicking leg extension (table 10.2; fig. 13.4). Quadriceps of the kicking leg work in concert with the gluteus complex to lift and rotate the upper thigh inward (medial rotation). Outward canting (lateral rotation) of the support hip counteracts momentum transfer by the kicking leg, while the contralateral gluteus complex compresses the support hip to minimize pelvic tilt.

Hip adductors (adductor brevis, adductor longus, gracilis) and abductors (gluteus medius, gluteus minimus) generate the highest torque force among the LPHC musculature.[36] Their functionality accounts for explosive hip actions that, together with the mass of the hips and the kicking leg, give rise to enormous kicking power.

The tremendous force delivered by a side thrust kick may also incur hip injury from overuse or due to joint misalignment from movement compensation. Fascia inflammation is a common injury of the iliotibial band that serves as insertion points for the gluteus maximus, the largest and outermost buttock muscle, and sartorius and tensor fascia latae, less powerful muscles that sustain hip-knee alignment. A weakened gluteus medius (dormant butt syndrome) causes knee injury when the destabilized hip brings excessive pressure to the knee joint and the anterior cruciate ligament (ACL) of the support leg.[37]

The single leg squat is a common exercise to evaluate quadriceps strength and the gluteus complex's hip stabilizing functions.[38] The exercise assesses hip adductor and knee abductor performance, core stability, and movement coordination.[39] Repetitions are useful for strengthening the participating muscle groups.

The athlete starts by balancing on one foot while lifting the other leg to hip level and straightened toward the front, parallel to the ground. Left and right hands and arms are fully extended in the same direction. The toes are pointed upward. The athlete then sinks down slowly by bending the support knee while maintaining balance and keeping the back upright. They slowly return to an upright position after reaching the limits of ankle flexibility, while keeping both arms and the leg extended. By alternating between the left and right legs for support, the exercise informs on core strength and stability of the left and right hips.

2. Hip Flexion and Extension

Hip flexion brings the body and thighs closer to one another, allowing the thigh to come forward when the upper body is stationary, or bending the torso forward when the thighs are fixed in position. Hip flexor muscles sustain posture by preventing arching of the lower back. They counteract forward pelvic tilt during the leg's forward-lunging motion while stepping or running.

The iliopsoas is the strongest hip flexor. It inserts at the lower spine, running downward to overlay the hip girdle at the front, then wraps to the backside of the femur (see table 10.2). This leveraged configuration enables posture control, core stability, and gait. Another flexor muscle, the rectus femoris, traverses the hip and coalesces as part of the quadriceps musculature at mid-thigh. It acts with the iliopsoas to achieve flexion of the thigh or pelvic tilt.

When executing a front kick (see fig. 13.4), hip flexors work with the quadriceps to bring the knee up toward the body's center. Improper hip flexor movements cause a loss of power in the kinetic chain at best, and at worst increase injury to the back or lower limb.

Hip extensor muscles maintain the upright posture and keep the upper body from toppling over when one leg pushes from behind, as in the case of Zenkutsu-dachi, or when the sprinter accelerates off the starting block. The primary hip extensors include the gluteus maximus, posterior head of the adductor magnus, and hamstrings.[40] Psoas major of the iliopsoas is a key hip extensor. It anchors the spine to the pelvis and maintains vertical stability. The hip abductor complex (gluteus medius and minimus) also plays a stabilizing role, when the back-leg thigh is required to lift and turn outward during the natural forward movement. This process activates the gluteus maximus, part of the adductor group, to limit pelvis sway as the center of mass moves forward or back.

Adductors and abductors contribute to strength of movement during hip flexion and extension, functioning alternately over each stride.[41] In particular, adductors are instrumental for quick directional changes to accommodate force delivery by the inside surface (medial aspect) of the foot, such as in swimming or kicking a soccer ball. The karate front kick (mae geri) or the roundhouse kick (*mawashi geri;* see chapter 13) involves adductor muscle activation of the kicking hip and thigh.

Hip flexion and extension are coordinately controlled by the upper thigh muscles (quadriceps and hamstrings). Adductors and hamstrings work together to modulate body momentum during forward movements as soon as the back leg moves past the torso to the front.

Lunging and squatting exercises call on the same muscular demands; hence, they are applicable for conditioning quadriceps, hamstrings, back muscles, and the gluteus complexes.

3. Hip Rotation

Hip rotation is the hallmark of wushu and budō arts, as exemplified by the counterpunch (gyaku-zuki) in Shōtōkan karate (see fig. 4.2). Medial rotation of the trailing hip transmits and amplifies the coiled spring compression force of the lower limbs, contributing to the punch's explosive power.

A number of the hip's medial rotating muscles work together to bring the trailing hip toward the midline of the body. They include the gluteus medius, gluteus minimus, and tensor fasciae latae as well as the hip adductors. Internal rotational torque is boosted by the corresponding movement of the opposite hip, which rotates externally away from the centerline. This action is engineered by the piriformis and gluteus minimus, which compress and stabilize the hip joint. Biomechanical studies show that a flexed leading hip (the thigh brought in forward position, such as in Zenkutsu-dachi) can increase rotational torque by up 50 percent.[42]

The karate-ka utilizes rotational torque and core strength to coordinate hip and shoulder movements, creating an upper body motion like that of a swinging door panel. Energy is transmitted into the arms through activation of the deltoids, triceps, and forearm muscles (fig. 4.2). This course of action differs from the decoupled hip and shoulder rotations of the baseball pitch but provides biomechanical stability when recoil shock waves return to the body after impact (see chapter 9).

Core Awareness and Conditioning

The LPHC cradles our center of mass, underscoring its importance in kinematic chain performance. There is an abundance of online resources on how to improve core strength and LPHC stability. Readers are directed to the "Further Reading" section for a sample of blogs for conditioning the abdominals, lower back, and hip muscles.

Sports science informs us that hip flexors and extensors critically sustain our balance and posture by interconnecting the lower back with the thigh and knee. Hip movement is coupled to diaphragmatic breathing through sharp flexor contraction that coordinately activates the abdominal core muscles (see chapter 11). The powerful gluteus complexes impact mobility by their stabilizing influence on the hips and spine. They steer directionality of torso movements, contributing to quick changes in body position in a major way. Accordingly, a heightened awareness of LPHC functionality enables the martial artist to overcome constraints that are imposed by subpar performance in stance and core.

Takeaways from "Hips, Core, and the Kinematic Chain"

- Our center core strength (lumbopelvic hip complex, LPHC) defines peak performance in track and field, court sports, and martial arts. Apart from postural control, the LPHC conveys biomechanical energy from lower to upper body.

- The hip is the largest joint in our body and is its central pivotal point of movements. Hip-connecting muscles enable mobility through torque and angular momentum, which are key sources of technique speed and power in martial arts.

- Exercise physiologists consider the feet and ankles, knees, hips and pelvis, shoulders, and head and neck as the five checkpoints of the kinematic chain. These pivotal joints transmit dynamic force through serial and overlapping skeletal-muscular activation from the legs, through the hips, and into the upper extremities.

- In keeping with "form follows function," movement aesthetics are predicated on stance dynamics, balance control, and proper performance of the LPHC.

Breathing and Martial Arts

Enlightenment Through Breathing

For centuries, sages in ancient China, Egypt, and Greece practiced controlled breathing for enlightenment and health improvement.[1] Early Greek philosophers considered humans to have two souls. *Psyche* was our immortal soul. We acquired our mortal soul, *thymos* (later termed *pneuma,* "vital spirit," "life-force," or "spiritedness") from the first breath. It resided in the chest and departed on the last breath. This life force that is tied to our breath is called *sugs* by the Tibetans, *mana* by Polynesians, *orenda* (Iroquois tradition) or *manitou* (Algonquian tradition) by Native Americans, and *od* by ancient Germans.[2]

Aristotle (384–322 BCE) first associated breathing with air intake and expulsion but considered its purpose was to expel heat produced by the heart. Another two millennia passed before Danish scientists August and Marie Krogh showed that respiration brings about balanced gas exchange and oxygenation of the body.[3]

Meanwhile, controlled breathing took root in religious and philosophical pursuits of spiritual transcendence. Abstraction of "the breath" as vital energy is recognized by cultures worldwide, not surprisingly, since key sensations of breathing bring a state of well-being, energy, and power.

Ancient Hinduism, Buddhism, and Jainism (1500–500 BCE) sought spiritual enlightenment through breath control. These religions believe in harmony with the universe by mastering one's own vitality (from Sanskrit: *prana,* "life force"), which is tied to the breath. Controlled breathing was "one's link to the divine."[4] An elevated awareness of the breath (pranayama, from *prana,* "life force," and *ayama,* "extension") served to promote the union of mind and body.

This "floating technology" among the Veda-based religions was subsequently co-opted by spiritual undertakings such as yoga (from Sanskrit: *yuj,* "union" with the divine).[5] Yogis and yoginis sustain regimented poses and controlled breathing to seek a relaxed mind. Advanced levels of pranayama aspire to heightened self-awareness and the unification of mind, body, and spirit.[6] A beginner focuses on the rhythmic process of the in breath and the out breath *(svasa-prasvasa)* and internal-external prana circulation. Later, regimented and monitored inhalation *(svasa)* brings consciousness of the body's filled *(puraka)* and residual *(kumbhaka)* spaces, and their emptying *(rechaka)* during controlled and monitored exhalation *(prasvasa)*.[7]

Qigong

The Chinese philosophy of qi dates back to 500 BCE.[8] Qi (breath, 氣, *ki* in Japanese philosophy) was considered the universal life force for a person as well as the cosmos (see chapter 16).[9] Chinese philosophy, traditional medicine, and martial arts are steeped in the belief of qi to the present day, and the dao as the path to harmonize human qi with that of the universe.

Qigong (craft of life energy cultivation, 氣功) is the Chinese martial arts practice of focused breathing. Qigong originated from the ancient practice of *daoyin* (guiding and directing internal qi, 導引), Daoist mind-body-unity exercises that date to 168 BCE.[10] Qigong adopts slow and flowing exercises and deep rhythmic breathing to aspire for a calm meditative state.[11] The practitioner directs their attention within, striving to channel their own qi throughout the body.

Qigong practice bestows a sense of well-being and elevates the spirit. It is now integral to both "internal" (taijiquan) and "external" (Shaolin quánfǎ) wushu practices (see chapter 1), albeit emphasized at differing levels. Contemporary qigong was standardized from no less than thirty different historic styles in 2007 by the Health Qi Gong Management Centre of China's General Administration of Sport.[12] Rudimentary forms are intended as health-promoting exercises for the aging public, bringing popularity worldwide. A small number of randomized clinical trials suggest that qigong may benefit chronic hypertension, arthritic disease, and stroke rehabilitation.[13]

Qigong exercises begin with a physically relaxed body, seated or standing upright. Correct posture is paramount. The mind is quieted and directed inward to explore the breath, accompanied by an attitude of patience and curiosity.[14] The three levels of qigong assign escalating demands on breathing control. The novice starts with natural breathing, progressing to yogic breathing (also called diaphragmatic breathing), ultimately to Daoist reverse breathing at the expert level, which is also called martial arts breathing.

Natural breathing focuses on relaxed inhalation through upper chest expansion and the unforced movements of the diaphragm. The process reverses during exhalation. Consistent practice for about three months achieves balance between mind and body, at which point students are introduced to yogic or Buddhist breathing.

Both Buddhist and Daoist teachings subscribe to deep breathing practices to focus the mind. The exercise starts with awareness of the conceptual energy centers in our body called dantians (field of elixir; also called *tan t'ien,* 丹田, see chapter 16), similar in concept but not identical to *granthis* (psychic knot) in yogic philosophy. The three dantians nurture intent (upper dantian), emotion (middle dantian), and physical vitality (lower dantian) (fig. 11.1).

The lower dantian is located conceptually below the navel at the center of the body (see chapter 16). Lower dantian exercises are called variously abdominal breathing, diaphragmatic breathing, or belly breathing. In simplistic terms, diaphragmatic breathing focuses on up and down movements of the diaphragm. It brings order and control to thoughts and emotions, ultimately toward total awareness.

One inhales by consciously contracting the diaphragm, which moves downward into the abdominal cavity. This action compresses the stomach area (epigastric region), bringing out the belly, and gives the impression that breathing originates from the lower abdomen. The diaphragm relaxes during exhalation, returning to its upward, resting state.[15]

Figure 11.1 Physical locations of the upper, middle, and lower dantians. The upper dantian centers between the eyebrows behind the forehead. The middle dantian corresponds to the location of the heart. The lower dantian approximates the center of mass position below the navel.

Daoist thought considers our constitution to replicate that of the universe in pattern, nature, and structure (the microcosm-macrocosm concept).[16] The dao of enlightenment aligns the human nature (mind, body, spirit) with the natural order of the cosmos (wu-wei: effortless action, 無爲). This is the purpose of controlled breathing during the Daoist meditative ritual of *ziao zhou tian* (microcosmic orbit; 小周天).

Ziao zhou tian seeks to cultivate and direct the flow of internal qi through the body's meridians ("energy highways"), in turn bringing awareness of, and sensitivity to, qi of the surroundings (earth qi, 地氣), and that of the universe (heaven qi, 天氣) (see chapter 16).[17] As the meditator focuses on diaphragm contraction to fill the lungs, they draw air qi (*zhen qi,* 真氣) into the middle dantian (located near the heart), where it blends with nutrient qi (*ying qi,* 營氣), then with the person's innate essence (*yuan qi,* 元氣) into human qi (ren qi, 人氣) (see chapter 16). Ren qi is directed to the upper dantian during inhalation by way of the Governor Meridian (du mai, 督脈) in back of the body (fig. 11.2). Upon exhalation, ren qi descends via the Conception Vessel (ren mai, 任脈), a key meridian at the front of the body, replenishing the lower dantian without hindrance.

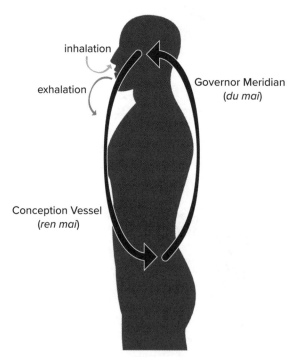

Figure 11.2 Ziao zhou tian orbit. The Daoist microcosmic-orbit exercise involves meditative breathing by way of the *du mai* and *ren mai* meridians within the body, in order to channel internal qi from the lower dantian to the upper dantian and back.

Having mastered ziao zhou tian, qigong experts incorporate this practice into more complex techniques, such as the macrocosmic orbit (*da zhou tian,* 大周天) that extend qi flow throughout the torso and into the arms and legs.

Chest Versus Diaphragmatic Breathing

The diaphragm is the primary physiologic motivator of respiration. Shaped like a hollowed-out flying saucer, it comprises contractile muscle fibers at the outer rim that encircle a tendinous, noncontractile dome. The diaphragm sits at the bottom of the rib cage, partitioning the viscera into thoracic (lungs and heart) and abdominal (gastrointestinal organs) cavities. It attaches to the base of the rib cage at front (xiphoid process) and sides (costal margin), as well as by multiple insertion points with the spinal column. The central upward-facing dome accommodates encroaching anatomies of the stomach and liver from below.

Restive breathing engages the chest muscles and the diaphragm. The process is controlled by the autonomic nervous system, which also governs other vital functions such as heartbeat, perspiration, and pupil dilation. Air intake involves contraction of the diaphragm, as it pushes down against the abdominal cavity. At the same time, external intercostal muscles of the rib cage contract while internal intercostal muscles relax, unfolding the rib cage upward and out. The enlarged chest cavity leads to negative pressure that draws ambient air into the lungs and alveoli.

The diaphragm recovers to its native position during restive exhalation, a passive process without effort or expended energy. The intercostal muscles reverse their actions as well, drawing in the rib cage downward. The reduced chest volume builds positive pressure, expelling air through the nose and mouth.[18]

Up and down movements of the diaphragm coordinately activate reflexive events of other respiratory tract organs (lungs, larynx, mouth, and nose). These processes are all subjected to conscious intervention. Starting and stopping breathing can be controlled at will, up to the limit of inflicting physiological harm to oneself.

Deep breathing seeks to regulate breath volume and tempo by consciously controlling diaphragmatic movements. The process offers physiologic and spiritual benefits and has been broadly adopted to elevate athletic performance in swimming, platform diving, weight-lifting, and track and field events from sprints to shot put. Vocal training and musical wind instrument performance also engage diaphragmatic control. Controlled air uptake and expulsion serve to modulate tonality and amplitude and to convey emotion. The conscious process of diaphragmatic forced inhalation markedly elevates peak intake volume, from 500 milliliters (tidal volume) to more than 3,000 milliliters (inspiratory reserve volume, IRV).[19]

Breathing control collaterally modulates our neurological responses. The natural movements of the diaphragm provide homeostatic feedback to the autonomous nervous system through the vagus nerve. Although we cannot control autonomous functions directly, we can influence its regulation through diaphragmatic movement. Inhalation naturally increases our heart rate to promote gaseous exchange as fresh air fills the chest cavity. This process represents a sympathetic up-tempo response of the overall physiology. Our heart rate decreases reflexively on exhalation, representing a parasympathetic response that promotes the restive state and a sense of well-being.

By focusing on diaphragmatic movement and the exhalation process, autonomous functions are shunted to parasympathetic control, as has been demonstrated with the deep breathing practices of pranayama. The deep breathing drills utilized by emergency first responders tamp down fight-or-flight responses and are based on the same concept. Also known as combat-tactical breathing, this practice lowers adrenaline shakes by curtailing sympathetic (excitatory) overexcitation, in turn stabilizing autonomic respiratory and cardiovascular functions.[20]

Scientists at Trinity College in Dublin have recently established the link between breathing and attention. Intense mental focus triggers reflexive functions that modulate breath volume and tempo, revving up cardiovascular performance. Pranayama moderates noradrenaline release by the brain (see chapter 5), which normalizes emotions and improves thinking and memory performance.[21] These findings provide clues on how controlled breathing calms the nerves and brings neuroadaptive benefits, apart from the benefits of oxygenating our body.

Breathing and Martial Arts

Martial arts workouts fall mostly within the anaerobic zone, characterized by short bursts of high-intensity strength and speed in between moderate exertions. The anaerobic zone occurs at 80 to 90 percent of the maximum heart rate as muscle-stored glycogen is metabolized to quickly supply energy without external oxygen input.[22] Sprinting, jumping, and powerlifting are other examples of short-exertion high-intensity anaerobic events.

Elite athletes and martial artists gravitate toward deep diaphragmatic oxygenation that fortifies the anaerobic state. This practice promotes gas exchange by maximizing differential chest volumes. Diaphragmatic breathing engages apex and middle as well as base of the lungs where air flow (ventilation) is 50 percent higher than at the apex. Diaphragmatic oxygenation shifts breathing control from the muscles of the shoulders and chest to the abdomen and the core musculature, a process that improves thoracic-spine mobility and upper-body movement control.

Deep, diaphragmatic breathing is mastered by meditation, as the student gains awareness and control of diaphragmatic movement. The process also draws attention to the coordinately activated abdominal core muscles.

Coordinative diaphragmatic–abdominal core movements are intrinsic to our autonomic physiology. During restive inhalation, downward diaphragmatic pressure activates the abdominal core to redistribute intra-abdominal pressure (IAP).[23] Conversely, an increased IAP that is initiated by the abdominal core triggers diaphragmic contraction, which is part of our reflexive physiology, such as vomiting, urination, defecation, and childbirth.

By consciously controlling the shortening and relaxation of abdominal core muscles, the martial artist acquires the skill of modulated breathing through IAP. IAP control is key to Daoist reverse breathing, which represents the highest form of controlled breathing. Daoist reverse breathing, also called martial arts breathing, should only be practiced after mastering the processes of natural and diaphragmatic breathing.

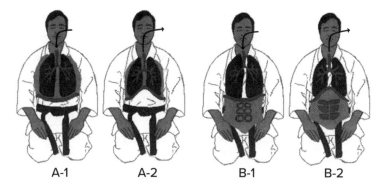

A-1   A-2   B-1   B-2

Figure 11.3 Natural breathing versus reverse (martial arts) breathing. Inhalation during restive breathing (A-1) involves external intercostal muscle contraction that lifts the rib cage outward as the contracting diaphragm moves downward. An expanded thoracic cavity volume draws air into the lungs. Air is expelled during exhalation (A-2) as the diaphragm and external intercostal muscles relax. Reverse breathing focuses on exhalation (B-2) by exercising core muscle tension and building of intra-abdominal pressure (IAP). An increased IAP prevents the diaphragm from moving downward. Its contraction draws in the rib cage instead. A reduced chest cavity volume builds internal pressure, expelling air. As abdominal muscles relax, IAP drops, allowing the rib cage to expand outward. Air is drawn into the lungs to achieve inhalation (B-1).

Daoist Reverse Breathing

Unlike restive breathing, martial arts breathing focuses on the process of exhalation (fig. 11.3). Forced exhalation is initiated by activating the rectus abdominis complex, the "six-pack abs" (also called "eight-pack" to account for the 4th pair below waist) that wrap the front of the abdomen (see table 10.1).[24] As these muscles shorten, the abdominal cavity draws inward like an hourglass from both sides, giving the sensation of filling the belly from built pressure in the intragastric (above the navel) and lower abdominal region below the navel (B-2 in fig. 11.3).

Muscle tension in the abdominal core elevates IAP, triggering the reflexive contraction of the diaphragm through the sympathetic nervous system. However, its natural downward movement is contravened by the squeezed abdomen. The diaphragm's dome is held in place from crowding of the stomach and liver. Contractile muscle fibers of the diaphragm resort to tug at the lower circumference of the rib cage accordingly, pulling the chest cavity inward.[25]

Internal pressure builds from a reduced chest volume, forcing air to be expelled. Then, as core muscles relax, a reduced IAP returns the abdominal cavity to its resting configuration. The diaphragm resumes its downward tack that is accompanied by expansion of the rib cage. Air draws into the lungs by virtue of an increased chest volume (B-1 in fig. 11.3).

Reverse breathing refers to the sensation of a tight belly while exhaling, where diaphragmatic movements play second fiddle to IAP control. The process increases efficiency of gas exchange, as the diaphragm's travel path during exhalation is doubled compared with the range of one to two cm in restive breathing.[26] It expels an additional 1,200 milliliters of air (expiratory reserve volume, ERV) over the tidal volume of 500 milliliters. The markedly increased exchange volume enhances aerobic capacity.

Forced exhalation is a deliberate process that expends energy. A neutral posture is important to clear the airway in martial arts breathing. Perineum integrity should be maintained through pelvic-floor muscle contraction (see table 10.1) in order to prevent herniation from unintended IAP overexertion. IAP modulation elevates awareness of core muscle tension, now the common agency for respiration and technique execution.

Ibuki and Nogare: The Yang and Yin of Controlled Breathing

Japanese philosophy and martial arts consider the *hara* (lower abdomen, center of essence, 腹) as seat of a person's vitality, and synonymous with the concept of tanden or dantian. Masters of Japanese calligraphy, tea ceremony, and martial arts supposedly draw inspirations from the hara. Hara is cultivated through *ibuki* (息吹), the practice of *ki* (qi in Chinese) awareness through breathing control.

Ibuki in budō practice is divided into *yō-ibuki,* hard and heavy abdominal breathing, and *in-ibuki,* also called *nogare* (smooth and firm abdominal breathing, 逃れ). Yō-ibuki exercises total-body tension to sustain strength and resolve, while invoking a deep gut-tural sound over the course of drawn-out exhalation. This is followed by short, sharp inhalation with a light rushing sound.

In-ibuki favors smooth, rhythmic, and unstrained air exchange through deep abdom-inal breathing, seeking to unify body and spirit.[27] *In* and *yō* breathing emulate the yin and yang (陰陽) polarity of the universe's forces, respectively (see chapter 10).[28]

Both approaches engage the reverse breathing practice. Gōjū-ryū, Kyokushinkai, and many Okinawan karate styles practice yō-ibuki to strengthen the core during kihon and kata training. The fortified abdominal wall during yō-ibuki enables the practitioner to withstand strong blows to the body. Nogare is adopted when sitting in zazen (meditative pose), or during kumite when rhythmic air volume exchange is desired to power the tran-sitional anaerobic phase during movements.

A review of the key aspects of breathing practice follows.

INCREASED VIGILANCE.   Proper breathing is predicated on correct posture, which clears the airways for inhalation and exhalation. An erect spine, lowered chin, and connectedness to the floor offer the proper frame of reference of self to the sur-roundings, thereby optimizing perception and preparedness (see chapter 9). The martial artist is relieved of anxiety by centering their mind to the tanden. The calming effects of deep breathing serve to strengthen their resolve during engagement.

FOCUS ON TANDEN.   Breathing through the chest predisposes one's mental atten-tion to upper body actions. In contrast, martial arts breathing attends to the abdom-inal core musculature. Core muscles of the skilled practitioner are conditioned to maintain hip stability and IAP control simultaneously, in effect consolidating upper and lower body connectivity.

The martial artist triggers their kinematic chain by contracting their abdominal muscles, which increases IAP, and the coordinative compression of their hips and lower body joints. Exhalation is coupled to the unitized core, which acts as conduit to augment and transmit the coiled-spring reactionary force from the lower body to the torso (see chapter 12).

KIAI.   IAP exertion (internal tension) peaks with kime, the application of total-body muscle tension from the hips and core to the limbs (external tension; see chapter 9). Kime is triggered by vocalizing an abrupt, guttural sound from the abdomen, called kiai (ki integration, 気合), which resembles a short reflexive grunt as we summon physical strength and attention to lift a heavy object.[29] This mental focusing technique aims to synergize mind and body at the very instant of energy transfer.

Through incessant practice, one learns to appreciate the pros and cons of different breathing patterns. Swift mental-shock techniques such as leading-hand straight punch (kizami-zuki) are accompanied by sharp exhalation and an abrupt kiai. Deep, drawn-out exhalation and a sustained kiai are used to accentuate powerful techniques such as yoko kekomi geri (side thrust kick) or the load-carrying judō one-arm shoulder throw *(ippon seoi nage).*

TIMING AND STRATEGY.   A thorough understanding of body integrity throughout the breathing cycle (strong during exhalation, weaker during inhalation) informs on timing tactics. Breathing rhythm is a key part of kinematic signature (see chapter 12) that betrays an opponent's intent and movement. Success is ensured if the defender learns to hack into the adversary's telltale breathing rhythm, then interrupts the rhythm with a preemptive strike.

The budō art of aikidō uses locks, holds, and throws to redirect the opponent's momentum in self-defense. The literal translation of *aikidō* (合気道) is "the way of unifying ki," often represented as a form of Moving Zen practices (see chapters 2 and 15). The aikidō-ka believes that inner calm and awareness transform physical ki into spirit, intent, and an exquisite balance of "bio-physio-psychological energies."[30] Most martial arts also seek to cultivate and unify ki (see chapter 14) to this end, albeit differing in approach and manner of practice.

This philosophical and operational common ground underlines the importance of mastering breathing control, which can be pursued in a traditional martial dōjō through due diligence in meditation and the trifecta of kata, kihon, and kumite exercises (see chapter 7).

Takeaways from "Breathing and Martial Arts"

- Diaphragmatic breathing promotes a sense of wellness and offers physiologic and spiritual advantages over chest breathing.

- Martial art workouts fall mostly within the anaerobic zone, characterized by short bursts of high-intensity strength, power, and speed in between moderate physical demands. These endeavors require deep diaphragmatic oxygenation that in turn sustains the anaerobic state.

- Martial arts breathing focuses on exhalation by modulating intra-abdominal pressure (IAP) through abdominal core muscle contraction and expansion. The martial artist exerts IAP to trigger their movement's kinematic chain of events. IAP exertion (internal tension) peaks with kime, coordinating hip and core muscle tension with that of the limbs (external tension).

The Physics of Mastery and Effort

I learned of Bruce Lee's passing on July 20, 1973. It was the typical hot and humid summer day in Hong Kong. I had returned to the city-state to visit my parents after having graduated from university. Their home was a mere block away from Lee's temporary residence at Lancashire Road, and no more than a ten-minute walk after Lee moved to his permanent home on Cumberland Road. It appeared, however, that I would not have the opportunity to meet Lee.

At first, I discounted the news as fake. Lee was in his prime at age thirty-two. Just two years before, he returned to Hong Kong to make martial arts movies and ascended to instant superstardom. It would be an unusual news day if Lee wasn't in the daily headlines, be it fact or fiction.

Tragically, his passing was soon confirmed. Autopsy reports attributed recurrence of brain edema as the cause of death due to an allergy to medication. Lee's funeral was attended by thousands. Tens of thousands more braved the heat and watched the funeral procession as a last tribute to their native son.

A year prior, the TV series *Kung Fu* had debuted on American network television. The series portrayed the adventures of Kwai Chang Caine, a "peace-loving Shaolin monk traveling through the American Old West."[1] Lee previously presented a parallel concept named *The Warrior* to the movie moguls. *Kung Fu*'s lead role went to David Carradine nonetheless. Carradine had no prior martial arts training but came from a family of actors. Lee had been considered for the lead role but was passed over. Though part Caucasian, his thick accent was deemed detrimental to audience appeal.[2] The *Kung Fu* series went on for three seasons, beguiling the American audience with Eastern philosophical musings and Hollywood's choreographed martial arts.

Lee returned to Hong Kong on the advice of Hollywood producer Fred Weintraub. The idea was for Lee to be cast in a feature film that would impress Hollywood executives.[3] His instant popularity far surpassed expectations. The three films produced in Hong Kong (*The Big Boss*, 1971; *Fist of Fury*, 1972; *Way of the Dragon*, 1973) enjoyed enormous box office success throughout Asia. Lee captivated the audience with his charisma and confidence on and off-screen, the diametric opposite of Hollywood's subservient stereotype for the Asian male.

Lee's last completed film, *Enter the Dragon* from Warner Brothers, was released a week after his untimely demise. Some forty years later, the film remains one of the most popular martial arts movies, having grossed more than $200 million worldwide. The theme of Lee's movies was all based on variations of the misunderstood martial artist who sought social justice. These portrayals cemented his legacy as a cultural icon across ethnic and generational lines. Larger-than-life statues of Lee have been erected in Hong Kong, Los Angeles, and Mostar, Bosnia-Herzegovina, in memory of his unquenchable spirit.

Mastery and Effort

To many, Bruce Lee's legacy symbolizes kung fu from the standpoints of personal triumph and empowerment of the underclass. The term *kung fu* (Wade-Giles romanization of 功夫 in the Cantonese dialect; in pinyin: gōngfu) likely originated as southern Chinese street slang, used so extensively through history that its origin was lost.[4]

The character *kung* (功) translates as skill, achievement, or craft. *Fu* (夫) commonly refers to an individual who asserts physical efforts. Hence *kung fu* describes someone who has mastered a craft through hard work, or an expert in the field who can apply their skill effectively.

Kung fu is now construed in the abstract as accomplishments in mastery and effort. This type of semantic shift is common as the historic Chinese language is replete with homonyms. Prior to the convention as an accolade for martial arts practice, master chefs, calligraphers, painters, sculptors, or craftspeople were also accorded the honor of possessing kung fu; in deference to their depth of knowledge (kung) and expertise to translate knowledge into outcome (fu). By the same token, martial arts expertise is as much about technical knowledge (kung) as its practical applications (fu).

Lee was exposed to many martial arts disciplines over his lifetime. He was well known as a student of the famed Wing Chun master Ip Man in Hong Kong. Lee also taught Wing Chun in Seattle and San Francisco. He trained in Southern Chinese Choy Li Fut Fist and Wu-style tai chi. Lee was well versed in Western boxing and took first in a Hong Kong (High) Schools Boxing tournament. As well, he was a strong advocate of rigorous conditioning to improve endurance, flexibility, and strength. His heavy emphasis on forearm development and grip was widely reported.[5] Lee was a student of Asian and Western philosophy and acknowledged influence from Daoism, Jiddu Krishnamurti, and Buddhism.[6]

Bruce Lee founded Jeet Kune Do, a fighting style that incorporates knowledge from many disciplines, which he called "the style of no style." This sentiment echoed the assertions of karate-dō master Gichin Funakoshi. In *Karate-dō: My Way of Life,* Funakoshi stressed that "there is no place in contemporary karate-dō for different schools.... The ultimate aim of both karate and sumō was the same," concluded Funakoshi, "the training of both body and mind."[7] These insights are in line with the common path of budō for a disciplined mind and body, without necessarily fixating on the beauty or uniqueness of the outside form.

Kinetic Energy, Force, and Momentum

Martial arts techniques take the form of kicking, punching, striking, and grappling. Effectiveness is predicated on the precision of imparting force and momentum to the target. The dynamics of these actions are best understood through the fundamentals of mechanics, a subdiscipline of physics that quantifies motion in terms of force, energy, and momentum according to Newton's laws of motion.

The mechanical outcome of two colliding objects depends largely on their intrinsic makeup. Consider object A that travels in a straight line when it collides into B, a stationary sphere of comparable size and mass. If A is a nondeformable solid steel sphere ("elastic" in physics terms), the impact results in the partial transfer of A's stored kinetic energy to B. If B is also nondeformable and free-moving, the transferred energy generates momentum in B. A and B now move in the same direction at a reduced speed, as previously stored kinetic energy is now shared between the two. This is a simplified illustration of an elastic collision, where both momentum and kinetic energy are conserved with no appreciable loss after the event.

However, momentum and kinetic energy are not conserved in an inelastic collision. When B, as an inelastic object, suffers structural damage during the collision, part of A's stored kinetic energy is dissipated in other forms. This is the case when a motor vehicle runs into a guardrail. When the car's forward motion is blunted on impact, its stored kinetic energy produces heat, sound, and mechanical force that penetrates and deforms the guard rail and the vehicle's crush zones (energy transferred into work, which equals force × distance) (see table 12.1).

The impacted areas of both car and guardrail are a key determinant of the extent of damage to both (force per unit area, or pressure; see appendix A). Force tends to create more penetrating damage when applied over a small area than when spread over a broad impacted surface.

Table 12.1 Collision outcomes.

TYPE OF COLLISION*	MOMENTUM TRANSFER	BLUNT FORCE PENETRATION
Elastic	primary	secondary
Inelastic	secondary	primary

*Elastic collision occurs when two nondeformable objects impact one another. Inelastic collision occurs when one or more deformable objects impact one another.

Collisions in real life rarely present as wholly elastic or inelastic, but a combination of both. The human body is by and large an inelastic entity that is prone to structural damage. The outcome of physical impact can best be understood in biomechanical terms, which examine biological structure and function in relationship to motion, and hence are highly instructive for quantifying force and speed in athletic performance. Let's consider the example of when a baseball catcher fields a speeding ball. The typical high school varsity pitch can achieve a terminal velocity of 55 miles per hour (25 meters per second). The kinetic energy stored in the five-ounce ball as it impacts the catcher's glove is about 1,016 foot-pounds. Part of this energy dissipates from compression of the baseball on impact. Heat (from impact friction) and sound (the "pop" of the ball) consume some more stored energy. The remainder is discharged as residual force onto the catcher's mitt that halts the ball's forward motion.

To accommodate impacting momentum, the catcher moves their elbow back about six inches, and if necessary, shifts their perch in line with the caught ball. The ball's kinetic energy is now converted to performing work (force × distance) onto the elastic mitt webbing and the backside of the glove. Assuming half of the baseball's stored energy remains, the impacting force (F) is calculated to be 1,016 pounds:

$$(50\%)(1,016 \text{ ft·lb}) = 508 \text{ ft·lb} = F \times 0.5 \text{ feet};$$
$$F = 1,016 \text{ pounds of force.}$$

Catchers routinely receive 150 pitches per game. Repetitive force penetration takes its toll on the catcher's glove hand. According to studies by Dr. T. Adam Ginn at Wake Forest University, a catcher's glove hand suffered from a much higher incidence of blood vessel injury than their throwing hand due to this penetrating force that is focused on a small area.[8] Over time, swelling and reduced blood flow lead to damaged nerves, giving symptoms of numbness and tingling, reduced sensitivity to cold, and blue-tinged skin.

These concepts apply equally to the physics of defending a side thrust kick (yoko kekomi geri). If the kick lands at the defender's upper arm with full force, its considerable kinetic energy may cause bruising or fracture the humerus bone. If the arm is braced against the body, tissue damage may be reduced, as force is broadly distributed over the defender's torso. The kick would likely lift the defender off the ground, sending them to the opposite side of the room.

Power of the Punch

Scientists have been intrigued by the tremendous punching force of elite boxers and karate black belts for quite some time. Consensus differs with respect to their peak speed and power due to divergent study conditions, methods of analysis, and the participants' level of expertise.

A study on peak punching force showed that flyweight to super heavyweight Olympic boxers generated 447 to 1,066 pounds of force.[9] Variances in the weight classes, the weight of the boxers' hands and gloves, the speed of their punches, and the rigidity of their wrists contributed to the broad range of values. Remarkably, flyweight boxers as a whole delivered more oomph than the majority of super heavyweights. Researchers were not surprised, though; based on laws of mechanics, the boxers' speed differential (Δv) contributed exponentially to the stored kinetic energy (KE), whereas mass differences only affect the final outcome linearly (KE = $\frac{1}{2}mv^2$, where m is mass and v is the velocity of the punch; see appendix A).

In another study of seventy boxers, the overall average of delivered force was 776 pounds. Elite fighters as a group punched twice as hard as novices, landing the strongest punch at 1,300 pounds of force.[10]

The analysis of a karate punch is somewhat easier, as it is delivered in a straight trajectory. Transferred force (F) is calculated by the formula F = $(m)(a)$, where m is the free mass of the arm and a the rate of change from peak velocity right before impact to zero (at impact), as determined by $\Delta v / \Delta t$, where v is peak punching velocity and t the duration of contact with the target (see appendix A).

In *Dynamic Karate,* Master Masatoshi Nakayama writes that advanced karate-ka (fourth-degree black belts) can generate 1,500 pounds of punching force with an ungloved gyaku-zuki (counterpunch).[11] The test subjects were top-flight athletes and graduate standouts of the JKA Kenshusei Instructor Program. Their expertise was reflected in their remarkable peak speed of 13 meters per second when executing a lunge punch (oi-zuki), as compared to 7 meters per second by a novice practitioner.[12]

Unlike the follow-through of a gloved boxing punch, the karate-ka completes their action with kime, the practice of total body skeletomuscular compression at the split second of impact (see chapter 9). Karate experts complete their kime in less than 10 milliseconds.[13] By sharply tensing the muscles of the entire body on impact, the terminal velocity of a karate punch quickly goes down to zero within inches into the target. The mechanical advantage of kime comes from transforming body momentum into a penetrating force (force = rate of change of momentum). This biomechanical feat has been analyzed in detail by Michael S. Feld and Ronald McNair.

McNair, a NASA astronaut and physicist, was already an accomplished karate black-belt competitor when he pursued his PhD at the Massachusetts Institute of Technology. His mentor, the late Professor Michael S. Feld, was well known for his scientific contributions in the fields of laser science and applied biophysics.[14] With the help of McNair, Feld delighted in using board-breaking demonstrations to impress his physics undergraduate classes, as he referred to the karate punch as "one of the most efficient human movements ever conceived."[15]

Feld and McNair coauthored the highly popular 1976 *Scientific American* article titled "The Physics of Karate." They used sophisticated mathematical models to analyze peak force output and reconciled these values with the energy required to shatter a pine board or concrete block.[16] Complex calculations revealed that an average peak force of 600 newtons was needed to split a pine board, whereas 2,500 to 3,100 newtons would shatter a concrete block.

According to high-speed stroboscopic measurements, even a beginning student can launch a downward karate hammer strike *(tatsui-uke)* at 6.1 meters per second with the aid of gravity. Experts like McNair achieved peak velocity of 14 meters per second with the same technique. His straight punch was clocked at 6–10 meters per second. The impact speed of his kicks was 7–15 meters per second.[17]

When McNair's hammer fist struck a concrete block at peak velocity, his hand made contact and transferred its energy in 5 milliseconds, in effect decelerating at a rate of 3,500–4,000 meters per second per second.[18] Without accounting for the mass of the upper arm or contributory vibrational energy, McNair generated 2,800 newtons of peak force with his 1.5-pound fist, which was at par with the force needed to shatter the 14-pound concrete block. The fist's high terminal speed, and McNair's masterful kime, accounted for the tremendous force imparted on the block.

For readers who are interested in learning more about the physics of the punch, a summary on mathematical relationships between force, velocity, and related concepts is provided as appendix A.

Maximizing Force and Momentum Transfer

The explosive power and speed of budō techniques is predicated on a self-defense mind-set, the ethos of unwavering situational control in spite of being outmatched in strength and in number, summarized below and in fig. 12.1. To survive unscathed, there may only be a singular opportunity to upend an opponent's mental and physical equilibrium. The martial artist strives for peak mental and physical performance on demand.[19] Critical elements for this caliber of execution have been discussed earlier and are summarized below.

RESOLVE.   Resolve is the mental commitment to pursue a course of action. Certainty of purpose, such as will for survival or protecting the young, galvanizes resolve and clarity of mind (see chapter 15). Resolve is the key to overcome fear, anxiety, or uncontrollable distractions due to adrenalized excitability.

Resolve impels the mind to sustain physical dexterity, bolstering psychological (confident decision-making, commitment to technique) and neurological (shortened reaction time) performance. A commitment to act is founded on certainty of solutions distilled from prioritized options.

STANCE.   Lower body musculature and core contribute more than half of the impact force in punching or throwing.[20]

Stance pressure compresses hip, knee, and ankle joints. Adjoining muscles and tendons work together to achieve a coiled-spring effect, propelling the speed phase of technique execution (see chapter 9). The ground reactionary force gives impetus to body and foot movements such as pivoting, stepping, or shifting. At the point of kime, stance stability serves as a foundation to channel biomechanical energy into the target, then redirects recoil energy back to the ground to produce secondary shock waves.

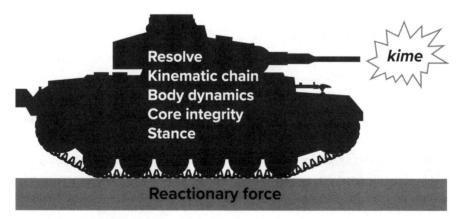

Figure 12.1 Critical elements for stationary technique execution.

BODY DYNAMICS.   The speed of body motion (kinematics) critically determines final outcomes of force and momentum transfer (see chapter 10). Controlled movements of the hips and knees effectuate torso rotation, shifting, or rising and dropping of the center of mass, dynamically transmitting lower-body power to propel upper-body techniques.[21] Consistency in body dynamics performance elevates force and speed outputs. Deliberate practice of complex, closed kinematic chain movements (see chapter 10), with or without the benefit of visual guidance, can heighten our subconscious proprioceptive feedback by up to 50 percent.[22]

CORE INTEGRITY TO SUSTAIN POSTURE AND BREATHING CONTROL.   Core strength sustains proper posture and relieves the spine and hip joints of unnecessary shearing force during movements. The concerted efforts of abdominal and back muscles anchor the spine and maintain pelvic joint stability, which together with the gluteus complex and hip flexors potentiate rapid multidirectional movements of the torso. In particular, gōngfu masters talk of "waist strength" (yāo-lì, 腰力), the importance of powering techniques through the hips and midsection.

Core awareness is critical for dynamic delivery of punches and kicks (see chapter 10). The core muscles' capacity to modulate intra-abdominal pressure (IAP) is key for spontaneous and unitized body movements.

IAP exertion works to synchronize internal tension (deep breathing) with external muscle tension (skeletomuscular actions) (see chapter 11). Peaked IAP couples exhalation to leverage power to the limbs (chapters 10 and 11). Tensed obliques and abdominis muscles buffer vital organs against impact force to the midsection (chapter 10).

Neural feedback from tensed core muscles heightens awareness. Controlled breathing through the diaphragm produces a calming effect that strengthens resolve and sustains mental clarity during engagement.

KINEMATIC CHAIN AND THE SUMMATION OF SPEED.   Effective force transfer requires coordinated, sequential coupling of body kinematics (see chapter 10). Each segmental movement adds power and speed. Precision in each segmental movement contributes to strength and directionality of the final biomechanical outcome.[23]

Punching, striking, and grappling actions are the end results of a multiplied biomechanical force that originates from the floor, then is transmitted through major joints of the body (fig. 12.2). Foot techniques similarly start with support-leg knee and ankle joint compression, leveraging ground reactionary force to coordinately activate the hip and thigh muscles of the kicking leg. Whether punching or kicking, the

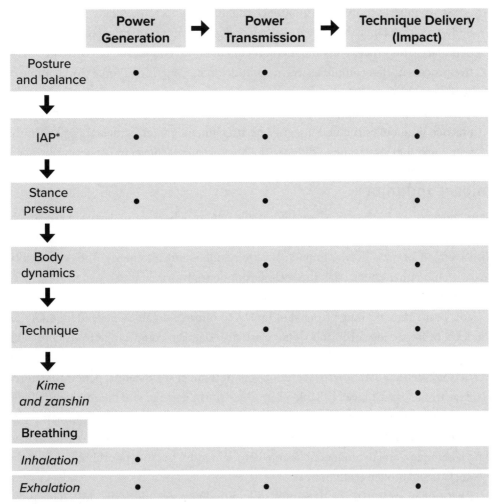

	Power Generation	Power Transmission	Technique Delivery (Impact)
Posture and balance	•	•	•
IAP*	•	•	•
Stance pressure	•	•	•
Body dynamics		•	•
Technique		•	•
Kime and zanshin			•
Breathing			
Inhalation	•		
Exhalation	•	•	•

*IAP: intra-abdominal pressure; *kime*: total body muscle compression; *zanshin*: sustained state of total awareness (chapter 8).

Figure 12.2 Physical components of technique execution.

coordinative action of each intervening segment adds successively to technical force and speed.[24]

KIME AND ENERGY TRANSFER.   A cross-body boxing punch is propelled by shoulder rotation and follow-through. The circular motion creates enormous angular momentum that drives back the opponent. In comparison, straight-line budō techniques are designed for penetrating force delivery onto an impact area.

Sharp muscle contractions that lock the major joints sequentially from the lower body, trunk, and then the extremities halt body momentum after penetrating the plane of impact. This process of kime is completed in milliseconds. The intensive decelerative process in effect amplifies the magnitude of the penetrating force (force = rate of change of momentum).

The compressed interposing joints also consolidate the effective mass on impact. This practice promotes structural integrity by strengthening intersegmental rigidity of the arm, torso, and lower limbs, adding to the biomechanical efficiency of energy transfer.

Impact and Injury

Skin, muscles, and soft tissues are deformable, whereas bones and cartilage are largely nondeformable. According to mechanical principles that define the outcomes from the displacement of mass during impact, the tremendous impact energy from a punch or strike would either knock back the recipient or be transformed into a penetrating force that fractures bones or disrupts soft tissue integrity.

Soft tissue traumas range from skin breaks to internal organ damage, depending on the level of impacting force and tissue elasticity. Damage could range from contusion (bruising from blood vessel damage below the skin); abrasion (an open wound caused by friction, such as a scrape); to laceration (cut or tissue tear). Residual penetrating force (such as from a thrust kick) is liable to produce organ damage and hemorrhagic shock. Blunt trauma to our nondeformable framework (bones, cartilage) leads to fractures or dislocations.[25] Blows to the head, whether from a direct punch or shock waves from snapping techniques, create brain concussion, disorientation, fracture of orbital bones of the eye socket, and broken teeth or nose.

I have been an instructor of Shōtōkan karate-dō for decades. The dōjō has graduated scores of black belts and trained national and international champions. We hold advanced training every Saturday, when everyone gives their very best and pushes the envelope during sparring. As a traditional dōjō, participants do not wear gloves, pads, or body armor. Over the years we have been fortunate for having sustained no injuries more serious than cut lips and sprained fingers or toes. Remarkably, members come in with casts from mishaps in soccer, baseball, or even slipping on the kitchen floor.

Rather than assigning our good fortune to sheer luck, I consider our safety record to reflect the participants' attention to tasks, mental discipline, and physical control. Precepts of self-defense specify awareness and mental focus that accompany physical performance. These elements also empower participants to master the consequences of their actions.

Takeaways from "The Physics of Mastery and Effort"

- The dynamics of martial arts kicking, punching, striking, and grappling follow Newton's laws of motion and are based on mastery in imparting force and momentum to the target.

- Kime, the practice of joint compression throughout the body, delivers impact force in less than 10 milliseconds. The rapid deceleration amplifies impact-force transfer by a karate punch.

- Technique execution involves the ordered events of power generation, transmission, and delivery. The final outcome requires proper posture and balance, stance pressure, IAP modulation, body dynamics, and technique impact with kime and zanshin.

Hand and Foot Techniques in Karate-dō

Hand Technique Execution

Martial arts techniques are evolved to favor the primacy of biomechanical outcome, given their original purpose for self-defense applications.

Karate hand techniques utilize the forearm, wrist, open hand or closed fist, or elbow for punching, striking, gripping, or blocking (fig. 13.1). Relative fragility of the hand favors soft tissue targets or cartilaginous connective tissues.

The most commonly practiced techniques are based on punching with a closed fist, with thumb and fingers wrapped tightly against the palm. They include the straight punch (choku-zuki) directed to the face or mid-level in a linear course, and the back-fist strike (*ura-ken,* fig. 13.1) or roundhouse punch *(mawashi zuki)* with a circular trajectory that targets soft tissues on the sides of the face or neck with the top knuckles (metacarpophalangeal joint, MCP).

Strikes are also executed with the open hand or the edge of the wrist, which leads directly into grappling after delivery of force. Follow-up techniques or takedown maneuvers can be applied to further disrupt the opponent's mental composure.

The knife-hand strike *(shuto)* is easily the best-known open hand technique, recognized as the "karate chop" to the side of the neck (carotid artery). *Shuto-uke* strikes with the lower side edge of the palm to inflict nerve and soft tissue trauma or to impart shock to pressure points on an opponent's forearm as a defensive move. The ridge-hand strike *(haito),* based on the same concept, lands with the top outside ridge of the palm with an over-the-top rotating trajectory. The technique is intended to inflict shocking force to pressure points such as the temple. In comparison, *teisho* uses the bottom edge of the palm as a forward strike on soft tissues (see fig. 13.1), or executed as a sharp downward thrust to blunt incoming, straight-line attacks.

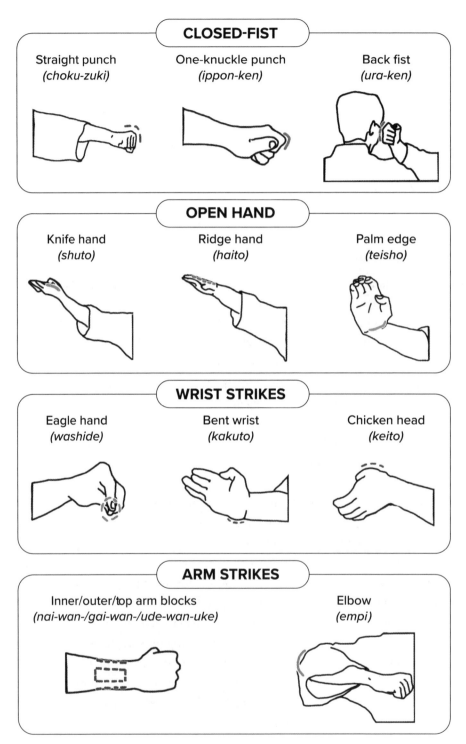

Figure 13.1 Karate-dō hand techniques.

Other wrist and arm strikes are part of the close-range defensive repertoire that incorporates the option of gripping the assailant's arm or wrist to control their mobility. Wrist strikes *(washide, kakuto, keito)* target pressure points of the opponent's wrist and lower arm, while arm strikes *(uke)* apply blunt force to deflect directionality of an impending attack (see fig. 13.1) or to disable an opponent by targeting nerve endings.

Open-hand techniques work best within arm's length. Linear punching techniques are effective from a farther reach and are governed by front-foot positioning in reference to the opponent.

The effectiveness of hand techniques rests on optimizing body dynamics that amplify speed and force delivery (see chapter 12). The following concepts are commonly practiced to improve consistency in outcome.

ARMS FOR SPEED, HIPS FOR POWER.   Studies show that the straight punch can generate anywhere from 4,500 to 6,670 newtons.[1] Our arm and shoulder musculature are not as strong as our legs and torso. Accordingly, the peak velocity of punching, striking, or blocking with the arm is best achieved by sequential acceleration of the lower body segments (knee, ankle), through the hips, converging at the elbow and wrist.

Similarly, maximized force delivery is dependent on the summation of lower-body forces (stance pressure and weight, segmental velocity and acceleration) and core strength. Locking the wrist immediately before kime promotes energy transfer into the target and reduces the likelihood of sprained joints.

UNLOCKED SHOULDER JOINTS.   Stiffening of the shoulders gives a false sense of strength. Chronic over-use of the shoulder joints leads to inflammation of key tendons (rotator cuff tendinitis).

The dynamic fluidity of the shoulders (glenohumeral connectivity) should be maintained over the course of punching, striking, or blocking. Deliberate tension to lock this joint at kime is counterproductive, running the risk of damage from impact recoil. The shoulder girdle is best stabilized by activating the lower trapezius and rhomboid muscle sets, which pulls the shoulder blades downward and medially toward the spine. Tension from musculature of the upper back and underarm (serratus anterior) works in concert with the lower back and front abdominal core muscles to fortify the scapula while bracing the spinal column on impact (see table 10.1).

ELBOW CONTROL.   Alongside core functionality and fluidity of the shoulder joints, proper elbow positioning elevates precision in hand-technique execution. The upper-back and underarm (serratus anterior) muscles work in concert to promote elbow control from the torso.

The process incorporates external rotation of the elbow joint, which together with the back muscles (latissimus dorsi) helps to guide the forward trajectory of the karate straight punch (choku-zuki). The punch starts with the fist tucked against the side of the body, corresponding to the anatomic position of where the elbow meets the waist. The back of the fist faces the ground. Fingers are tightly wrapped and braced by the thumb. As the fist drives toward the target, the elbow stays tucked and in close contact with the side of the body. Sharp outward rotation of the elbow (pronation of radioulnar joint) turns the fist by 180 degrees as soon as the elbow clears the front of the body (fig. 13.2).

This rotational action, also called "elbow snap," works to achieve a spiraling course as the forearm maintains a linear course toward the target. Rotation of the elbow reinforces directionality and helps to sustain forearm control, similar in concept to a twirling projectile speeding toward the target. The snap of the elbow generates angular momentum, which together with forward thrust of the forearm contributes to peak velocity at the time of impact.

Figure 13.2 Elbow snap in oi-zuki. Starting with the technique arm tucked against the side of the body, the fist drives toward the target with the elbow in close contact with the side of the body. The elbow rotates sharply outward for 180 degrees as it clears the front of the body. At the same time, the opposite elbow turns inward as the arm draws back toward the side of the body *(hikite)*.

For applications of cross-body strikes with a looping trajectory (shuto, haito, *empi*), the elbow functions as a biomechanical fulcrum between the upper arm and the forearm. A sharp elbow snap produces axial rotation of the forearm and wrist, delivering shocking power immediately before impact.

These movements are powered by pivoting of the hips, which transmits waves of kinetic energy onto the shoulders to launch the upper arm's circular trajectory. Maximum torque transfer to the forearm is achieved by a well-positioned and compressed elbow joint. In contrast, a wandering elbow requires constant recalibrations of shoulder and upper arm positions, which detract from proper kinematic transfer and add to the margin of error in execution.

CONCENTRATED POWER DELIVERY.   The top knuckles (MCP joint) of the index and middle fingers are used to focus impact force delivery in choku-zuki (straight punch) or ura-ken (back fist strike). Single-digit punches, such as ippon-ken, engage the center knuckle (proximal interphalangeal or PIP joint) of the index finger as the strike point (see fig. 13.1), intended to focus force penetration onto fragile soft tissues (pressure = force ÷ unit area).

HIKITE.   Musculature from both sides of the body is engaged when executing a hand technique. As one arm drives toward the target, the opposite arm pulls back sharply toward the torso. This reciprocal action, called hikite (pulling hand, 引手), specifies reciprocal elbow rotations by both arms (see fig. 13.2) and is commonly utilized in all budō arts.

A coordinated executed hikite promotes mechanical efficiency and heightens the awareness of back muscle engagement for proper arm control. Imagine our two arms as rigid segments tied to the opposite sides of a horizontal pulley (our spine and hips). A directional impulse from the opposing segment (the pulling hand) activates the rotational process of hips and shoulders, adding to the mechanical advantage of the technique arm. Bilateral activation of the shoulder, core, elbow, and wrist also heightens neural feedback.

Our open hand with inwardly curled fingers is a natural implement for grappling. Apart from biomechanical advantage, hikite likely originated from simultaneous gripping applications that restrict the attacker's options as the defender repels with a striking technique. The adversary's balance is severely compromised when movement of their attack arm or torso is disrupted. Hikite's pulling action also serves to upend the attacker's mental equilibrium (fig. 13.3).

In judō grip and throw techniques (seoi nage, 背負投), coordinative two-handed push-pull motions of *tsurite* (lifting or gripping technique hand) and hikite lead into the explosive hip turn and leg sweep. Reciprocal movements from both shoulders and arms improve center-of-mass leverage to complete the throw (fig. 13.3).

Foot Technique Execution

Our naturally stronger hip and leg muscles contribute to kicks that are more powerful than hand techniques. Kicking techniques can achieve farther reach, especially when the karate-ka drives in with their hips. Unlike punching actions, however, it is more difficult to sustain balance and power with kicks in rapid succession.

Kicking places a higher demand for hip stability when balanced on one leg. Both hips should be aligned at right angles to the direction of the front or back kick (median plane alignment) to maximize stability on impact. The support knee is aligned in the direction of technique trajectory. This orientation helps to mitigate hip sway from side to side. A 90-degree geometry is maintained in side kicks between the support knee and the hips, as the kicking leg drives toward the target in parallel to both hips (frontal plane alignment) (fig. 13.4).

Hip stability, peak velocity, and kime at impact are key to power delivery. The tremendous momentum of kicking techniques and their impact recoil can rattle our sense of balance. To this end, the martial artist is mindful of the following considerations to maximize momentum transfer.

Figure 13.3 Hikite applications. Left: karate gyaku-zuki. Right: judō *eri-seoi otoshi.*

Figure 13.4 Karate-dō foot techniques.

KICKING INVOLVES CENTER-OF-MASS TRANSITION.   To consolidate balance and force transmission, the karate-ka focuses on their center of mass to power the trajectory through the hips (core musculature). Muscle tension of the upper and lower legs should only contribute secondarily to the kick force.

The front snapping kick (mae geri) is one of the most common kicks in Shōtōkan karate practice. Its snapping motion resembles that of a whip. The bent (flexed) kicking knee is anchored above the waist in front of the body center, acting as a fulcrum between the thigh and the lower leg. An explosive drive and return action are powered by corresponding pendulum movements of the hips (hip flexion), followed by return to neutral posture in One-Legged Stance. The sharp contraction of the thigh quadriceps, in coordination with the muscles of the shin (tibialis anterior, soleus) and ankle (shank peroneal), powers the lower leg toward the target at angular velocities of up to 13 meters per second.

The reciprocal kicking action of forward penetration and instantaneous pull-back requires explosive contraction of the hamstring and calf muscles of both legs. The sharp pull-back hip movement in effect doubles the rate of momentum transfer, intended to maximize force delivery and to create shock waves on impact (see appendix A).

To maximize force transfer, the kicker locks the kicking leg ankle, drawn toward the shin (dorsiflexion) at approximately 90 degrees. This practice of delivering impact with the ball of the foot (*koshi;* see fig. 13.4) reduces the likelihood of fracture to the complex bone structures on top of the foot.

The Shōtōkan side-thrust kick (yoko kekomi geri) starts at the same preparatory position, with the kicking knee tucked up front (in flexion) of the hips. Activation of the hip abductors (see table 10.2) opens the hip joints. A sharp tilt of the pelvis toward the target launches the kick, effected by differential activation of the hip gluteus medius at the kicking and support hips. Sideward extension of the kicking leg (thigh adduction) instantly follows this center-of-mass pivot (see fig. 13.4). The kicking foot is angled with its outside edge facing the target (eversion). Medial ankle rotation (toes pointing in) focuses impact to the lower outside edge of the foot (*sokuto;* see thrust kicks, fig. 13.4).

BALANCE IS KEY.   The mechanical impulse created by the kicking motion tends to pull the lower body in the direction of the kick. Center core, hip flexor, and thigh muscles of the support leg are tasked to prevent the hips from submarining in the same direction. The kicker would otherwise tilt over, losing their balance and ending up on the ground.

The mishap is preventable through strengthened abdominals and lower-back muscles, which work in concert with hip abductors and support-leg musculature to sustain downward compression of the support-leg hip, knee, and ankle. An appreciation of this infrastructure brings control to kicking dynamics and improves precision in force delivery.

A well-executed kick also takes into account the reactionary force that returns on impact. Unless properly addressed, impact recoil upends the kicker's balance and may produce joint injuries. The following are safeguards to preserve structural integrity during this process:

1. Properly align leg and hip positions, as discussed earlier;

2. Be conscious of support-hip and leg adductor muscle activation for balance and stability;

3. Bend and compress the support knee to lower the center mass, which reduces side-shearing force to the knee joint and channels recoil impulses to the ground;

4. Maintain balance on the heel of the support foot with full-sole contact to the ground. The practice creates a strong base for weight support, also serving as a conduit for recoil energy dissipation, thereby reducing the likelihood of knee and ankle injury.

CORE REMAINS CRITICAL.   As the kicker perches on one leg, their core strength is critical for sustaining hip stability and kinematic performance. The martial artist focuses on exhalation by modulating their IAP. This downward pressure is initiated at the tanden, followed by sequential compression of the hip, thigh, knee, and ankle joints, from the moment of drawing up the kicking knee to completion of the kick. This complex motor routine is predicated on proper alignment of the spinal column and hips (see chapter 9).

KNEE DEFINES THE TRAJECTORY.   As counterpoint to hand technique execution, precision in kicking rests on the trifecta of correct posture, fluidity of the kicking hip joint, and a proper knee position. Reach and height of the kick are founded on correct alignment of the kicking knee and hip flexion control.

Coordinative activation of segmental joints is needed to achieve a proper course of kicking. A front kick is propelled by sequentially flexing the hip joint, then kicking-knee extension that launches the lower leg in a pendulum arc. The side thrust-kick trajectory is guided sequentially by sideways tilting of the hips, movement of the upper thigh away from body center (adduction), then inward (medial) rotation of the kicking knee joint (see fig. 13.4).

RECOVERY IS PART OF TECHNIQUE.   Technique recovery is an integral part of kicking. Recovery for a snapping kick is an intuitive process, simply by reversing outbound motion from the flexed hip and knee. The same concept applies to thrust kicks. Through hip adductor and upper thigh muscle control, the tilted hip, then the knee of the extended kicking leg, is drawn back to preparatory position, allowing the kicker to recover balance before launching follow-up techniques.

A kicker's lack of control during kicking and recovery is easily exploited by an experienced opponent. The kicker's center of mass shifts rapidly as they launch their technique while supported by one leg. A loss in balance could land them on the floor, placing them in a highly vulnerable state.

Readers are referred to the many excellent reference texts listed in "Further Reading" for style-specific treatises of technique execution. Expertise invariably requires perseverance, deliberate learning, and exhaustive trial and error to arrive at mastery (kung) and focused effort (fu). This enduring process is best carried out in a traditional training environment under the proper guidance of a knowledgeable teacher and reinforced by encouragement and support from one's peers.

The Floating Board Break

In the early 1970s, I started formal karate-dō training under Robert Graves sensei. I first met Graves sensei while an undergraduate, when he conducted beginner and intermediate courses on campus.

Graves often pulled me in front of the class for technique demonstrations, as I was reasonably well coordinated. Though forewarned of his actions, I never ceased to be surprised by his quickness. "Do not move," warned Graves sensei. Then this *whomp* of energy would charge in at an unsafe speed. Just as abruptly, he came to a dead stop in his tracks, completing his oi-zuki at less than an inch from my face. Columns of shock waves enveloped my entire being as I forgot to breathe. His inimitable spirit, technical mastery, and the snap of his *gi* at kime were enough to put me in awe, motivating me to pursue Shōtōkan and JKA training that has continued to this very day.

Graves is the stereotypically robust 6-foot-plus outdoorsman. He moved (and still does) like greased lighting. His impeccable techniques were delivered with amazing go-no-sen timing. That, combined with his stealthy speed and mastery in sweeps, made him close to unstoppable in sparring.

Graves sensei trained in Wadō-ryū karate-dō while stationed in Japan after the Korean conflict. Wadō schools were few and far between when he returned stateside in the late 1960s. In light of the parallels with Shōtōkan practices, he turned to Hidetaka Nishiyama sensei for further guidance, at the same time growing the Pacific Northwest Karate Association to the formidable presence that it is today.

I stayed in Oregon on and off for over eight years until I completed my postgraduate education. By then I had attained nidan (second-degree black belt) in the Nishiyama system. When Susan Radtke and I moved to Dallas in the early 1980s, we started a karate club within the Recreation Sports Division at the University of Texas at Dallas.

Soon after, we invited Graves sensei to give a training seminar in Dallas. Graves sensei loves an audience and is famous for recounting historic anecdotes. One evening after dinner, he started telling Nishiyama stories. I consider the following to have bested them all.

Nishiyama sensei often gave karate demonstrations after he came to the United States. One that fascinated most was the gyaku-zuki floating-board break.

In run-of-the-mill breaking demonstrations, wooden boards are braced firmly from behind by fellow students. Backside support reduces board movement when struck, hence ensuring force penetration that splits the target.

A floating board is held only on its top side, with no other support. It is far more difficult to rupture because of the lack of leverage, to the point of defying common sense. Utmost mental concentration is needed, together with an explosive kime to impart shocking force. The board would otherwise fly off intact from the fingers of the holder.

Yet Nishiyama sensei shattered the floating boards unceremoniously and without fail. For him it is no more challenging than routine training at the dōjō.

In a private conversation with Nishiyama, Graves sensei asked about the secret to the floating board break. Nishiyama sensei stared into the distance. He turned to Graves after a pensive moment. "There can be no doubt in your mind," he said. Having known and trained under both of them for decades, I have no doubt regarding the veracity of the story.

The enlightened seeks transcendence in mind and spirit, having perfected physical execution.

Takeaways from "Hand and Foot Techniques in Budō Karate"

- Consistency in hand technique execution requires proper body dynamics, the use of the arms for speed and the hips and legs for power, unlocked shoulder joints, directional control with the elbows, reciprocal motion of opposing arms, and concentrating power delivery with the top knuckles of the fist or the edge of the open hand.

- Kicking generates appreciably more power because of the muscle mass of the hips and kicking leg, and it has a greater reach, but it places a higher demand for balance and hip stability.

- Consistency for foot-technique execution requires strong balance support from the support leg, hip-stabilized center-of-mass transition, strong core strength, directional control with the kicking knee, and proper recovery to sustain balance after the kick.

Shén

Have the courage to use your own understanding.

IMMANUEL KANT, *WHAT IS ENLIGHTENMENT*

It is not death that a man should fear, but he should fear never beginning to live.

—MARCUS AURELIUS, *MEDITATIONS*

Where there is great power there is great responsibility.

—WINSTON CHURCHILL

PART IV

The Spirit

The Spiritual Journey of Martial Arts

Of Soul and Spirit

We humans are likely the only living beings capable of introspection: the capacity to reflect on our thoughts and feelings. Our awareness of time's passage is counterweighed by the desire for permanence. Since the beginning of culture, our existence is defined by the physical body and the metaphysical soul and spirit.

Current philosophy and theological teachings consider the soul and spirit to be inter-twined, yet separable. Latin for *soul* is *anima,* inferring attachment to the physical living body and its innate natures (id and libido). Each of our souls is unique, manifesting as emotion, desire, affection, and response to the senses.[1]

The word *spirit* traces back to a mid-thirteenth-century usage meaning "animating or vital principle in humans and animals." The English word came from the Latin *spiritus* ("breath," "breath of a god") and its Proto-Indo-European origin, *peis,* synonymous with "life," "character," "courage," and "vigor."[2] The belief of an animating life force or stream of consciousness pervades ancient cultures.[3] The Hebrew word *ruach* (mysterious power of the wind or breath, חור) relates to "spirit" whose essence is divine.[4] Similar concepts are found in Hindu (prana), Chinese (qi), and Japanese (ki) philosophies.

According to Judeo-Christian beliefs, the spirit is transcendent as a formless intel-ligent energy, individualized yet connecting us to a higher being. Eastern philosophy subscribes to a broader view—that spirit is within people, beasts, earth, and the cosmos. There is broad consensus across cultures that the spirit embodies a deep-rooted bio-logical intelligence protecting us from disease. We fall sick, age, or die when our spirit subsides.[5]

Teachings abound that our soul is individual whereas spirit is universal for all lives. Thomas Moore, the eighteenth-century poet and songwriter, long considered Ireland's national bard, spoke of "movement of the spirit and attachment of the soul." Hence, the lexicon of "when the spirit moves me" refers to doing something that is apart from habit.

Spirituality

In accepting the concept of spirit, one aspires for spirituality, a desire to connect with something beyond ourselves and to contemplate the meaning of life. It is a universal human condition, its practice profoundly affected by personal experience and social mores.

Early Christian religions pursued spirituality to recover people "in the image of God."[6] Contemporary spiritualism, with its infusion of Eastern ideas, is more nuanced. Many regard spirituality to be the foundation of happiness and inner peace.[7] Secular purveyors follow spiritual paths without worship of supernatural forces to attain moral character (love, empathy, patience, tolerance, forgiveness, contentment, responsibility). Christian author Kristina Kaine considers that spirit moves within our soul, making it more accepting of others and of the future.[8]

Spirituality frees the soul from habits and prejudices that confine one's outlook in life. The spiritual journey is transcendent through introspection, redirecting a materialistic sense of well-being to one based on positive experience. Enlightenment comes from embracing a larger reality. One moves from the realm of "me only" values to welcome family and friends (love, friendship, mutual nurture), community (local, regional, national), global humanity (respect for other people), nature (compassion for all living things, the environment), the cosmos (order versus chaos), to the divine.[9]

Martial Spirit

The secular pursuit of martial spirit in wushu is grounded in *li* and *neijing*. Li (strength, effort, endurance, 力) reveres physical prowess that is epitomized by Shaolin Wushu (see chapter 1). Neijing (inner contention, 內競) seeks inner awareness of qi, its cultivation, and oneness with nature in order to achieve enlightenment. This practice is emblematic of the southern Wudang martial arts (see chapter 16).

The wushu spirit is tethered to a code of conduct called *wudě* (martial morality, 武德), which imparts "morality of deed" and "morality of mind."[10]

Regardless of style or origin, morality of deed has been the lingua franca of the military, law enforcement personnel, and martial artists. This code of conduct was grounded on tenth-century neo-Confucianism with the infusion of Buddhist and Daoist ideas, and it embodies humility, respect, righteousness, trust, and loyalty.[11] *Yi* (moral rectitude, 義) is highly valued, reflecting the complex Confucian concept of justice, tolerance, integrity, restraint, actions founded on strong moral values, or simply the conviction to do the right thing.[12]

This righteous attitude is epitomized by the lore of Guan Yū (關羽), a military general in the warring Third Kingdom Period (220–280 CE). Guan Yū was revered through the ages as a brilliant tactician, a great warrior who "slayed thousands." His legend is held in high esteem to this day across Southeast Asia, symbolic of the Confucian social order for his sense of duty, loyalty, and righteousness. Guan Yū is deified in Chinese Buddhist and Daoist religions.[13] His statue permeates martial arts schools, police stations, as well as triad societies (which have roots as rebels against foreign invaders of China), symbolic of the moral compass for those who assert power over others.

Morality of mind guides the path of discipline and intensity in one's martial spirit. It cultivates the inner balance of intellect (*hui,* 慧) and emotions (xīn, 心), and encompasses will, endurance, patience, and courage.[14] The martial spirit also engenders the tenacity to rise above adversity, inculcated from the strength of resilience from overcoming setbacks in training. The goal is to achieve wu-wei (see chapter 2), a Daoist concept on the mastery of self, to be discussed later in this chapter.

The Japanese martial spirit has long been synonymous with bushidō (the warrior's way, 武士道) and its contemporary interpretation as budō (the martial way, 武道). These codes of chivalry trace back to the nation's war-torn heritage and an idealized notion of the samurai as the model warrior.[15]

The literal meaning of *samurai* (侍) is "one who serves." They were men of noble birth assigned to protect the imperial court. The service ethic spawned roots for the samurai nobility caste, who distinguished themselves with their social standing and spiritual authority.

Contrary to common belief, the unified bushidō code was a relatively modern nineteenth-century convention.[16] This ideological treatise on martial ethics and moral aspirations came from Inazō Nitobe's English publication titled *Bushidō: The Soul of Japan* in 1899.[17] While there were codes of duty and honor prior to this era, values differed greatly by time period, region, and individual.[18]

Bushidō unquestionably drew influence from seventeenth-century Chinese Confucianism and Daoism, giving recognition to the concept of dō and emphasis on social order.[19] Nitobe's portrayal of Japan's indigenous "feudal knighthood" also echoed the gallantry of the medieval European knight and attitudes of nineteenth-century English "gentlemanship" (*jentorumanshippu* in Japanese literature).[20]

As Japan implemented Western attitudes and technology after the Meiji Restoration (1868–1912), the success of the reformation was credited to the indigenous samurai spirit and accompanying virtues such as nationalism, diligence, and *giri* (duty or obligation, 義理). The popularity of bushidō reached its zenith after Japan's victory in the 1895 Sino-Japanese War.[21] These core beliefs, however, were soon tempered by social unrest from the elimination of the samurai class and notoriety from Japan's imperialistic aggression in the early and mid-twentieth century.[22]

With the loss of the samurai class, the bushidō era gave way to budō, founded on the Zen-influenced belief that the martial spirit is directed at personal enlightenment through betterment of mind, body, and spirit. Budō aspires for honor, discipline, fidelity, rectitude of judgment, and enlightenment through constant improvement of one's craft, embodying moralities of mind and deed that transcend history and time.

Moving Zen: Perpetuating the Martial Arts Spirit

The Annual AAKF Summer Camp, later renamed the International Traditional Karate Federation (ITKF) Summer Training Camp, ranked as one of the most attended martial arts training events from the late 1970s to the early 2000s. Spearheaded by Hidetaka Nishiyama (Shōtōkan/JKA) and Richard Kim (Okinawan kobudō, Shōrinji-ryū), the camp initially was modeled after the grueling Kenshusei (Instructor Intern Training) program that Nishiyama codeveloped while at JKA Honbu Dōjō in Tokyo. Karate-ka of all levels of training attended, although instruction was geared toward the majority black-belt participants. The program also incorporated competition training for athletes at national and international levels.

The weeklong camp has developed a reputation for its high caliber of instruction over the years, and to a lesser extent, the appeal of the locale and the climate. Held in balmy La Jolla, California, at the University of California San Diego campus, the weather rarely drifts below 50 or above 90 degrees Fahrenheit, conducive to extreme workouts and recovery of sore muscles.

The camp started with about twenty-five trainees in the early 1970s that included the famed Ray Dalke and Frank Smith.[23] By the 1980s and 1990s, international attendance ballooned to more than three hundred. Nishiyama sensei and Kim sensei personally conducted the full week of vigorous training. Attendees also learned from the stellar teaching roster of who's who in traditional karate. The distinguished guest teaching faculty over the years comprised renowned instructors worldwide, including Masatoshi Nakayama (Japan), Kenei Mabuni (Japan), Tomoharu Kisaki (Japan), Keinosuke Enoeda (Britain), Hiroshi Shirai (Italy), Takeshi Oishi (Japan), Masaaki Ueki (Japan, currently chief instructor of JKA), Yutaka Katsumata (Canada), and Masoe Kawasoe (Britain).

The participants' spirit energized the six-court gymnasium during daily workouts, and its wood floor rumbled like an earthquake. Each session began with mokusō (meditation). Two long rows of black belts sat seiza, stretching from one end of the gymnasium to the other. It was a sight to behold.

Training incorporated credentialing lectures from 1990 on. The didactics were part of Nishiyama's vision to seek IOC recognition of traditional karate as an official Olympic sport. But the primary focus has always been about hands-on training. Everyone was expected to adhere to the timetable, unless waylaid by sickness or injury.

Daily training started at 6 a.m. with a one-mile run at the beach (fig. 14.1), followed by limbering drills. The morning sessions continued after breakfast and a brief rest, with three hours of kihon focusing on body dynamics and technical fine points. Three hours on kata and kumite followed in the afternoon, while select athletes took part in separate, even more rigorous competition training.

Enthusiasm from the first day was soon dampened by the extraordinary demands on the body from seven hours of strenuous daily workouts. Most hit a wall of fatigue by the end of the second day. Yet campers enjoyed a heightened spirit by the fourth day, the body having acclimated to the physical exertion. They felt rejuvenated by newfound stamina and the camaraderie of fellow participants; the mind now had been purged of extraneous thoughts other than *train, eat, rest, repeat!* Enduring friendships developed among fellow karate-ka, who came from all corners of the earth to share this unforgettable transformative experience.

Martial arts practice for the betterment of mind, body, and spirit is often considered "Moving Zen," meaning the way of attaining enlightenment through action. Its attribution to martial arts was popularized in the West by martial artist and author C. W. Nicol. In his autobiography of the same title, Nicol recounted his transcendent experience while studying Shōtōkan/JKA karate-dō in Japan for two and a half years. Nicol considered

Figure 14.1 Beach training. On the closing day of the ITKF training camp, participants waded into the shallows of the Pacific Ocean and practiced punching and kicking against the forces of nature.

this apprenticeship as an initiation to the gentleness of the soul and enlightenment of the spirit.[24]

Moving Zen (動禪) has been part of Zen (禪) Buddhism since the thirteenth century CE, brought from China to Japan by the Japanese monk Eihei Dogen. Zen Buddhism originated from Chinese Chán (Zen in Japanese) Buddhism, a fusion of Daoist philosophy and the Mahayana Buddhist concept of dhyāna (training of the mind through meditation).[25] Dhyāna is to "contemplate," "meditate," or "think," centering Chán and Zen practice on "the meditative state."[26]

Meditation seeks awakening of the mind and spirit, although each Zen Buddhism school differs in emphasis and practice.[27] A key doctrine is to achieve kū (emptiness, 空), the absence of self from physical and mental elements of existence, and busshō (Buddha Nature, 佛性), the natural and true state of the mind.

Soon after the popularization of Zen Buddhism in Japan, the Rinzai school came into prominence. Rinzai Zen focuses on *kenshō* (seeing one's true nature, 見性); and *mujōdō no taigen* (free functioning of wisdom, a state of spontaneity without judgment in daily activities, 沒有約德諾的表現) as a gateway to enlightenment. In addition to zazen (seated meditation), Rinzai Zen proactively engages in *kōan* (paradoxical anecdote, 公案) and *samu* (physical work done with mindfulness, productive service, 作務) to attain busshō.

Other Zen schools considered these approaches as overt and intrusive. Nonetheless, Rinzai Zen's goal-oriented practices were broadly endorsed by the thirteenth- to seventeenth-century shogun and samurai castes. The bonding of Zen, particularly Rinzai Zen, and bushidō peaked by the late nineteenth century.[28] Introspective Moving Zen practice became a staple of martial arts training, revered for sharpening the mind as well as its interpretation of karma by aspiring to higher moral values. Following the abolition of the samurai caste, technique-oriented bushidō and bujutsu (techniques of the warrior, 武術) evolved as the aspirational budō attitudes to guide mind, body, and spirit toward realizing one's full potential.

Given the Daoist influence on Zen Buddhism and Southern wushu (see chapter 2), it comes as no surprise that Moving Zen tracks parallel to the concept of gōngfu, broadly interpreted as a depth of knowledge and expertise to translate knowledge into outcome (chapter 12). Chinese master chefs, calligraphers, painters, sculptors, and craftspeople were also accorded the honor of having mastered gōngfu well before the term's contemporary exclusive reference to martial arts. These experts in their field were regarded to have achieved the natural state of *ziran* (self-organization, 自然), the central Daoist value of spontaneity and creativity that emulates native forms of the cosmos.[29] Ziran assigns process to the attitude of wu-wei for Daoist martial arts (qigong, taijiquan, xingyiquan, baguazhang), aspiring for unencumbered actions, even when under duress, that are free from conscious thought.

Likewise, the flowering of Japanese culture after the Meiji Restoration was attributed to Moving Zen practice, including calligraphy, painting, literature, tea ceremony, and martial arts.[30] Zen (inner peace) practice covets the mushin (no mind, 無心) mind-set, abbreviated from *mushin no shin* (無心の心), or "being free from mind-attachment" that parallels the attitude of wu-wei.[31]

In particular, "mind of no mind" is highly sought in martial arts practice. Through cultivation of the spirit, one achieves mushin that focuses on the very moment, casting out extraneous feelings of inadequacy, anxiety, fear, and ambition that weigh the spirit. The mind is agile and far-reaching, freed to act and respond without hesitation, and is unhindered by emotional longings.[32]

Releasing the Spirit: Bushidō and Seppuku

America's fascination with the samurai and bushidō culture came decades after the abolition of Japan's warrior class. While critics scoffed at Nitobe's romanticization of servitude in feudal Japan, the author's discourse took hold to signify the Eastern rectitude of personal and professional chivalry.

Apart from Zen Buddhist and Confucianist influences, the unified bushidō code also embodied Shintoism and the era's socioeconomic trends.[33] The Eight Virtues of Bushidō (rectitude and justice, courage, benevolence and mercy, politeness, honesty and sincerity, honor, loyalty, character and self-control) defined the proper discharge of giri and other moral obligations.[34] Through Buddhist beliefs, the samurai accepted the cyclic inevitability of life and death. This elite military caste was committed to the ultimate sacrifice in order to maintain public order and to properly serve under their master's charge.[35]

Bushidō was officially abandoned after World War II but enjoyed a resurgence in the 1970s. It came to represent the spirit of Japan during that period's rising economic growth and post–World War II nationalist movement.[36]

Yukio Mishima was the pen name of Kimitake Hiraoka. He was a highly acclaimed poet, playwright, actor, and film director of the twentieth century, and a nominee for the 1968 Nobel Prize in Literature. Mishima committed seppuku (ritualized suicide, 切腹) on November 25, 1970.

Mishima's body of work focused on inner turmoil between the ideals and the desires of contemporary Japan. His personal life was equally conflicted. Mishima's later works obsessed with blood and death, increasingly gravitating toward a self-destructive persona.[37] As an ideological right-wing nationalist, Mishima detested the modern materialistic Japanese society.[38] He formed the Tatenokai (Shield Society; 楯の會), an unarmed civilian militia of over a hundred members, seeking to reinstate the emperor as Japan's de facto head of state.

The scandal that ended Mishima's life would later be known as *Mishima jiken* (the Mishima incident). On that fateful day in November 1970, Mishima and four Tatenokai members gained an audience with the Tokyo commandant of the Self-Defense Forces (Japan's military after World War II) under false pretenses. They barricaded the office and held the commandant hostage. Mishima then stepped out onto the balcony and implored the troops to join his coup d'état. Soon realizing the futility of his cause, Mishima turned back into the office and proceeded to kill himself, an act that he may have planned months before.[39]

In twelfth- to nineteenth-century Japan, seppuku was performed on the battlefield by the dishonored samurai or by edict of judicial punishment. In accordance to bushidō, seppuku served to redeem honor from cowardice or defeat in battle, shame over a dishonest act, or loss of sponsorship or death of a *daimyō* (great lord). It also may be an extreme protest over policy clash with one's superior. With ultimate sacrifice, the samurai preserved his reputation and his entire family's standing in society.[40]

In the highly planned and witnessed ritual, the samurai prepared himself by eating a final meal, then bathed and dressed in all-white *shini-shōzoku* (funeral garb). He knelt in formal position as his *kaishakunin* (appointed second, 介錯人) stood behind in waiting with a sharpened katana. The samurai might choose to read a final poem that he composed for the occasion. Collecting his resolve, the samurai held either his own wakizashi (companion sword) or a provided dagger wrapped in white cloth. He then opened his kimono.

With both hands on the knife, he plunged the dagger deep into his left abdomen. He dragged the dagger across from left to right, ignoring the extreme pain, and sliced open the belly. He turned the blade upward toward the diaphragm to ensure a lethal wound.[41] To mitigate prolonging agony, he signaled his kaishakunin to perform a skilled downward cut (*kaishaku,* 介錯), which partially decapitated the samurai without having the head roll onto his lap.[42]

In the case of Mishima's grisly act of defiance, the petrified kaishakunin was unsuccessful with the kaishaku cut after three attempts. Another member had to step in to bring Mishima's trauma to its end.[43]

The antiquated practice of seppuku rarely occurred after the abolition of the samurai class. Misguided critics nevertheless condemned this practice as barbaric and indicted bushidō as a culture of death. This couldn't be any further from the truth.

The concept of a "good death" perpetrates all ancient cultures, dating back to Solon, the sage of Athens, who argued that one couldn't judge a person's happiness until one was privy to the manner of his death. The Greeks recognized that we are all destined to die; the best we can hope for is a death that benefits family or humanity.

In the eighteenth-century monograph *Hagakure Kikigaki* (hidden leaves, 葉隱聞書), author Tsunetomo Yamamoto considered that, "the Way of the Samurai is founded in death." Contextually, however, this opinion was not to encourage the samurai to actively seek an end to life. A dead retainer served no purpose to his master. A more studied interpretation is that in seppuku, the samurai accepted "honorable death" as the ultimate sacrifice. A samurai was expected to constantly place his life in the balance. By making final reparation as befitted his station, he gave his life to preserve honor, integrity, and fidelity.[44]

Ancient warriors considered seppuku as resolution to one's inner conflict between mind (duty, honor) and spirit (passion, will). Other than the Mishima jiken, well-known incidents of seppuku in the twentieth century included General Nogi Maresuke, who disemboweled himself in 1912 out of loyalty to the deceased Meiji Emperor. Numerous military personnel and civilians also chose the sword over surrender during World War II.[45]

Zen Buddhism subscribes to the belief that one's spirit resides within the hara (lower abdomen below the navel; see chapter 11). The conscious act of seppuku releases one's spirit, liberating it to a better place.

Contemporary budō values have served to uphold order and ceremony in martial arts practice after the Meiji Restoration. While bushidō tempers the soul through the morality of loyalty and obligation, budō seeks to elevate the spirit through an understanding of self and the maximizing of one's potential. An enlightened citizenry lends credence to the cultivation of benevolence in society. The collective protocol was now centered on *wa* (harmony, togetherness, 和), which sustains the tradition of service and self-sacrifice and manifests as cooperation and preserving the common good. Pierre Teilhard de Chardin, the French philosopher, paleontologist, and Jesuit priest, once said, "We are not human beings having a spiritual experience. We are spiritual beings having a human experience." In the next chapter, we will examine manifestations of the spirit in human endeavors and the nurturing of the human spirit through the paths of martial arts.

Takeaways from "The Spiritual Journey of Martial Arts"

- Our actions are driven by our spirit, the essence of our character, passion, courage, energy, and aspirations.

- Morality of the mind cultivates the inner balance of intellect and emotions, forging discipline and intensity of the martial spirit.

- Martial arts training places high demands on our physical body and psyche. As students of the art, we recognize that progress is built on learning from failure. Sustaining a strong spirit is essential for this path.

- "We are not human beings having a spiritual experience. We are spiritual beings having a human experience."

Manifestations of the Spirit: The Will and the Way

Until the Battle Is Won

On Monday, August 14, 2017, the 1,217 cadets of the West Point Class of 2021 marched for twelve miles from Camp Buckner to Central Post of the Military Academy in full military gear.[1] Led by the class banner, inscribed with "Until the Battle is Won," the ceremonious March Back was affirmation of having survived the initial curriculum, known as "Beast Barracks."

Beast Barracks, or simply "Beast," marks the cadet's first basic training hurdle in four years.[2] The seven-week summer curriculum comprises code-of-conduct indoctrination that teaches cadets how to act, speak, and present themselves in a soldierly manner. Operational trainings cover military tactics, weapons, rappelling, land navigation, and the use of gas masks.

Many consider the grueling schedule to be the hardest transition from civilian to military life. Reveille comes at 0500 (5 a.m.), followed by running or calisthenics before a shower and breakfast. Formal training ends at 1700 (5 p.m.), after which cadets take part in competitive group sports (mass athletics). There are more briefings and additional training after dinner at the squad level. Dismissal falls at 2130 (9:30 p.m.), and lights-out at 2200 (10 p.m.).[3]

Cadets admitted to the U.S. Military Academy have passed rigorous selection comparable to that of an Ivy League University. These men and women represent the top 10 percent of all applicants. Their grades average 3.64 and accompany equally stellar standardized test scores. Applicants are also evaluated for physical and leadership potential, which, together with their academic achievements, form the basis of an aggregate Whole Candidate Score (WCS).[4] This metric is used by the academy to predict the candidate's likelihood of surviving the rigors of West Point training.

Ninety-seven percent of the class of 2021 completed Beast Barracks, but it must be heartbreaking for the thirty-plus cadets who didn't. The preadmission WCS score did not effectively reflect their fitness to overcome this formidable challenge.[5]

Grit

In 2007, a research study titled "Grit: Perseverance and Passion for Long-Term Goals" was published by Angela Lee Duckworth, noted psychologist and professor at the University of Pennsylvania. Duckworth showed that perseverance and passion were key contributors of success for young overachievers, in addition to their native intelligence.[6] The social psychology community was instantly enthralled by the formal validation of these commonly held beliefs.

Professors Duckworth and Chris Peterson considered psychological parameters called "grit." They developed a Grit Scale that measures a person's attentional focus and inclination to follow through (consistency of interests and perseverance of effort). Respondents evaluated themselves by answering twelve simple questions on a scale of 1 to 5. Their score serves as an indicator of their determination to complete near-term and long-term goals.

Duckworth's study enlisted five thousand participants, including twenty-five hundred West Point cadets from the classes of 2008 and 2010. The test was administered before the cadets entered Beast Barracks. Their confidential responses were unsealed after the seven-week curriculum.

For the classes of 2008 and 2010, dropout rates were around 5 percent (71 cadets or 5.8% of the 2008 class, 61 cadets or 4.7% of the 2010 class). The Grit Scale turned out to be highly effective at predicting success at Beast Barracks, garnering a higher level of accuracy than WCS, intelligence quotient, or grade point average.[7] It was as effective as the WCS in forecasting a cadet's overall performance through the first year of training.

Duckworth modeled this study on her own life experience as a late bloomer. While her peers achieved success in their twenties, Duckworth did not commit to a career until her mid-thirties despite top flight academic credentials.[8] She theorized that achievement requires both talent and commitment, taking a page from the century-old dissertations of Professor Catharine M. Cox that, IQ aside, lifelong success can be predicted by childhood traits of "persistence of motive and effort, confidence in their abilities, and great strength or force of character."[9]

Beast Barracks call on a person's fortitude by deliberately challenging physical, emotional, and mental presumptions. The Grit Scale represents self-assessed views of the candidate's mental focus, determination, and tenacity. Duckworth's retrospective analyses found that follow-through was the key contributor to survive Beast Barracks.

Duckworth defines this attribute as consistency in one's long-term goals and the stamina by which one pursues these goals over time.[10] Grittiness engenders the confidence to overcome the unanticipated from moment to moment.[11] Unless the cadet is prepared to confront the dawn-to-night challenges head-on, they succumb to the urge to abandon the program. Her subsequent studies showed that follow-through was also the single best predictor for success (apart from intellectual talent) in science, art, sports, and communications, to name a few. While some are born with a high level of grit, Duckworth considers grit as a trait that develops through experience, enabling sustained interest and effort toward long-term goals.[12] Those who devote time and effort in martial arts training will find that this pursuit is also conducive to forming the habit of grit.

Nurturing the Martial Arts Spirit

A person's grittiness falls within five common traits—namely, courage (true to one's belief), conscientiousness (vigilance, attentiveness), follow-through (perseverance, focused efforts toward long-term goals), resilience (optimism, toughness, confidence), and passion (deep sense of purpose, will, enthusiasm).[13] Psychologists consider that these "noncognitive" attributes reflect a person's patterns of thought, feelings, and behavior.[14] The traits continue to evolve over one's lifetime and do not peak until late adulthood.[15]

Spirit is defined in the *Oxford English Dictionary* as "the quality of courage, energy, and determination." As applied to martial arts, spirit is synonymous with confidence, fortitude to overcoming adversity, and the conviction to sustain the moralities of deed and mind (see chapter 14); in other words, it's the possession of grit.

These parallel concepts infer that nurturing of the martial spirit also fosters success in everyday life. Circumstances dictate that we cannot be the strongest, quickest, and in control at all times. The martial spirit aspires for an attitude of sustained equanimity and confidence under duress. To achieve morality of the mind, the martial artist focuses on exercising will, discipline, patience, and courage.

The expectation is to perform at the highest level in order to sustain duty, honor, and dignity. Through disciplined training, one learns to embrace setbacks and fatigue to rise above the fray. The following summarizes the attitudes and practices that sustain the martial spirit.

1. Purpose

Having secured physical comfort and security, our mind craves a higher calling, often termed purpose. In its most abstract, we seek the purpose of life—the reason why anything is done, is created, or exists. Purpose guides life decisions and gives a sense of direction.[16] A sense of purpose motivates us to get up in the morning. Purpose enriches our spirit by giving meaning to day-to-day activities, instilling motivation and tenacity.

To enrich the spirit, social influencers preach to experience as many things as possible in life.[17] Asian philosophies such as Zen Buddhism take an introspective path, believing that spirituality starts with an understanding of the enlightened self.

Budō seeks to achieve *kenshō* (seeing one's true nature) through samu (mindful physical work), the paths (dō) of Zen that elevate spirit by melding mind and body (see chapter 14). The martial artist immerses themself in deliberate practice, refining techniques and then testing them in kata performance and sparring exercises. The technical toolbox in any martial art is far from bottomless. To perform each technique correctly is only the beginning of learning. There's always room for improvement as the student seeks to elevate power output, precision, and consistency. One accepts setbacks as part of growth and relishes scaling higher technical planes.

The enduring process builds competence and mastery, fortifying the spirit. It gives confidence to overcome adversity. The realized skill set spurs the desire for constant improvement and instills passion for lifelong pursuit. Over time, the spirit of dedication and discipline spills over to everyday endeavors, soon becoming a way of life.[18] From increased confidence comes the urge to shape our own destiny and the mettle to steer daily outcomes in a positive way.

2. Positivity

Psychologists consider the need to be perfect as detrimental to success.[19] Perfection is subjective. To declare that "My manager, Josh, is a perfectionist," is not necessarily a compliment. It implies that Josh is more invested in crossing t's and dotting i's than in fulfilling overall objectives. An obsession to abide by an arbitrary standard of flawlessness comes from insecurity, which invites anxiety, low self-esteem, and clinical depression.[20]

In contrast, the martial spirit aspires for an attitude of excellence and consistency in outcome. One accepts that circumstances evolve rapidly and relies on proven judgment to arrive at effective solutions. These virtues pave the way to confidence and determination to fulfill purpose and function, and the tenacity to accommodate setbacks and emotional struggles as part of personal growth.[21]

An aspiring martial artist passes eight ranking examinations before reaching shodan (first-level black belt, 初段) in the Shōtōkan-style karate systems. They devote more than two years, likely three, in diligent uninterrupted training. Each rank promotion incentivizes by the realization of incremental learning goals. The student's standing is the recognition of their progress through the ranks. Having endured in-the-face challenges, risk-taking, and setbacks, the young yūdansha (black belt, 有段者) owns their elevated mental strength and buoyed spirit through perseverance and determination. Like accomplished experts in athletics, chess, music, or mathematics, a black belt confronts the psychological challenges of failure as stepping-stones for success and personal growth.[22]

Empirical evidence suggests that 1 to 3 percent of beginners eventually reach the rank of shodan. This watershed recognizes the student's skill, efforts, and perseverance and represents an endorsement of their command for the fundamentals of the art. They have earned the privilege to embark on a more personalized learning experience that further shapes their identity as a martial artist. The skill set and understanding of the art are melded by the martial artist's unique psyche. This transcendent journey requires the martial artist to wade into new terrain as teacher and arbiter of the art, reexamining and interpreting each technique that incorporates their personality when applied under different circumstances.

The enlightened martial artist is the modern-day stoic who recognizes that real life is imperfect.[23] The key is to set goals for the best possible outcome through continuous improvements. Along with progression in rank, guidance and feedback from the instructor bolster their confidence. The student accepts that perfect form is predicated on proper function, and beauty in outside form is secondary to correct execution. Proper body dynamics supersede stilted actions that feel powerful but are not.

This maturational process calls for the utmost dedication and passion, earmarked by corresponding ranking promotions from shodan to *jyūdan* (tenth degree, 十段). The entire process engages the individual for over fifty years, by all accounts a substantial investment of one's lifetime.

3. Community

As social beings, we crave fellowship with others who share commonalities in attitudes, interests, and goals.[24]

I've had the privilege to train in numerous traditional dōjō over the last several decades, either as a member or a visitor across the United States, Canada, South America, Europe, and Asia. These experiences of meeting, training, and interacting with sensei and fellow martial artists have never ceased to enlighten and inspire.

Budō dōjōs commonly display the dōjō kun (training hall rules, 道場訓) at the entrance or the dōjō facade *(shomen)*. Dōjō kun is a code of conduct that defines personal behavior. To instill mutual respect (table 15.1), its origins traced back to the precepts of practice as promulgated by Gichin Funakoshi and Ankō Itosu. The guiding principles mirror morality of deed in wushu arts, to be observed at all times by members as well as visitors.

Table 15.1 Shōtōkan dōjō-kun.*

一、人格完成に努めること
First: Seek perfection of character
一、誠の道を守ること
First: Be faithful
一、努力の精神を養うこと
First: Endeavor to nurture spirit
一、礼儀を重んずること
First: Respect others and observe etiquette
一、血気の勇を戒むること
First: Refrain from hot-blooded behavior

*The five guiding principles of Shōtōkan karate-dō, according to Gichin Funakoshi.

Students come from all walks of life, ethnicities, and genders. The adult class may comprise professionals drawn to the discipline of training and mental and physical diversions to offset demands in their day jobs. Varsity athletes and fitness enthusiasts come for the well-rounded curriculum, while adults of all ages may seek the allure of being a martial artist, then stay for the training and camaraderie. Some joined at grade-school age and still attend as adults. Thrill-seeking bad apples, whose main concern is self-gratification, can be found in any dōjō. Yet, when fully engaged in the intensity of the workout, everyone's ego takes a back seat in order to keep up with demands. A wise sensei understands that exhilaration is earned through hard work and mental focus and hence organizes each class accordingly.

The apprenticeship system of martial arts traces its origin as far back as the art itself. The historical Chinese martial arts title of *sifu* (or *shī fù,* "master and father," 師父) assigns authority and respect to the head of the school, who is expected to impart knowledge as well as moral and philosophical guidance. Japanese budō arts take after Zen Buddhist influence and accord the title of sensei (one who is born before you, 先生) in deference to the master's wisdom and expertise.

Students are given titles of senpai (senior student, 先輩) and *kōhai* (junior student, 後輩) based on rank seniority. The hierarchy of sensei and senpai functions as the backbone of a dōjō. This is similar to the wushu order of shifu, *shī xiōng/sihing* and *shī jiě/sije* (teacher's older son and daughter, 師兄-師姐) and *shī dì/sidai* and *shī mèi/simui* (teacher's younger son and daughter, 師弟-師妹).

Each school or dōjō is a social microcosm of the martial arts community at large. Members are bound by the common core of purpose and mutual respect as stipulated by the dōjō kun. By exercising vigilance, all are assured of relative safety despite training intensity and the spirit of competitiveness. In turn, members have the collective responsibility of etiquette and proper regard for one another, hence contributing to the collective spirit and well-being of the dōjō.

Over time, seniors earn respect through dedication, hard work, and mastery, serving as examples to younger students; otherwise, they would not be encouraged to stay. The senpai teaches by example when paired with kōhai in sparring drills, challenging the younger student with speed, power, precision, and control. Senior students may also take on the role as junior instructors, hence assuming leadership as the sensei's proxy and nurturer of younger students.

In a thriving dōjō, sensei and senpai are each teacher and student at the same time. They mark the kōhai's path of development by striving for a deeper understanding of the what, why, and how of the art. By illustrating the possibilities of dedicated training in person, accomplished members serve as role models for kōhai to venture out of their comfort zone, inspiring confidence to face uncertainties. Many senpai attend the same dōjō for decades. Remarkable life-long friendships are forged from pushing one another to the limits.

4. Contemplation

The meditative discipline of zazen (seated contemplation, 座禅) has been central to Zen Buddhism since inception. Buddha was believed to reach nirvana (enlightenment, state of total liberation, 涅槃) while in zazen. Zazen is a simple idea: just sitting and letting our breath focus on the present. Yet taming the wandering mind takes a lifetime to master, while each session offers a unique experience.

Budō arts subscribe to mokusō (silent contemplation, 黙想) to nurture the spirit and prepare the mind. Mokusō, which translates as "the silencing of conscious thoughts," is the act of meditation on the dōjō floor without assuming the zazen pose. It is carried out at the beginning and end of training.

Mokusō seeks to recenter oneself to the present, liberating the spirit from burdens of anxiety, regret, anger, and fear. The practice summons awareness without analysis or emotion. By focusing on the task at hand and at the very instant, we fully attend to ourselves and our surroundings.

This is also the concept of mushin (no mind, 無心) (see chapter 2). The martial artist takes the meditative path to achieve clarity of mind even when under pressure. It is a mind without emotional hindrance, more open and responsive to subtle sensations, intuition, and capable of spontaneous response.[25]

In mokusō, the assembled class lines up by rank on the dōjō floor, facing the shomen. Everyone follows the lead of the most senior student to assume seiza (proper sitting, 正座), kneeling first with the left knee, then the right. One's backside is settled on the calves and upturned heels (see fig. 15.1).

Proper posture is key to the process. The spine is erect, and the head and neck are aligned with the shoulders on the same frontal plane (see chapter 9). The chin is slightly tucked, and the external muscles are relaxed, other than to maintain posture and core integrity (fig. 15.1). The hands rest on top of the thighs in a state of preparedness, instead of the zazen half-clasped *hokkai-join* (cosmic mudra, gesture of reality) held below the navel. Eyes are held at level gaze. As the senpai snaps the command of "mokusō!" everyone's eyes become half-closed. Gaze is fixed to the floor about three to five feet ahead. The tongue is relaxed.

Mokusō begins and ends with gentle, controlled breathing (see chapter 11 and fig. 11.3). The calm and deep rhythmic cycle is centered on controlled exhalation from the diaphragm while putting aside all other concerns of daily life. Attention turns inward to attend to one's ki of the very moment, although all senses are open to the surroundings. Martial arts breathing is preferred through the modulation of IAP (see chapter 11). This process also primes the core muscles for the impending workout.

Figure 15.1 The author holding mokusō at beginning of practice.

Exhalation is held for a longer interval (five or six seconds) than inhalation (three or four seconds). Mokusō is generally conducted for three to five minutes before physical training, and for the same period or longer thereafter.

Over the course of inhalation, veterans of mokusō channel the in-breath (qi) through the middle and upper dantian, which collects at the lower dantian. To complete the breathing cycle, exhalation centers on the lower dantian, expelling air through the modulation of intrabdominal pressure. By maintaining proper posture, the student seeks connectedness to cosmic energies at the crown of the head, and earth's energies at their feet (see chapters 11 and 16).

Spirit and Meditation

His Holiness the Fourteenth Dalai Lama, Tenzin Gyatso, has long been a vocal advocate for understanding Buddhist spiritualism through the lens of modern science.[26] In 2003 the Dalai Lama took part in the scientific conference titled "Investigating the Mind" with Nobel laureates Daniel Kahneman and Eric Lander at the Massachusetts Institute of Technology.[27] This high-profile event helped to launch the research subdiscipline known as contemplative neuroscience. The new area of study makes use of modern scientific tools, such as functional magnetic resonance imaging (fMRI), to better understand brain function and consciousness in "introspective, first person" experiences, such as Buddhist meditation.[28]

The year 2005 was the Dali Lama's seventieth birthday. The Dalai Lama paid a ten-day visit to Washington DC to meet with President George Bush. He also lectured at the 35th Annual Meeting of the Society for Neurosciences. His address to the world's largest gathering of brain scientists was not without controversy, given the Dalai Lama's limited scientific background. As he took the podium, he quickly disarmed the audience with a loud sneeze. His usual charismatic self, the Dalai Lama complained of a sudden affliction from modern-day stress.[29] He professed an interest in science from an early age, then, in a mix of halting English and Tibetan, expounded on parallels between Buddhist spirituality and contemporary science.[30]

The genesis of this lecture dated to 1992, when he first met the neuroscientist Richard Davidson in Dharamshala, India. "You've been using the tools of modern neuroscience to study depression, and anxiety, and fear," commented Gyatso. "Why can't you use those same tools to study kindness and compassion?"[31] Dr. Davidson initially was taken aback by his forthrightness, but from this gathering sprang a lifelong friendship and an enduring scientific collaboration between the fourteenth incarnation of the Living Buddha of Compassion and the professor from Madison, Wisconsin.

Some twelve years later, Davidson published the first of a series of scientific findings in the *Proceedings of the National Academy of Sciences,* one of the world's most prestigious peer-reviewed scientific journals.[32] These insights paved the way to a scientific understanding of the effects of meditation on brain function.

There are more than twenty forms of contemporary meditation practiced today.[33] They can be divided into three major categories—namely, focused attention, mindfulness (open monitoring), and loving kindness meditation, as exemplified by the practices of the Tibetan monks in Davidson's study.[34]

Focused attention is the starting point for the novice meditator. They are required to concentrate on an object or an event, such as the in and out cycles of breathing. This deliberate effort keeps the mind from drifting. Mokusō before and after martial arts training is a form of focused attention. It nurtures the spirit by steering innate awareness to our breath (ki in Zen Buddhist thought, qi in Daoist thought). Mokusō is also part of the Zen practice called *joriki* (stabilized strength, 定力), referring to a mind that has been tethered to a singular task.[35]

Mindfulness (open monitoring) is the second level of meditation, to be practiced after the practitioner learns to focus their mind. The mindfulness concept originates from the ancient Indic word *sati,* translated into the Chinese Buddhist concept of *nian* (in Japanese *nen,* 念), meaning having achieved the presence of mind. Whereas focused-attention meditation is directed at a single object or event, mindfulness (open monitoring) practice requires the meditator to attend to all external or internal matters of the very moment, but without being burdened by them. This harks back to the Daoist meditative process of *neiguan* (inner vision, inner observation, 內觀), of turning the mind's eye inward to embrace elements of nature within the body, such as to mentally channel the circulation of qi through the body's meridians (see chapter 11).

Loving and kindness meditation represents the highest level that seeks the spiritual state of *anatta* (non-self, 無我). The practitioner develops positive feelings of love and compassion by forsaking negative emotions, then extends the same benevolence to family and friends, people within the community, all humankind, and ultimately to all living beings.

The three types of meditation produce common as well as specific neuroadaptations.[36] Neuroscientists regard all forms of meditation as "cooling off the brain." It orders thought processes, centers our stream of consciousness, and diverts negative mental energy that saps the spirit. There is strong evidence that meditation streamlines the neurological network as well as modifies brain structure.[37] As we learn a new skill such as a novel kicking combination, brain areas that control the activity enlarge progressively as we master

the skill. Similar neural adaptations (neuroplasticity) are evident over the course of prolonged meditative practice.[38]

Participants can realize markedly reduced stress, anxiety, and fatigue after only five
twenty-minute sessions of integrative body-mind training (IBMT).[39] IBMT is a form of
mindfulness meditation based on qigong exercises of relaxation, breathing, and centering postures. The participants enjoyed measurable gains in attention, along with lowered
heart rate and oxygen consumption and modified brain-wave patterns. They learned to
temper their cognition skills and emotions and moved away from the impulse to simply
react to external stimuli.[40] The key body functions of meditators are shifted toward
parasympathetic, autonomous regulation and hence are less susceptible to emotional
fluctuations.

Coordinative brain functions generate rampant bursts of electrical signals among a
large number of neurons. Their synchronized activities produce electromagnetic waves
of varying frequencies, detectable by EEG sensors that are placed on the scalp. Mid-
frequency (3–8 hertz) theta brain waves predominate during learning and the forming
and recall of long-term memory. Theta waves reflect subconscious awareness and are
prevalent during vivid dreaming and intuitive thinking (see chapter 3). The brains of
IBMT participants displayed highly prominent theta brain waves as they underwent the
meditative process, indicative of internalization of their senses and a state of well-being.

Brains of the Tibetan monks in Davidson's study emitted even higher-frequency
(25–70 hertz) far-reaching gamma brain waves as they entered deep meditation. These
seasoned practitioners observed loving and kindness meditation, as is common in the
Tibetan culture (*dmigs med snying rje,* "benevolence and compassion pervades the mind
as a way of being"). Gamma-band waves represent neural-synchronizing signals during
extremely high brain activities, reflecting the neural-interactive process that aligns attention, working memory, learning, and conscious perception. High gamma activity is tied
to heightened cognition and social and emotional experiences, which was confirmed by
Davidson with the monks' brain imaging profiles.[41]

The monks had far higher gamma activities even at rest. Their gamma pattern spiked
to five times the level of control subjects and were sustained at the heightened state over
the course of meditation. Davidson concluded that meditation elevates attentional functions and can even impart feelings of affection.[42]

The attributes of love and compassion meditation align with the martial spirit of duty,
honor, mutual respect, service, and seeking the common good. Equating the goals of
meditation with those for martial arts practice may seem disingenuous. Nonetheless,
the activism of Shaolin martial monks was well known throughout China's history. Their

martial spirit manifested as yi (moral rectitude) through efforts to intervene on behalf of a downtrodden citizenry (see chapter 14).

In Zen Buddhism, to become one with the world is considered key to authentic, durable happiness.[43] A person's certainty of purpose is balanced on awareness of self and the altruistic non-self. The non-self perspective maintains that empathy, compassion, and consideration for others counterweighs self-gratifying egotistic pursuits.[44] An enlightened spirit, acquired through meditation or budō practice, achieves a state of "non-duality" or oneness with the world, a sense of unification between person and environment.[45] Individualism is defined by having reached the heights of mind and spirit potentials, yet being cognizant of one's accountability to society.[46]

Over the long term, the meditative process rewires neural networks that govern awareness, stress, and empathy. Meditators show heightened activity in anterior insular and anterior cingulate cortex areas of the brain. These areas' connectivity with the limbic system (see fig. 5.2) is key to promoting the translation of sensory information to complex, cognitive functions. Their overdevelopment represents a heightened capacity to transform novel events into judgment, feelings, and impulse control.

The hippocampus, tasked for learning and memory, is enlarged among meditators.[47] Forebrain regions that connect focused attention and rational thinking (the anterior prefrontal cortex) to cravings and uninhibited impulse control (the posterior cingulate cortex) are also highly developed.[48] The size of the amygdala, our brain's emotional center for fear and anger, is reduced.[49]

Experienced meditators can better tolerate negative sensations of pain. They accept the experience rather than interpret, change, reject, or ignore it. Brain imaging studies revealed reduced activity in the person's insular cortex, which together with the amygdala are regions that trigger anxiety.[50] There is reduced connectivity between the frontal cortex and ventral striatum, which dampens spontaneous recall of prior experience. With reduced memory intervention on current experience or second thoughts, the meditator was less prone to snap judgments.[51] These physiologic changes all contribute to a healthier sense of fellowship and empathy.

Moving Zen practices in budō arts involve total immersion in a highly cultivated skill (see chapter 14) and are regarded as a path toward achieving nirvana.[52] By focusing on the task at hand, the mind is calmed and relaxed. This state of awareness is highly valued, as the practitioner discerns current experiences of bodily sensations, thoughts, and emotions without being overly invested or distracted by them.[53] In *Hagakure,* the seventeenth-century practical and spiritual guide for the samurai, Tsunetomo Yamamoto described the mind-set as to "enter into nothing else but go to the extent of living single thought by single thought."[54]

Kata (*tàolù* in Chinese martial arts) is often considered the soul of martial arts, its practice an essential component in the lifelong pursuit of karate-dō, judō, or tai chi, to name a few. Individual movements frame a uniquely balanced moment of self and surroundings, yet the practitioner is not transfixed by either (see chapters 10 and 11). One cultivates calm and collected thoughts and feelings as opposed to frantic actions out of fear and anxiety. Mindfulness, in concert with affirmation from mastery of skill, elevates the spirit as inner confidence. These attributes will be examined further in the last chapter of the book.

Takeaways from "Manifestations of the Spirit"

- Grittiness engenders confidence and fortitude to overcome the unanticipated from moment to moment.

- The martial spirit epitomizes qualities of grit. It is guided by purpose, positivity, shared values of community, and the process of contemplation. These qualities help to sustain our outlook in times of self-doubt or apathy.

- Meditation nurtures the martial spirit. The contemplative process orders our thoughts, centers our stream of consciousness, and diverts negative mental energy.

- Meditation is an important aspect of traditional martial arts practice. Introspection invigorates our inner strength, casting aside fear and anxiety. Meditation switches on neuroadaptive processes in the brain, giving long-term benefits in mental focus, self-awareness, and emotional health.

Nature of the Spirit

Human beings have been in awe of nature's forces since the beginning of time. Our ancestors considered tornadoes, thunder, lightning, life, and death as the handiwork of the gods, who asserted their will with forces that overpower our sense of proportion.

By the late seventeenth century, our ancestors commenced to detect, measure, explain, and ultimately harness the forces of nature. Mystic rituals and philosophical beliefs persist nonetheless as tradition, emotional solace, and paths for spiritual enlightenment.

Qi and the Cosmos

The belief that the breath equals "life force" has endured over millennia. This "vital principle" called qi (氣) in 500 BCE Chinese philosophy represents an all-pervading ether that envelopes self and non-self, permeating the universe. Chinese traditional medicine (CTM) holds that strength and vitality depend on the balanced flow of qi through the body. This concept persists today as perhaps the most enigmatic concept in spiritual practice and martial arts.

The forces majeures in Chinese philosophy are tian qi (heaven qi, 天氣) and di qi (earth qi, 地氣) that coexist with ren qi (human qi, 人氣).[1] Tian qi sustains balance of the cosmos, the order of the planets, and heavenly events such as the weather, gravity, and energy in the stars. Di qi permeates all earthly lives and is controlled by tian qi. It flows in a grid pattern, according to Daoist thought, and is modulated by earth's magnetic poles and its molten core. All living things thrive when the yin and yang of di qi are in balance. Disasters (earthquakes, volcanic eruptions) happen when they are not.[2]

Ren Qi

Ren qi is the essence of our being. It is influenced by heaven and earth (tian qi, di qi). Physical health and mindfulness rest on the yin and yang balance of ren qi, its unimpinged flow throughout the body, and its free interaction with tian qi and di qi.[3]

According to Daoist philosophy, ren qi comprises yuan qi (innate qi, 元氣), derived from *jing* (essence, 精), the congenital essence that we inherited from our parents, putatively stored in the kidneys, and exogenous *zong qi* (gathering qi, 宗氣), stored in the chest, sourced from the energy of fresh air (true qi, or zhen qi from the lungs, 真氣) and nutrients (nutritive qi, or ying qi from the stomach and spleen, 营氣) (fig. 16.1).[4]

Zong qi resides near the heart. It guides circulation and breathing. Zong qi is fortified by the yin nature of ying qi (nutritive qi) that circulates in the blood and bodily fluids and nourishes the organs. The yang nature of ying qi, called *wei qi* (protective qi, 衛氣), shields us from infection, synonymous with the body's immune system.

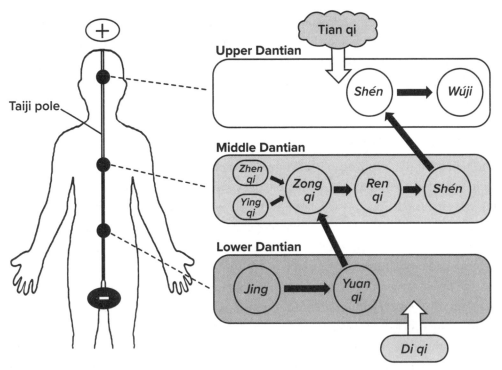

Figure 16.1 Inner alchemy of jing, qi, and shén (simplified).

Fields of Elixir: Alchemy for the Spirit

Medical qigong practice advances that good health and longevity hinge on balanced yin (−) and yang (+) energies. A person can start to purify their nature after having balanced these polar life forces through an internal process called "inner alchemy" (*naidan*, 內丹). This feat starts with the assimilation of ren qi by blending our innate nature (yuan qi) with postnatal constituents (zong qi), ultimately transforming into shén (spirit, 神), our spirit that seeks enlightenment (fig. 16.1).

Qi, alongside jing and shén, constitute the "Three Treasures" (sānbǎo, 三寶) in Chinese traditional medicine. Jing is the seed for growth, development, and procreation. Qi is the medium to sustain strength and vitality. Shén embodies blossoming of consciousness, emotions, judgment, and our transcendent spirit that connects to a higher level of consciousness.

Inner alchemy takes place at the three "inner crucibles," energy centers in our body called the dantians (fields of elixir, 丹田) (see fig. 16.1). At the lower dantian, jing transforms into yuan qi, the prenatal vital energy. The middle dantian catalyzes the blending of yuan qi and zong qi into ren qi. Ren qi is refined into shén. As shén gathers in the upper dantian, it is exposed to the Order of the Cosmos (Tian Dao, 天道). Union with Tian Dao transcends shén to *wújí* (limitless, nothingness, the infinite, 無極), the spirit's most radiant and unblemished form when the person's nature is perfectly balanced.[5]

The dantians align at the midline of the body between left and right. They are located at the center of the brain (upper dantian, 上丹田), near the heart (middle dantian, 中丹田), and within the abdomen below the navel (lower dantian, 下丹田) (see fig. 16.1). The dantians are linked by a central energy channel called the Taiji Pole (太極寶), which runs longitudinally from the yang (+) pole on top of the crown (Bai Hui, 百會), to the yin (−) pole at the perineum (Hui Yin, 會陰).

The lower dantian is the primal elixir field, seat of physical consciousness, and root of our life force, wisdom, and emotions. It is referred to as "the dantian" without further designation. Daoists also named the lower dantian "Golden Stove" (金炉). They believed that when jing processes into yuan qi, heat is emitted to sustain body temperature.[6]

The lower dantian is located about three finger-widths below the navel and two fingers into the lower abdomen, coinciding with our center of mass for physical balance (see chapters 10 and 12). Daoist medical qigong attributes our "muscle sense" to emanate from the lower dantian.[7] This awareness for movement and limb-positions is distinct from sight, hearing, smell, touch, and taste (see chapter 4).

The tanden (Japanese for dantian) is synonymous with the hara (lower abdomen) in Japanese culture, regarded conceptually as the wellspring for mastery in an endeavor (see chapter 11). It serves as the focal point in Zen Buddhist meditation and martial arts breathing.

As qi depletes from stress, exhaustion, or illness, di qi is drawn subconsciously into the lower dantian as a survival instinct. This process is enhanced by meditation or qigong practices.[8] The lower dantian also nourishes and replenishes the middle and upper dantians through connectedness with di qi.

The middle dantian is the seat of our emotional consciousness and attitudes of empathy and compassion. These "earthly" attachments are collectively referred to as our *shàngling* (aspirational spirit, 上靈).[9] Located in the upper chest, the middle dantian sustains the health of the internal organs by dispensing zong qi (qi acquired from fresh air and nutrients) first to the heart and secondarily to the lungs. Ancient beliefs considered our thoughts and feelings to emanate from the heart. Accordingly, our emotional stability depends intricately on the well-being of the middle dantian.

Table 16.1. Daoist concept of the three dantians.

	LOWER DANTIAN (下丹田)	MIDDLE DANTIAN (中丹田)	UPPER DANTIAN (上丹田)
Association	yuan qi	zong qi	shén
Connectivity	earth qi wei qi I*	earth qi heaven qi wei qi II*	heaven qi wei qi III*
Transformative Role	jing → qi**	qi → shén**	shén → wúji**
Associated Organs	kidney	heart, lungs	brain
Manifestations	physical	emotional	spiritual
Realm of Function	physical and kinesthetic consciousness; survival instincts; subconscious reflexes	emotional consciousness; respiration; circulation; empathy	spiritual consciousness; intellect

*Protective qi levels I, II, and III.**Jing: essence; shén: spirit; wúji: Daoist spiritual state of limitless.

The ancient martial arts axiom of "qi follows yi" conveys that our physical action (qi) is predicated on mental intent (yi, 意). Qigong speaks of qi as the horse, yi as its rider, and *jīngluò* (meridians or energy-flow channels, 經絡) as metaphorical roads within our body that qi travels and asserts its influence on.[10]

The upper dantian is the seat of our spirituality and the focal point of intent. It is centered between the eyebrows from the front, often referred to as the third eye or the mind's eye. The upper dantian houses our spirit, the product of jing and shén. *Jing shén* (stamina and intellect, 精神) is our binary state that brings positive attitude in welcoming day-to-day challenges, awareness of self and surroundings, and our aspirations.

The upper dantian acts as an inner crucible to further distill shén to wújí by drawing qi from the lower and middle dantians. Wújí is the all-knowing mental state after having achieved wu-wei, the Daoist attitude of effortless action or flow, hence becoming one with nature in deed and in spirit. It extends from the Zen ethos of "no-mindedness" (mushin; see chapter 14), and the budō concept of mushin no shin (unhindered mind, 無心の心), to be uncluttered from thoughts and emotions, extending our mental reach to all that is possible.

Mushin no shin is heightened awareness without conscious thoughts. To "know without knowing" is to open our mind's eye, allowing the subconscious to control our actions.[11] Unencumbered by distractions or emotion ties, the martial artist's awareness is heightened to the vibes of the surroundings. Our stream of consciousness enables spontaneous response without self-doubt.[12]

The three dantians collectively radiate wei qi (protective qi) that putatively protects against infections. The first layer of invisible defense emanates from the lower dantian. It hugs the skin and permeates muscles throughout the body, extending outward to about two inches from the skin. Wei qi from the lower dantian corresponds roughly to the contemporary understanding of our innate immune system, where our skin and mucous membranes act as physical barriers, alongside resident white blood cells that patrol, survey, and tackle foreign invasive pathogens.

A second layer of wei qi connects to the middle dantian and manifests as the aura. The field extends from two inches to about a foot outside the body. Our aura is more pronounced when ren qi fills the middle dantian. Given that the middle dantian is tied to emotions, mood fluctuations change this energy field, and hence color of the aura.

The third level of wei qi is controlled by the upper dantian. This field envelops the body, extending out to several feet. Qigong experts consider its reach to correspond to the person's spiritual power and psychic perception.[13]

Qi Cultivation

Ren qi cultivation starts at the lower dantian. The process requires proper posture and focus of the breath into the abdominal area, directing the body's cyclic energy flow from the lower dantian, through the middle dantian, onto the upper dantian, then back to the abdomen (see chapter 11).

Advanced practitioners focus their qi to the middle dantian to unburden the heart from psychoemotional tensions. Only medical qigong experts can refine the workings of the upper dantian. By drawing in tian qi (heaven qi) through the Bai Hui of the Taiji Pole (see fig. 16.1), healers transmit their own wei qi into patients through the practice known as external qigong.

These ancient beliefs reinforce traditional martial arts practices of posture, connectedness to the ground, core integrity, and focus on breathing. By sustaining soundness of the body (lower dantian) and positive emotions (middle dantian), the martial artist achieves clarity of mind and transcendence of the spirit (upper dantian).

Fostering Intuitive Thinking: The Gut-Brain Axis

The conceptual lower dantian is enveloped by our gut. "Guts" have been associated with courage and determination since ancient times, in line with beliefs that our physical consciousness resides at the lower dantian.

Humans have long equated sensations in the belly with instinct or premonition. "Going with my gut" is to act by intuition. When former President Donald Trump claimed that "My gut tells me more sometimes than anyone else's brain can,"[14] he considered that his intuitive thinking far surpasses the informed decisions of his advisers.

We suffer from a "gut-wrenching" experience when we are very upset. "Butterflies in the stomach" accompany light-headedness when we have performance anxiety, as blood flow is diverted to muscles in a fight-or-flight response. Fear and discomfort give rise to nausea that is felt in the "pit of the stomach."

Scientists have revealed the physiological connection between feelings, emotions, and our sense of well-being.[15] In the United States, 5 to 15 percent of the adult population are afflicted with irritable bowel syndrome (IBS). They suffer from chronic abdominal pain, bloating, and abnormal bowel functions in the large intestines.[16] Another three out of a thousand people are diagnosed with inflammatory bowel disease (IBD), manifesting more severe symptoms from chronic inflammation of the digestive tract (Crohn's disease, ulcerative colitis).[17]

Causes of these ailments are diverse and hard to pinpoint. The patients are prone to anxiety or depression. Conversely, those suffering from anxiety or depression are more likely to develop IBS and IBD.[18] Scientists describe the association of gastrointestinal dysfunction with emotional distress as the "gut-brain axis."

Proper performance of our gastrointestinal track is essential for survival. A multilayered system monitors and regulates these processes, enlisting hormones, immune mediators, and a complex meshwork of nerve cells that keep tabs on the alimentary canal.

Some five hundred million neurons populate the gut's nervous system (the enteric nervous system, ENS), five times as many as in the spinal cord. It monitors gut motility and blood flow and communicates gut health status to the brain by the vagus nerve, the body's longest cranial nerve, which runs from colon to brainstem.[19] Although this nonspinal nerve also relays motor commands, 90 percent of its fibers are tasked with apprising the brain of the operational status of internal organs, in particular the gastrointestinal tract. This communicative path coordinates digestive control with other vital functions, such as heart rate and respiration, through the parasympathetic nervous system.

The ENS functions without intervention by the brain, representing our body's singular peripheral nervous system that operates independently. Professor Michael Gershon refers to the ENS as "the second brain"[20] in deference to its autonomy and far-reaching influence on other bodily functions, from appetite and immune response to our emotional state.

Our body reacts to news and risks faster than to conscious thoughts.[21] "Gut feeling" is part of the collective interoceptive (inner body) sensations triggered by emotional distress. These sensations extend to the heart, circulatory blood flow, and various brain areas.[22] A "punch in the gut" stands out through connectedness of the gut brain axis, the high nerve density of the alimentary tract, and the exquisite sensitivity of gut muscles to neurotransmitters (such as serotonin) that are released locally and systemically.

What is commonly portrayed as intuition represents subconscious memory recall and outcome prediction that stems from prior risk-taking. The widely known Pavlov's response illustrates this conditioned response, when dogs automatically salivate by associating food with the sound of a bell.[23] Through trial and error, our body aligns confidence in outcome with a particular thought or emotional state, which translates into gut feelings.[24] When applied to martial arts, gut feelings likely represent a subconscious response to subtle external stimuli in touch, pressure, vibration, or temperature that have previously been associated with risk or danger (see chapter 4).

Diet habits that affect gut health profoundly influence our sense of well-being.[25] Healthy diet habits can help reduce false alarms in gut feelings, hence improving our confidence in intuitive response. Both Western medicine and Chinese traditional medicine attend to our physical health as prerequisite to nurturing the spirit. Chinese traditional medicine preaches that the spirit (shén) is profoundly affected by zong qi, the acquired essence from fresh air and nutrients.[26]

Open space and a light breeze invigorate and lighten our spirit in everyday life, whereas tight quarters and pollution stifle and contribute to depression, anxiety, and psychosis.[27] Comfort food fondly reminds us of kinder and gentler times, and a great meal can be "life-transforming." On the other hand, high-calorie, high-fat diets invite lethargy after the initial high from spiking bursts of dopamine, the "feel-good" neurotransmitter. Poor diet habits are linked to psychological distress such as irritability, anxiety, and even depression.[28]

When deep breathing is carried out as part of sitting meditation (mokusō) or Moving Zen (kata), it calms our spirit by clearing the mind (mushin) (see chapter 14). Habitual diaphragmatic breathing expands the gas exchange volume, leading to more efficient respiration. Controlled breathing reduces sympathetic (excitatory) nervous activity from arousal and stabilizes autonomic respiratory and cardiovascular functions (see chapter 11).

For proper nutrition, diet experts recommend a portion-appropriate, minimally processed diet that also fulfills our mineral needs. There should be a balanced intake of carbohydrates (for energy and performance), proteins (the building blocks of tissues, muscle mass, and strength), and a moderate amount of fat (for available and stored energy, and brain health).[29] The strength and vitality of Shaolin monks and their vegetarian habit are testimony to the nutritional capacity of plant-based diets. Plant-derived fibers aid in digestion. They moderate food sugar absorption, reducing sugar rushes and crashes.[30]

Martial artists are already inclined toward vigorous exercise that promotes circulation, builds endurance, and boosts the spirit. Proper hydration is necessary to support a demanding workout. Peak physical and mental performance relies on robust cardiovascular output. It is advisable that aerobic exercise (swimming, running) be integrated in the training regimen, which can be conducted outside the dōjō. As described previously, core strength is essential in multiple aspects of performance biomechanics (see chapter 10) as well as in digestive health. Running and calisthenics such as push-ups, sit-ups, squats, and leg-lifts help to build core endurance to anchor the hips and spine.

Figure 16.2 Northern lights (aurora borealis) in Norway.

Forces of Nature

In a cycle of eleven years or so, Arctic and Antarctic residents are rewarded with nature's spectacular aerial lightshows. Haunting patterns appear to swirl from the ground to the heavens (fig. 16.2). Ancient Roman mythology attributed these boundless fluorescent green and pink formations to the descent of Aurora, goddess of dawn, as she traveled from east to west, announcing the coming of the sun.[31] Inuit people of the Arctic believed that the northern lights were the spirits of hunted seals, salmon, deer, and whales ascending to heaven.[32]

The northern lights (aurora borealis) are nature's most haunting yet benign display of force. Explosive nuclear fusion within the sun's core produce charged electrons and protons. These subatomic particles escape through holes in the sun's magnetic field. Some arrive at earth's atmosphere by the solar wind, the result of the sun's rotation and the earth's gravity pull. They enter earth's atmosphere at the north or south pole, where the magnetic field is weaker.

Visible light is emitted as these charged particles ionize and excite gas molecules in earth's atmosphere, appearing as dancing swirls of aurora borealis (Arctic) and aurora australis (Antarctic). Ionization of thermosphere oxygen from sixty miles up creates ethereal yellowish-green images. Blue and purplish red auroras appear when nitrogen ionizes at lower attitudes.[33]

The sun's multistage fusion process converts hydrogen atoms to helium at a rate of five hundred million metric tons per second.[34] Tremendous energy is released as gamma rays, which sustain all life forms in our solar system. The nuclear force unleashed by the sun, also called the weak nuclear force, is crucial to the nuclear fusion and radioactive decay that power star formation and create new elements. This weak nuclear force, together with the strong nuclear force, electromagnetic force, and gravity, are the four fundamental forces that sustain the fabric of the universe.

Unlike mechanical forces, such as muscular force, impact force, and friction, which require physical contact to assert an effect, the forces of nature persist as oscillating waveforms and exert their effect from a distance.[35] Like different countries' disparate monetary currencies, each force requires unique subatomic force-carrying particles (bosons) to achieve energy transfer.[36]

The strong nuclear force sustains the fundamental atomic nuclear structure of protons, electrons, and neutrons through the workings of gluon particles (massless bosons that "glue" quarks together). This structure becomes the building block of solids, liquids, and gases. The strong nuclear force is a hundred times stronger than the electromagnetic force, about a million times stronger than the weak nuclear force, and 1,038 times stronger than gravity.

Electromagnetic forces define most of our day-to-day experience, from Wi-Fi and electric cars to diverse medical applications. Electric current, contact force, and elasticity are all outcomes of charged or neutralized particle interactions. Electromagnetic forces are produced when charged particles swap photons—massless force-carrying bosons that deliver energy as radiation. The strength of the energy carried by the photons determines the type of radiation, ranging from low-frequency, low-energy infrared radiation to visible light and high-frequency, high-energy ultraviolet radiation, X-rays, and gamma rays.

Biomagnetism, Qi, and Biofield

In the mid-eighteenth century, electrical science and technology were the raves of the era. Luigi Galvani accidentally uncovered "animal electricity" in 1791 from the twitching muscle of an injured frog leg. He attributed this phenomenon to electricity that is produced in living tissues. Western societies became instantly enamored by this electrical "life force." Mary Shelley published the tragic Gothic novel *Frankenstein* within the same period, its narrative centered on a scientist who brought life to a corpse by infusing it with electricity. Yet another 150 years passed before the widespread electrification of industrial processes and broad-based generation and power distribution for public consumption. The relationship of electricity and magnetism became well established by the late nineteenth century, giving rise to the unified concept of the electromagnetic field (EMF).

Animals have long been known to detect and respond to electromagnetic forces.[37] Sharks home in to minute bioelectrical signals of their prey by highly developed biosensors (ampullae of Lorenzini). Honeybees and migratory birds navigate their flight paths through innate magneto-receptive mechanisms in alignment with the earth's magnetic poles.

The human body gives off electric and magnetic energy. The average human produces around 100 watts of power at rest. This output is equivalent to around 2,000 kilocalories of food energy, a point of reference in the recommended daily caloric intake.[38] Trained athletes can ramp up to peak force output within minutes and comfortably sustain power outputs of 300 to 400 watts for hours.[39] Outputs of up to 2,000 watts have been recorded in short bursts of assertion such as in competitive sprinting.[40]

The living cell maintains an electrical polarity across its membrane barrier. This charge gradient is critical for cell-to-cell communication through ionic exchange. Nerve cells fire off brief pulses of electric current called action potentials to achieve neurological signaling, which triggers contraction and expansion of muscle fibers. The small but measurable membrane potential of our body's cells generates a corresponding magnetic field. The resulting EMF resonates continuously within atoms, molecular structures, cells, and organs.

Although all matter resonates, each tissue has a unique resonance signature according to cellular composition, molecular makeup, and the tissue's electrical activity. Among the different organs, the heart registers the highest level of electricity (2.5 watts), forty to sixty times higher than the brain, and it emits the strongest EMF.[41] The body's electromagnetic profile can be revealed by subjecting a patient to a strong external energy field. Malfunctioning tissues and cells display an abnormal profile after absorption or deflection of the external energy, represented as nontypical biophoton emissions.[42] As disease-causing free radicals interfere with energy-expending metabolic processes, weak but detectable energy-carrying light particles are released, enabling diseased tissue detection by diagnostic imaging from X-rays to magnetic resonance imaging (MRI).[43]

Similar to the force-carrying photons in quantum physics, biophotons mediate measurable biological outcomes from embryonic tissue organization, enhanced wound repair, tissue regeneration, and reduction in pain and inflammation.[44] These findings foster the intrigue of information-carrying biophotons that could be harvestable to influence health, broaching the reality of what could only be previously ascribed to science fiction or ancient philosophy.

In 1992 the Office of Alternative Medicine (OAM) was established within the U.S. National Institutes of Health, its scientific mission to "facilitate study and evaluation of complementary and alternative medical practices," undoubtedly as acknowledgment of the emerging public interest in New Age holistic medicine.

An ad hoc committee proposes to focus on promoting scientific rigor in the study of "energetic therapies," with the hope that these holistic approaches can one day complement biomechanics- or biochemistry-based treatments.

The committee defines biofield as "a massless field, not necessarily electromagnetic, that surrounds and permeates living bodies and affects the body." This hypothetical "human energy field" would conceivably transmit information-carrying energy that regulates and sustains our physical and emotional equilibrium. The OAM also seeks to better understand external close-proximity healing by energy transfer, such as by Reiki, therapeutic touch, and external qigong.[45]

Modern science shows that our body incorporates redundant regulatory networks. Apart from anatomically linked highways (the skeletomuscular framework, blood circulation and blood cell migration, neurological signaling), our endocrine and immune networks produce soluble effectors that ramp up local as well as systemic "spur of the moment" responses. The concept of biofield, whether at the level of biophotons, patterns of cell membrane resting potentials, or response to a geocosmic energy field, may possibly be another unexplored layer of governance that interconnects mind and body and orients our identity of self with space and time.[46]

For over forty years now, theoretical physicists have pondered the theory of everything (TOE). Starting with nature's four fundamental forces, the TOE seeks a single unifying concept to link all physical aspects of the universe. Thus far, the three nongravitational forces (strong nuclear force, electromagnetic force, weak nuclear force) are well framed within the standard model of quantum mechanics, best illustrated by subatomic particle interactions. The gravitational force remains the exception to the rule, as theoretical physicists and cosmologists continue to ponder its workings that bind massive stars and galaxies over infinite distances. In the meantime, mortals like us may be more enthralled in arriving at how all these forces fortify our very existence.

The reason for these discussions, dear readers, comes from the vexing parallels between qi, ki, prana, and the modern interpretation of biofield.[47] Biophysicists, like their quantum physics colleagues, consider energy and matter as one and the same. They are now faced with the challenge of defining how complex bio-information can be embedded in the everyday electromagnetic force, or biofields that leverage specific biologic functions.[48] Newfound knowledge may hopefully explain how ancient beliefs, like qi or prana, can serve as vehicle to traffic physical and emotional flux.

At this juncture, neither philosophy nor modern science fully explains the wonders of the universe or the destinies of humanity. Our sense of purpose and the meaning of existence can only be realized through unique personal beliefs when fortified by scientific facts.

The practice of martial arts offers a path for self-examination in the physical realm and nourishes and refines our spirit at the same time. Budō arts utilize meditation and Moving Zen practices to achieve this reflective process. Moving Zen is enlightenment through doing. This is represented by kata training, which is discussed in the last chapter of this book.

Takeaways from "Nature of the Spirit"

- The quality of the universal vital energy (qi, ki) within us (human qi, ren qi) and its unimpeded flow defines our strength and vitality according to Daoist and Zen Buddhist beliefs. It is nourished by earth qi (di qi) and cosmic qi (tian qi).

- The martial artist gains control of their physical actions (lower dantian), emotions and feelings (middle dantian), and intellect (upper dantian) through heightened awareness of qi flow through the three main energy centers (dantian or tanden). While qi remains a conceptual abstraction, focus on the lower dantian and controlled breathing produces unquestionable benefits to physical performance and mental focus.

- Confidence in our gut feelings adds to overall awareness. Our gut feelings are not prescient. Previously categorized as intuition, gut feelings are learned, now acting as subconscious emotional triggers by the highly sensitive enteric nervous system.

- Nature's electromagnetic forces permeate day-to-day experiences. Scientific studies continue to study the role of biomagnetism in maintaining our body's homeostasis, and the biophoton and biofield phenomena as potential candidates of informational energy.

Apollo, Dionysus, and the Dō of Kata

Apollo and Dionysus: From Dichotomy to Synergy

Apollo and Dionysus are the sons of Zeus in Greek mythology. Apollo is the god of the sun and master of truth, logic, light, reason, and healing. Dionysus is the god of wine, intoxication, sexuality, and ecstasy. These antithetic urges embody our primal nature, according to philosophers Plato, Ayn Rand, and Nietzsche. One is compelled to confront these driving forces when embarking on a course of action.[1] Stephen King, the horror novelist, similarly alludes to "Apollonian" as a metaphor for reason, restraint, and power of the mind, and "Dionysian" for emotion, sensuality, and impulsive behavior.[2] We are inclined toward structure, order, and harmony in times of peace and prosperity. When society is in flux and chaos, one is given to impulses that are driven by fear and anger (table 17.1).[3]

Our judgments are swayed by emotionally driven media excess from day to day. TV ratings are vaulted as much by accurate in-depth coverage as by overwrought human drama. Incessant news of murder and mayhem upend reason and dispassion. Today's despairs quickly fade as yesterday's events, while more of the same brings anxiety the next day, challenging our psyche to sustain equilibrium and control.

Table 17.1 The Apollonian-Dionysian dichotomy.

TRAIT	APOLLONIAN	DIONYSIAN
judgment	objective	subjective
decision-making	evidence-driven	emotionally driven
actions	deliberative	impulsive
social tendency	constructionistic	consumeristic
pop-culture icon	Mr. Spock	Captain Kirk

The year 2020, during which much of this book was written, will likely end up being the most challenging year in our lifetimes. A previously unknown virus, COVID-19 (coronavirus disease 2019), caught governments woefully unprepared, paralyzing communities worldwide.[4] The American way of life was also upended by high unemployment alongside racial justice unrest, political extremism, and an unprecedented level of natural disasters.[5]

By late fall, the 2020 U.S. presidential race unfolded as a referendum to President Donald Trump's autocratic populism—and a more vicious replay of the 2016 campaign. The Dionysian appeal of the Grand Old Party was pitted against Apollonian rhetoric of the Democrats. Uncertainties lingered well after election day. The pandemic ravaged unabated despite the promise of vaccines over the horizon. By year's end, twenty million were infected in the United States, and more than 350,000 people had died from COVID-related causes. Economic recovery remained tenuous, rankling public confidence in 2021.

Historians may yet recognize that our nation overcame previous challenges of a similar magnitude. From 1918 to 1920, more than one-third of the world's population were infected by the H1N1 influenza virus. The disease spread to the United States when World War I veterans returned from Europe. Reported pandemic deaths exceeded the total fatalities of the war, numbering more than 50 million worldwide and 675,000 in the United States.[6] The nation went into a deep recession. High unemployment and inflation spurred pro-Bolshevik movements (the Red Scare), major strikes, and bloody riots in Seattle, New York, Cleveland, and Boston.[7] Yet doom and gloom started to pass in the spring of 1920 with the emergence of herd immunity, civil rights advocacy, and the election as president of Warren G. Harding, who called for "return to normalcy."

Our social fabric cannot sustain everyone just fending for themselves in chaotic times. Democracy rests on public trust and an enlightened citizenry with a well-developed sense of community. The luxury of personal freedom hinges on acceptance of civic order and empathy for others. Traditional martial artists are gifted with the path to aspire for a clear mind and strong spirit. Clarity in thinking sustains equanimity and stable emotions, accepting that human problems in most cases can be solved by human solutions that lie within and without. Training affirms personal values and the physical and spiritual self, the latter resting on well-balanced Apollonian-Dionysian tendencies and consideration for others. These ideas of accountability are deeply ingrained in traditional martial arts.

By observing morality of deed and mind (see chapter 14), the martial artist refrains from uncontrolled urges that stem from fear, anger, and anxiety. The impulse for snap judgment is now tempered by disciplined actions. A heightened capacity to endure uncertainties benefits our daily life as we navigate through personal challenges and chaos in our nation and the rest of the world. Self-assuredness enables us to advocate mutual respect and dignity when engaging friends, family, community, and even adversaries.

Tàolù

Equanimity and heightened awareness in martial arts are fostered by mindfulness meditation (see chapter 15) or moving meditation, as exemplified through kata practice.

Performing movement sets to condition mind and body dates back to the early fifth century CE. Hand forms and poses (Luohan's Eighteen Hands, 羅漢十八手) introduced by Bodhidharma were companion to qi-circulating exercises (Sinew Metamorphosis Classic, 易筋經).[8] Between 1260 and 1368, Bai Yufeng (白玉峰), the Chinese martial artist turned Shaolin monk (see chapter 1), merged Luohan's Eighteen Hands with indigenous Chinese martial arts as the 173 techniques in the *Essence of the Five Fists* (*Wǔ Xíng Quán,* 五形拳). These movements are recognized as the foundation of Shaolin quan as we know it today.[9]

The *Essence of the Five Fists* incorporated instructions for individual techniques as well as tàolù (roadmap to movement sets, 套路), martial art forms of serial offense and defense combinations. Each emulates the fighting spirit of an animal. As wushu diversified over time, northern schools gravitated toward mimicry of dragon, tiger, snake, panther, and crane while southern wushu took after tiger, crane, snake, monkey, and mantis.[10]

Tàolù continues today as an important part of Chinese martial arts practice. Each style has its own tàolù, previously held in secrecy to guide the learning of proprietary techniques. Ancient tàolù engendered unarmed as well as weapons applications performed individually or in paired training (*dui lian,* 對練), many of which are still practiced today.

Between the eighteenth and twentieth centuries, Chinese expatriates and influential Okinawan martial artists brought tàolù from China through cultural and commercial exchange. These were modified over time and transformed into Okinawan and Japanese kata.

What Is Kata?

The word *kata* has been widely adopted. It now applies to the practicing forms of almost all Asian martial arts, including Japanese budō arts, Okinawan martial arts, Korean martial arts with Japanese influence, and Chinese quánfǎ. Contemporary kata commonly refer to unarmed forms that are performed individually in karate-dō practices.

The Japanese *kanji* (character) for kata, 型 or 形, translate as "form." The characters also denote model, style, template, or mold. A totality of these concepts is a better representation of its purpose. Kata is often referred to as a language or living texts of the art. It specifies the precise manner that techniques are executed sequentially and is believed to be the instrument for conveying practical knowledge through generations. At a fundamental level, kata portrays prearranged sequences of defensive and offensive techniques. The movements are imbued with additional attributes of rhythm and tempo to convey strategy and psychology, and hence are subject to broad interpretations and academic analyses.

Each kata is mastered at multiple levels, starting with proficiency of the constituent techniques. The student then integrates these techniques systemically within the context of the form, now assembled in an ordered array, much like links of a chain.

Mastery requires deliberate practice measured in years if not decades. After achieving balance and physical coordination from start to finish, the student then incorporates bio-mechanical know-how to build power and speed. By acquiring "muscle memory," they achieve fluid movement transitions and the rhythmic execution of serial techniques. This process demands repetitive processes of self-examination, discipline, and perseverance. In turn, the student gains an appreciation of their own strengths and the wherewithal to rise above weaknesses.

Kata practice sits opposite to the mind-set of sparring training. Free sparring is predicated on the visceral interpretation of an adversary's actions that demand spontaneous responses, whereas kata performance offers the bandwidth of contemplation. By seeking within, one reaches for the steeliness in mental resolve and precision in execution, depicting each movement with the best of one's ability.

Master Hidetaka Nishiyama, the late Shōtōkan great, compared kata mastery to an in-depth understanding of a core principle that is portrayed within the movements, such as strength versus speed, soft versus hard, penetration versus redirection, rootedness versus agility, and so on. The intent of the kata is often reflected by its name, as discussed in subsequent sections.

The Spirit of Kata

Once upon a time in ancient China, there was a man who grew up near a tall mountain. He often pictured the spectacular view from the mountain's peak in his mind. One day, the man decided to realize his dream. He reached the foot of the mountain after a half day's journey. "My, the mountain is much higher than I thought," he bemoaned. "It would take forever to get to the top."

He soon saw a traveler coming down from a mountain path. "What was it like to be at the mountaintop, and how did you get to the peak?" he asked. The traveler gushed with enthusiasm as he related the challenging but worthwhile experience. Sadly, his insights only served to discourage the man.

"Surely there's an easier way," the man ruminated. He started chatting up others who came down different trails. The man collected feedback from ten or twelve travelers after some time. And yet, he was increasingly demoralized. It appeared that each trail carried unique challenges.

"Well," he consoled himself, "now I know exactly what it'd be like to be at the mountain top. There's really no point to spending all day getting there." And so the man returned home, satisfied that he had fulfilled his mission without much effort.

This ancient Shaolin Temple parable, now part of Chinese folklore, speaks of the manner that different individuals tackle life's aspirations and challenges. On a smaller scale, the story addresses the process for mastering kata.[11]

Each of our lives is unique, our outlook shaped by community, personal experience, and values. The path (dao, dō) to fulfill one's aspirations is beset with challenges, dictated by circumstance within and without. Knowledge from others only contributes so much. Paradoxically, personal satisfaction is the fruit of one's labor, more cherished if preened from hard-earned successes and setbacks. The way that we overcome challenges molds our persona, enriches our life, endows confidence, and broadens our sense of reality.

Zen Buddhism seeks spiritual fulfillment by coming to terms with inner feelings and aspirations (one's "Buddha Nature"). Patience, endurance, energy, and effort are the cited virtues for cultivating the Zen spirit. These qualities can be realized with kata practice, part of a Moving Zen experience that is valued by wushu and budō arts.

Each kata represents a unique roadmap for structured learning, a template for focused practice to attain expertise.[12] Kata strives for constant improvement beyond mimicry. Executing complex techniques by memory recall requires intense mental concentration, and determination to overcome intrinsic limitations—a wandering mind, bad posture, a disobedient core, and overreaching techniques, to name a few. The practitioner must limit their attention to the present, elevating awareness of the physical self as they strive for precision in every movement (see chapter 4).

Expertise enjoins the performer's persona in accordance to their physical attributes, skill level, and temperament, as is the case when differently skilled musicians cover the same song. At a certain point in time, the martial artist realizes that they no longer perform the kata learned from the sensei but one that they own from their considerable investments of time, energy, heart, and soul. This deliberative process fosters spiritual growth through insights of self. Hence, kata practice is referred to as Moving Zen, as the student seeks enlightenment through action (see chapters 14 and 15).

Ichi-Shén Ni-Waza (Spirit First, Technique Second)

Master Gichin Funakoshi advised that martial arts follow the axiom of *ichi-shén ni-waza* (spirit first, technique second, 一神二技).[13] The prioritization of inner strength ahead of physical clout applies equally to kata and kumite training.

The kata's spirit embodies its defining elements in concept and practice, often referred to as the kata's dō. Its tenor guides a practitioner's skill-set development, as katas that demand strength and power fortify resolve, while katas of speed and quick changing movements promote situational awareness and agility of mind and body. Rising above the collection of techniques, the kata's spirit rouses our emotional state and guides our attitude as a martial artist.[14] It taps into our subconscious and channels our feelings, much like music, fine arts, or calligraphy.

Shōtōkan Kata: Nomenclature and Practice

In 1922, Master Gichin Funakoshi moved to mainland Japan to introduce karate to the elite citizenry. Funakoshi brought sixteen kata that showcased Okinawa Prefecture's karate community in his capacity as emissary. These included eight foundation or training forms of five Pinan (now called Heian, 平安) and three Naihanchi (also known as Tekki, 鐵騎). Eight advanced or application kata completed the roster: Kūsankū Dai (Kankū Dai, 觀空大), Kūsankū Sho (Kankū Sho, 觀空小), Seisan (Thirteen, 十三, later changed to Hangetsu or Half Moon, 半月), Patsai (Bassai, 拔塞), Wanshū (Empi, 燕飛), Chintō (鎮東, changed to Gankaku, 岩鶴), Jutte (Jitte, 十手), and Jion (慈恩). Other Shōtōkan kata practiced today include Sōchin, Chinte, Nijūshiho, Gojūshiho, Unsū, Wankan, and Meikyo.[15] As karate-dō flourished in Japan and worldwide, around twenty-six advanced kata are now practiced by various traditional styles.[16]

Kata nomenclature varies in attributes due to their mixed origins, many of which were lost in history.[17] A kata's spiritual aspirations are often reflected by its name, when presented as metaphors for the attitude of practice (strength and resolve, quickness and agility) or the mimicry of an animal's combat spirit. However, many kata names continue to be an enigma.

1. Names That Portray the Ethos of Execution

KANKŪ DAI.   The Kankū kata were formerly called Kūsankū, namesake of a Chinese diplomat and martial arts expert (公相君) who introduced Chinese quánfǎ to the Ryūkyū Kingdom. The original kata was created in 1700 CE by Kanga Sakugawa in memory of his teacher. Contemporary versions are practiced in many karate styles, including Shōtōkan, Wadō-ryū, Budōkai, Shitō-ryū, and Shūkōkai.

Master Funakoshi changed the kata's name to Kankū (Beholding the Emptiness, 觀空) to call on the Buddhist concept of emptiness (kū, 空), opening oneself to all possibilities. Separate versions assigned suffixes of *dai* (big, 大) and *sho* (small, 小) to emphasize power (dai) or speed (sho) in movement execution.

The Kankū kata are long known to be Funakoshi's favorite, and they are showcases of the repertoire Shōtōkan karate techniques (fig. 17.1). Closed-fist punches in tall stances alternate with swooping, open hand strikes in low, deep stance. Quick directional changes and whole-body repositioning (tai sabaki; see chapter 9) build agility. Powerful wrist-locks couple with smooth and graceful 180-degree body turns to portray catch-and-throw applications. These features are consistent with the quickness and power of Shaolin White Crane quánfǎ.

BASSAI.   The Bassai kata (formerly Patsai, 拔塞) portray a strong and unyielding spirit (fig. 17.2). Bassai Dai is one of two contemporary versions. The kata emphasizes powerful stances, robust forearm striking-blocks, penetrating counterattacks that are coupled with grappling and follow-through techniques to upset an opponent's balance. The kata fosters the toughened imagery of "Capturing a Fortress" with an attitude of physical strength and mental resolve against all odds. By comparison, Bassai Sho incorporates circular, parrying techniques that are followed with strong, powerful striking actions.

SŌCHIN.   Sōchin (壯鎭) translates as Strength and Stability by the display of overwhelming power, reinforcing the mind-set of a samurai who dedicated himself to preserving order and peace. The kata is made up of slow but intense movements that evolve into powerful explosive strikes (fig. 17.3). Unsurprisingly, the Sōchin Stance, also called Fudō-dachi (Immovable Stance, 不動立; see fig. 9.3) is used throughout. Rootedness of Fudō-dachi reinforces mental resolve to stand one's ground. Sōchin conjures the imagery of a martial artist who overcomes chaos by sheer will and authority.

JITTE.   Jitte strives for powerful techniques that overwhelm an adversary, hence the name, Ten Hands (十手). Slow and deliberate transitions that portray calm and imperturbability (fig. 17.4) alternate with sharp explosive attacks that breach the distance between self and opponent. Its name is symbolic of all-embracing defense and offense actions that equal the strength of ten fighters.

Jitte is one of many kata with the character 手 (*te* or *su*) in their names, denoting "hand" in Chinese (su) and Japanese (te). It references effort, workers, weapons, or techniques, as in Luohan's Eighteen Hands.

Figure 17.1 Kankū Dai.

Figure 17.2 Bassai Dai.

Figure 17.3 Sōchin.

CHINTE.   By the same token, Chinte (Precious Hands, 珍手) conveys uniqueness of techniques and finesse in execution (fig. 17.5). Rather than applying brute force, the movements seek opportunities not immediately obvious and engage body repositioning to penetrate an opponent's personal space. Close-range blocks and counterattacks favor someone of smaller stature but who can generate superior speed.

UNSŪ.   Unsū (Cloud Hands, 雲手) ranks as one of the most challenging kata, entailing a mid-level kick (mikazuki geri) that is followed by a full rotating 360-degree jump in the air (fig. 17.6). The form aspires to emulate multifaceted, constantly evolving cloud formations, the unpredictability of benign stasis quickly transforming into a blindingly fast and forceful skirmish. The broad range of multilevel offensive and defensive techniques is performed with lightning-fast transitions, encompassing joint locks, fingertip strikes, and kicks and sweeps after the performer drops onto the ground. Quick changes in tempo in various segments rally the emotional range of the martial artist over the course of elaborate offensive and defensive sequences.

Figure 17.4 Jitte.

Figure 17.5 Chinte.

Figure 17.6 Unsū.

JION.   Jion (慈恩) traces its origin to nineteenth-century Okinawan Tomari-te (see chapter 1). The kata was named after the Buddhist concept "Grace of Kindness" (慈恩) or possibly a Shaolin Temple by the same name.[18] Powerful linear actions and precise stance configurations are emblematic of the Shaolin quan influence (fig. 17.7). Jion is performed with the attitude of steadfastness as befits Zen Buddhism.

A collection of kata (Jitte, Ji'in), including Jion begins with the formal upright posture, with feet together and elbows tucked alongside the torso with straightened wrists. The hands meet at chin height in front of the body. Bent, open fingers of the left hand brace the clenched knuckles of the right fist.

This pose emulates the common greeting or salutation in Chinese martial arts called the *wushu baoquan li* (wushu wrapped-fist courtesy, 武術抱拳禮). The centuries-old gesture conveys respect, presentation of reason or open-mindedness (the open left hand) in parity with force (the fisted right hand), or union of the physical (earth and human qi, the right fist) and spiritual (heaven qi, the open left hand) aspects of the art (see chapter 16).

Figure 17.7 Jion.

2. Names That Mimic the Animal's Combative Spirit

GANKAKU.   Chintō (鎮東) was the original name of the kata Gankaku. According to legend, a Chinese seaman (or pirate) named Chintō was shipwrecked on the Okinawan coast in the early nineteenth century. He hid in caves and survived by pillaging local villages. Sōkon Matsumura, then Commander of the Ryūkyū Palace Guards, was sent by King Sho to arrest the sailor.

Matsumura was unable to defeat Chintō, according to a 1941 newspaper article by Gichin Funakoshi.[19] The seaman evaded Matsumura's attacks repeatedly by quick spinning movements. His powerful counterattacks put Matsumura on the defensive. Matsumura eventually came up with a compromise after repeated stalemates. He offered Chintō food and clothing for the next few months until he was able to leave the island. In exchange, Chintō taught Matsumura his unusual tàolù, which was to become the kata named Chintō.[20]

Funakoshi sensei renamed several kata when he introduced them to Japan. Chintō was renamed as Gankaku (Crane on a Rock), and Wanshū as Empi (Flight of the Sparrow) to portray different aspects of the avian combat spirit.

The crane (*tsuru* or *kurēn,* 鶴) is a mystical creature with supernatural powers in Japanese culture and is symbolic of longevity and good fortune. Gankaku (Crane on a Rock, 岩鶴) conjures the image of a crane that is balanced on one leg (Sagiashidachi, also called Tsuruashi-dachi, fig. 17.8). Extended arms present a menacing pose, mimicking the tall creature that spreads its wing to mesmerize a predator, then quickly launching its attacks with swipes of its wings, pecking, and clawing. In the kata, defenses in Sagiashi-dachi are followed by quick transitions to in-place or full-distance kicks and strikes. Graceful circular arm motions generate speed and angular momentum when coupled with sharp body turns. These practices are emblematic of Shaolin White Crane gōngfu, a highly popular southern Chinese style created by the seventeenth-century female martial artist Fan Qiniang (方七娘) (see chapter 1).

EMPI.   Empi (Flight of the Swallow, 燕飛) incorporates sharp snappy movements and quick body transition to highlight speed and momentum (fig. 17.9). The kata contains repetitive sequences of an upward bent-wrist face punch (furi-zuki) that is followed by a clinch and takedown movement. Wanshū (Refined Wrist, 腕秀), the original name of Empi, may be an attribution of this unusual technique. Wanshū may also be the pronunciation of the name of a Chinese diplomat and Fujian White Crane quánfǎ master (汪輯) who arrived in the Ryūkyū Kingdom Village of Tomari in 1683 CE.[21]

Japan witnessed rising nationalism after the imperial navy's limited but successful forays in World War I. After having occupied the Chinese-ruled Liaodong Peninsula since 1894, the Japanese army moved to annex Manchuria under false pretext in 1931. Manchuria fell under Japanese rule, though it was nominally still part of the Chinese Republic.

Anti-Chinese sentiments ran rampant when Master Funakoshi arrived in mainland Japan in 1922. Funakoshi referenced Zen Buddhist philosophy by changing the connotation of *kara-te* from "Chinese hand" to "empty hand" (see chapter 1), elevating the discipline to a budō art form. Revisions of kata names from Chinese historic figures (Kūsankū, Chintō, Wanshū) to Zen concepts of "void" and animal spirits were likely more accommodating to the ruling class. Given that Japan is located to China's geographic east, this sentiment was particularly relevant for Chintō (鎮東), whose literal translation is "Vanquishing the East." Similarly, renaming Wanshū to Empi aspires to the flitting spirit of the swallow (*tsubame,* 燕), which portends good luck and fidelity in Japanese culture.

Figure 17.8 Gankaku.

Figure 17.9 Empi.

3. Names from Numeric Codification

NIJŪSHIHO.   Nijūshiho (Twenty-Four Steps, 二十四步) originated from Niseishi (Twenty-Four), an Okinawan Shitō-ryū kata that traced back to Kenwa Mabuni and his son Kenzo Mabuni in the early twentieth century. It is one of many kata with a numerical designation. Others include Gojūshiho (Fifty-Four Steps) and Seisan (Thirteen). Nijūshiho techniques are most effective at close distance through the generation of "shocking" power (forced vibration). A series of combinations start with smooth and flowing parrying that leads into sharp and explosive strikes at close range (fig. 17.10). The form focuses on diaphragmatic breathing in coordination with contraction and expansion of the abdominal core, as well as sharp application of stance pressure.

Martial arts scholars believe that Nijūshiho originated from the Shaolin Dragon-style fighting, famed for harnessing qi as "internal force" to bring power and agility to kicks, hand-locks, and strikes.[22] The stylist subscribes to the mystical nature of the dragon, which attacks like the wind (swift and boundless), defends like the cloud (illusive and unpredictable), stops like iron (strength), advances like a tiger (ferocity), and retreats like a cat (agility).[23]

Figure 17.10 Nijūshiho.

GOJŪSHIHO DAI.   Formerly called Useshi (Fifty-Four), this kata reflects the heritage of the historic Shaolin Essence of the Five Fists (see chapter 1).[24] Gojūshiho (Fifty-Four Steps, 五十四步) corresponds to the Tiger-style tàolù by the same name. The kata was believed to have been brought from China by Sōkon Matsumura sensei of Okinawa Shuri-te, then passed to Kenwa Mabuni of Shitō-ryū.[25] Gojūshiho Dai, one of the two contemporary versions practiced by Japanese karate styles, is a study of advanced techniques such as the back-hand block (keitō-uke) (fig. 17.11) that swiftly transitions into explosive fingertip strikes (ippon nukite) while supported by a weight-leveraged cat stance (Nekoashi-dachi) (see table 9.1). Quick tempo changes condition the mental dexterity of the practitioner.

HANGETSU.   Hangetsu (Half Moon, 半月) is a modified version of Seisan (Thirteen, 十三), an ancient Naha-te kata practiced in multiple versions today. Seisan focuses on controlled breathing, the hara, and ki circulation throughout the body (see chapter 16). Funakoshi assigned Hangetsu for Shōtōkan practitioners for the conditioning of core muscles through controlled breathing, while other styles practice the Sanchin form toward the same end.[26] Hangetsu breathing is not as sharp and intense as in Sanchin, and it better aligns with the gentler White Crane practice of southern Wǔ Xíng Quán (Essence of the Five Fists).

The Hangetsu stance is a longer version of the Sanchin stance (see table 9.1). Both focus on inside thigh muscle tension to favor body-shifting or quick directional changes (fig. 17.12), enabling the martial artist to close in and control the opponent with strikes and takedowns.

Slowed and deliberate movements in the first half of the kata attend to connectedness to the ground, aspiring to absorb earth qi to consolidate body integrity (see chapter 16). Rhythmic breathing is adopted to synchronize joint compression and foot movements. The practitioner looks inward to harmonize mind, body, and spirit. A measured pace proffers the opportunity to study tsuki no kokoro, the unperturbed mental attitude of "mind like the moon" (see chapter 2).

The kata's tempo quickens to accompany sharp hand and foot movements in the second half, turning one's attention outward for technique applications. Stance pressure propels quick center-of-mass transition to within attack distance or serves to elude attacks through positional changes. The quick foot movements coordinate with the contraction and expansion of core muscles to render speed and power.

Kata names represented as numeric designations have confounded martial arts historians to this very day. Contemporary versions incorporate the character 步 *(ho)*, which means "steps" or "techniques" (Nijūshiho, Gojūshiho), whereas the originating versions did not (Seisan = 13, Niseishi = 24, Useshi = 54, Suparinpei = 108, and so on).

Figure 17.11 Gojūshiho.

Figure 17.12 Hangetsu.

Many, including myself, have attempted to reconcile the designations with the number of techniques, foot movements (steps), sets of combinations, and so on, and have come up short; in most cases, there are way too many. These notations also do not correspond to the number of directional encounters against multiple opponents.[27]

Intrigued by the embedded implications from Zen or Daoist thought, I posed the same question decades ago to the late Hidetaka Nishiyama sensei. Sensei was a scholar of Asian cultures, apart from his depth of knowledge in karate-dō. His interpretation was straightforward. He considered the kata numbers to represent their order of development within the originating style, similar to opus number of musical compositions (or the *L* designation for British weapons).

Sensei's advice to me? "Don't overthink everything, Alex!"

Kata Training Concepts

Master Masatoshi Nakayama, late chief instructor of the Japan Karate Association, advanced that, "A man must first be the master of his own house before becoming the master of another."[28]

The aphorism aligns with a series of Confucian wisdoms, starting with "to cultivate the body, (one must) first right the heart (mind, intention, passion)" (*yù xiū qí shēn zhě, xiānzhèng qí xīn,* 欲修其身，先正其心). A rectified heart and body empower the decision to act, sustaining purpose, then consideration for family, community, and ultimately, the nation.[29] The dictum as applied to martial arts specifies that intent and ability are essential for confidence in execution. It behooves us to put our own skill set in order through kata practice. Having attained total awareness and expertise, the same attributes can then be applied to control the opponent.

Common concepts underscore kata development regardless of training style. These ideas are summarized as follows.

1. Consistency in Practice

Kata trains for sustained peak performance throughout the series of defensive and offensive combinations. The physical demands to achieve precision and control are brought to bear under the purviews of one's own psychology.

In absence of an imminent threat, one is tempted to modify movements to accentuate physical prowess or to highlight personal preference in technical applications. Improvisation is discouraged when performing traditional kata. A deeper appreciation of nuance in movement and tempo can only come from consistency in practice. Ultimately, the martial artist gains deeper insights that have been embedded within the movements. As well, accomplished musicians adhere to the dynamics, pitch, tempo, key, and mood of classical compositions in order to plumb the depths of the composer's aspirations.

Faithful and consistent practice instills discipline in mind and body. "Muscle memory" aligns preconditioned brain automating functions (see chapters 4 and 7) with proprioceptive adaptations (chapter 8). A controlled and sustained performance (chapter 4) is predicated on having attained the mind state of "flow" (chapter 6).

Kata continues to be guided by long-held principles of kime (physical focus; chapter 9) and zanshin (sustained mental focus; chapter 12). The outside form is subjugated by intended purpose of action. Proper positioning is governed by awareness of stance functionality and specifications of the *embusen* (designated line of movement, 演武線), both required to maintain concise movement transitions.

2. Correspondence of Techniques

The kata's sequential movements are modeled after scenarios of self-defense and are presented as a study of pulsed executions for power and speed. Sparring techniques involve accommodations for distance, timing, or the strategy of the opponent. Kata performance does not suffer from these hindrances. The performer is afforded the opportunity to persevere within the kata's framework and to maximize output by untangling inner conflicts of mind, body, and spirit.

The unfettered setting of kata performance fosters body dynamics optimization, bringing insights on how muscles and joints work together to create the summation of speed. Accordingly, Master Hidetaka Nishiyama advised that the fundamental elements of kata techniques should align with those in kihon (basics) and kumite (sparring).

3. Autonomous Execution

A minimum of three months is allotted for beginner and intermediate students (white and green belts) to commit a designated kata to memory, then to perform it with competency. This is the first stage of kata study. To achieve a deeper understanding, black belts spend two years or more on a particular kata. The practitioner is expected to achieve the state of flow, sustaining autonomous execution (see chapter 3) where complex sequences are reproduced with minimal conscious intervention.

Motor function output is rapid, smooth, and effortless by delegating neuromuscular control to procedural memory, which expends minimal attentional resources.[30] The practitioner now directs attention to heighten their awareness as they fine-tune their movement transitions, guiding tempo with their emotional state, and invoke the mental attitudes of tsuki no kokoro (mind like the moon) and mizu no kokoro (mind like water) (chapter 7).

4. Quantum of Execution

Having mastered the required technical skills from start to finish, the practitioner seeks quantum of execution by attending to minutiae of the moment. Damage is done from what has already transpired. There is also no need to fret about what's to happen next, so long as action at the particular juncture can successfully control the opponent.

This is the mind-set of fudōshin (不動心), an unshakeable belief with the spirit of equanimity. Our physical actions are totally invested within this mental state. One is clinched to the present, without anxiety of fear for the future or anger and regret for the past, negative feelings that unnecessarily weigh our spirit.

5. Imagery

Kata performance differs from a gymnastic floor exercise. Each movement has an intended purpose and an expected outcome as portrayed by the martial artist. Imagery is projected from the mind's eye, directing what's to be accomplished by the body. Movements are otherwise no more than the flailing of arms and legs. Instead, the practitioner brings the audience along on their mental voyage as they paint vivid scenes with confidence and feelings of engaging virtual attackers with blocks, strikes, and throws.

Imagery is a mental process that calls on all the senses, enabling the martial artist to strive for peak performance.[31] Image projection is driven by intent and resolve. This visualization process is key to developing the necessary mental focus and physical strength in an actual conflict. Intent (yi) solidifies motor execution goals and rallies neuromotor activation to complete the physical tasks at hand (qi). Intent is guided by eye direction, which reinforces attention from start to completion of the technique.

6. Bunkai

Kata techniques revolve around the universe of self-defense applications, including striking, nerve-point shock, hand trapping, joint locks, chokes, and sweeps and throws. Mental imagery projection requires a firm grasp of the ramifications of each technique, which are studied as bunkai.

Bunkai (分解) is commonly annotated as "application," although a literal translation alludes to "dissect and explain" or "dissect and illustrate." Insights on potential applications solidify intent and the visualization process. Different levels of bunkai refer to technique applications as actually performed (*ōyō*, 応用), actions that represent variations (*henka,* 変化) in order to optimize positioning (body shifting, turning) or to augment the effectiveness of execution (changing a body blow to a face punch, adapting a body-turning into a throw).

Kakushite (hidden techniques, 隠して) is the third level of bunkai, extended to elective techniques that are not specified in the repertoire of movements. These added elements are incorporated without interrupting the overall flow of the application sequence, yet they give the expectation of significantly enhanced outcome when applied as self-defense. Examples include an opening move that adds an element of surprise, or a follow-up takedown technique to ensure complete incapacitation of an opponent.

The range of bunkai interpretations is mind-boggling. Some karate-dō styles have fundamental rules for their development, such as *kasai no genri* (rules based on principle of execution, 解裁の原理) in Gōjū-ryū. In the past, developed bunkai were practiced in earnest, held in confidence, and passed only to the most senior students of a school.[32] A fundamental understanding remains in that there is no singular definitive bunkai for a given technique, as application invariably hinges on circumstances of execution, while technical fixation leads to an inflexible mind.

Bunkai designations should address the kata's spirit and its dynamics. Correspondence of body dynamics that are part of the movement is a good starting point, as the application should sustain technical flow (hip rotation, dropping of the center of mass, and so on). For example, it is reasonable to add a grip and takedown action after the open-hand knife-hand strike (shuto; see fig. 13.1), as all these movements engage the same biomechanics from hip and core. The incorporation of kakushite techniques also serves to accentuate the performer's insights in self-defense or kumite strategies.

Ultimately, the martial artist is urged to exercise realism and restraint. International championships are filled with aerial flips and flying kicks rationalized as kakushite. The defender carries one attacker over their shoulders as they spin and deliver more kicks to another. The audience roars in approval for the impressive athletic prowess. Crowd-pleasing antics are not practical in self-defense, nor do they offer a realistic portrayal of the kata's spirit.

7. Attitude of Finality

Kata, like a musical composition, is storytelling in its most abstract. The protagonist encounters a challenge, navigates through entanglements, and finally commits to achieving resolution. The cycle of events repeats to address another quandary, and so it goes until the final reckoning (todome waza; see chapter 3).

A vivid performance reflects the performer's total investment of their physical and mental resources. Different segments engage the performer's emotions through tempo change and technical challenge. The ebb and flow of movements evoke contrasting feelings such as anticipation or steadfastness, prompting the practitioner to temper their mind-set (attack versus defend, strength versus agility, rootedness versus evasion).

Advanced kata such as Unsū or Gojūshiho incorporate capricious transitions and expansive technical demands, further commanding mental concentration.

We as individuals are unique in our ways of confronting challenges and arriving at solutions. Although movements in a kata are clearly delineated, physical attributes and psychological makeup uniquely color interpretation. This personalized patina is discernible from afar. Expertise captures and enthralls, whereas limited emotional involvement comes across as "flat" and uninspiring. Overindulgence in technical details gives the impression of being "clinical" or dull.

Each performance takes on a slightly different feel and outcome, given fluctuations in mood and physical well-being at a given juncture. By investing years of practice, the martial artist leans on their physical consistency without conscious thought so that they can devote their attention to interpret the kata's spirit through total immersion.

I often ask students to perform their kata for one last time at the conclusion of a kata study. They are expected to have utmost precision and total mental commitment and to apply speed and power without hesitation. This is the attitude of finality, to give your best because there may not be another chance. It is an understanding that one pushes the envelope in order to reach a higher plane.

This attitude conveys the Zen concept of ichi-go ichi-e (one lifetime, one encounter; see chapter 7). As practiced diligently in *chadō* (tea ceremony), participants completely immerse themselves in the ritual. Each party focuses on the subtlety and precision in every physical action, savoring moments of utmost tranquility. They are expected to fully treasure uniqueness in the encounter, as the exact circumstance will never happen again.[33] The process fosters depth of mental focus and aspires for an unperturbed spirit.

Performance Criteria

Competition is not the ultimate goal in kata training. Nevertheless, this forum tests one's mettle in front of an unsympathetic audience. The process adds perspective when witnessing another competitor who performs the same kata. Competition checks confidence, brings rapport among peers, and provides opportunities for feedbacks from judges, who would have mastered that same kata years before.

Competitors of national and international championships come from a diverse pool of athletes. Judges are tasked with ranking their caliber of performance across twenty-six kata.[34] The overall score has to reflect a competitor's mastery regardless of style or practice, as well as their insights on the kata's spirit.

The AAKF is the oldest U.S. nongoverning body for traditional karate competition. To ensure consistency in scoring, judges assign a cumulative score that is based on four equally weighed criteria.[35] These overlapping elements are worthy of deeper analysis to guide skill development.

BODY DYNAMICS. This scoring category assesses an athlete's display of skeleto-muscular biomechanics. Proper body dynamics, such as transition of the center of mass or hip rotation, require correct posture and balance, as well as the ability to harness stance pressure (see chapter 9) and hip actions to maximize kinematic outcome (see chapter 10).

POWER. The power criterion examines an athlete's force output. This score takes into accounts the competitor's physical attributes, then weighs their expertise to transform natural ability into technique power.

A small athlete can generate the same body power as a bigger competitor if they have a better biomechanical skill set (see chapter 10), in particular by synchronizing muscle contraction and expansion with core-based rhythmic breathing and an understanding of kime. These attributes will bring a higher power score than the bigger athlete deemed to have a lower level of mastery.

FORM. Form assesses the capacity to sustain physical integrity through mental concentration and body control. Stance structure, core integrity, and coordination contribute to proper form, which is also anchored by the athlete's confidence and emotional stability. Proper form requires that mental focus be aligned with directionality of physical action, which in turn portrays effective technical application (intended purpose) for the movement.

Form scores default to a mistake-free performance, as judges rank competitors according to overall physical competency and mental focus. An athlete's strength in spirit brings a more favorable score, whereas an agitated performance that precipitates in errors and disorientation would be downgraded.

TRANSITION. Transition score reflects the skill to sustain power, speed, balance, and mental focus within movement sets, from one set to the next, and when changing directions.

Athletes performing at the "flow" state of mind (see chapter 6) achieve fluidity in movement, where termination of one action leads naturally to the next with minimal windup or loss of energy. The mind, body, and spirit are melded as one and sustained through changes in tempo and rhythm. Through the skill of moderating intra-abdominal pressure, skeletal muscular actions are closely tied to the breathing rhythm in serially executed techniques.

Conversely, stilted movements belie a lack of technical competence or mental composure. Common errors include dropping the hand guard from one move to the next, lapses in attention that lead to loss of mental focus when changing directions, or being out of breath from nervousness. A lack of mental and physical continuity translates to *kyo* (breach) that would be exploited in an actual confrontation. The athlete's transitional score would be lowered accordingly.

Kata has been used for centuries to convey the essence of the art. Recently, the business community has also appropriated kata concepts toward improving everyday workforce performance. In particular, the Toyota Motor Corporation implemented repetitive training practices to spur worker skill development alongside ordered scientific-thinking drills. These processes have proven to be highly effective in bolstering self-confidence, and they help to promote progress in business practices.[36]

Discussions in this chapter serve to impress that kata is more than patterned learning and structured practice. Mastering a kata is an extraordinary and highly personal experience. Kata performance transforms our sense of reality, taking the karate-ka on an imaginary journal that brings respite from the intense materialistic demands of everyday life. This recurrent practice never ceases to challenge one's limits over time.

Martial artists over the ages realized that traditional practice is more than physical fitness and self-defense know-how. One can expect to maximize physical and mental performance by incorporating contemporary concepts in these areas of practice. Yet true purpose hinges on immersion in the martial spirit, which in turn shapes our way of life. These aspirations are best illuminated by the meditative process and through the practice of kata.

Takeaways from "Apollo, Dionysus, and the Dō of Kata"

- Kata, often referred to as a language or living text of the art, fosters physical competence, dexterity of the mind, and spiritual growth through gained insights of self.

- Techniques and tempo are but the cusp of learning, as kata also conditions the mind. In accordance to "qi follows yi" (energy follows intent), enduring practice cultivates heightened vigilance of the surroundings and directionality of mental focus.

- The kata's spirit embodies defining elements that rouse our emotional state and guide our mental attitude. It taps into the subconscious and channels our feelings, much like music, fine arts, or calligraphy.

- Body dynamics, power, form, and transition embody the fundamental elements in kata training.

Postscript: A Sack of Rocks

Even the stubborn rock nods its head

(wán shí diǎntóu, 頑石點頭)

—BUDDHIST EXPRESSION

In the 2007 film *The Drummer* (*Jin Gwu,* Battle Drum), Sid Kwan is a gifted but reckless young musician who runs among the triad gangs of Hong Kong.[1] Sid commits a fatal error and is exiled to the countryside of Taiwan for his own safety. There, he chances upon a Zen drumming *(taiko)* troupe in the midst of practice. Sid is taken by their intense focus and skill. As he approaches the sensei to join, his attitude and impetuousness are not well received.

To calm Sid's spirit and aid his development, the sensei instructs Sid to gather ten or twelve palm-size rocks from the forest and place them in a small burlap sack. Sid is to carry the sack full of rocks over his shoulder on the troupe's daily runs through the forest trails.

Sid becomes more grounded over time, gaining the skill and trust of the other members. He earns his place in the performing team that went on international tour. While passing through Hong Kong, he visits his older sister, his confidant. Now in his more collected demeanor, Sid offers his sister a personal gift. He opens his burlap sack and hands her one of the forest rocks. These rocks have lost their jagged edges over time from jostling inside the sack. Their surfaces are as smooth as river stones, with a fine sheen that belies their origin.

According to legend, the Buddhist monk Zhu Dao Sheng once spoke directly to a pile of rocks while seeking nirvana. His sermon on the Buddhist scriptures was so persuasive that even the insensate rocks nodded their heads in agreement.[2]

The martial arts learning process is equally quizzical. It is often compared to forging a samurai sword. The native steel is fired repeatedly, hammered, and folded into a near-perfect tool. By subjugating the body to harshness, one acquires an exquisitely responsive instrument that is primed for all physical activities. By ignoring pain, one gains a heightening of all the senses. In accepting setbacks as a way of life, a student sustains steadfastness and the inner strength to overcome. From newfound confidence comes generosity in spirit, bringing forth equanimity and happiness that is founded on an identity of non-self. For those who choose to test their mettle by pursuing traditional martial arts, the journey is enduring, at times frustrating, but ultimately exhilarating.

Like Sid, I carried a collection of unformed ideas for the longest time, jostling in my mental burlap sack. The completed work attempts to put to order knowledge, observations, and introspections gleaned through the lenses of traditional martial arts and biomedicine, two fascinating worlds that I traversed throughout my adult life.

This book took over three years to complete. It has been an astounding experience, not without the numerous distractions and detours. I learned about proprioceptors from my physical therapist, and about neuroscience from my karate students who were also PhD candidates at the University of Texas at Dallas. A firsthand appreciation of the fluidity of forces came from guiding Big Sucker, a suctioning device that draws leaves from the bottom of a swimming pool. The writing process was mind-boggling, as seemingly unconnected ideas merge with practice concepts.

I have been a student of the late Hidetaka Nishiyama sensei for over thirty years. This renowned authority of karate-dō constantly reminded that a sensei's role is synonymous to the *sina'n* (south-governor, 司南), the epic south-pointing navigating compass invented in China's Han Dynasty (2nd century BCE–1st century CE).

The role of a sensei, according to Nishiyama sensei, goes beyond spoon-feeding information. Instead, they make use of personal knowledge and wisdom to mark the path for the student's development, each according to their temperament and innate talents. In keeping with the concept of *shu, ha, ri* (comply, breakthrough, depart, 守破離), the sensei is also tasked with familiarizing students with all aspects of the art toward balancing mind, body, and spirit. The acquired skills are the student's instrument to take in any direction of their own choosing.

"For what is the use of new knowledge if it doesn't lead to novel behaviors?" argued Yuvan Harari in the epic tome *Homo Deus*.[3] At first glance, there appears to be no imminent need to incorporate scientific knowledge into traditional martial arts. I am inclined to disagree—so long as these proven facts augment learning and performance, at the same time affirming the true meaning of the art. I urge those who are more enlightened to contribute to the effort of assimilating new knowledge and insights into our incredible time-tested practices, and in doing so, propel the art to a higher plane.

Appendices

Mathematical Concepts of Motion and Impact

Physicists tend to be perturbed when martial artists toss around terms like *momentum, force, impulse,* and *energy* interchangeably, and rightly so. The contents of this chapter are an attempt to reconcile these terms by their definitions, and their mathematical relationships as applied to the study of mechanics.

Taking a simplified example of an incoming baseball traveling at speed, its momentum (P) is defined by the formula:

$$P = mv$$

where m is the mass of the ball and v is its linear velocity.

When a player is struck by the ball, the impact force (F) depends on the mass of the ball *(m)* and dissipation of the terminal velocity *(v_t)* over the duration of impact *(t)*, assuming that the player remains stationary and absorbs most of the impact energy. Impact force, F, is represented as:

$$F = ma$$
$$= m \left(\Delta v_t / \Delta t \right)$$

where $\Delta v / \Delta t$ is the change in velocity per unit time, and equivalent to linear acceleration a, in this case, deceleration. Accordingly, the delivered force (F) also equals the rate of transferred momentum, i.e.,

$$F = m \left(\Delta v_t / \Delta t \right)$$
$$= (m \Delta v_t) / \Delta t$$

In other words, momentum is converted to impact force on an immovable surface when no other external net force acts on the traveling ball. A shortened time for momentum transfer equals a higher impact force.

The same scenario applies in the context of energy transfer. Physicists broadly categorize energy (E) into kinetic energy (KE) from motion and potential energy (PE), which embodies mechanical energy, stored energy (electric charge, a chemical reactant, stresses within the object, the muscle coiled-spring effect), and energy gained from the object's relative position (such as height from the ground). Mechanical energy for movement in the human body is generated through metabolism and stored energy in compressed striated muscles.[1]

For a traveling baseball, the stored kinetic energy (KE) is expressed as:

$$KE = \tfrac{1}{2}mv_t^2$$

where v_t is its terminal velocity.

Energy equals the ability to do mechanical work (W). Work (W) is defined as force (F) applied over a distance *(d)*, i.e.,

$$W = F \times d$$

In a perfect world, where KE is transferred in its entirety into W,

$$KE = \tfrac{1}{2}mv_t^2 = W = (F)(d); \text{ and}$$
$$F = W/d$$

For example, the stored kinetic energy of a regulation Major League baseball of approximately five ounces (0.15 kilograms) that arrives at 55 miles per hour (25 meters per second) is:

$$KE = \tfrac{1}{2}(0.15)(25)^2 = 46.9 \text{ joules}$$

Damage from the impacting force would be amplified if it were dissipated within a short distance *(d)* after it hits the body. If the energy is dissipated over about an inch (0.025 m), the transferred force is:

$$KE = W = (F)(d) = 46.9;$$
$$d = 0.025 \text{ m}$$

$$F = 46.9/0.025$$
$$= 1,876 \text{ newtons}$$

From the standpoint of the pitcher, their power (P) is defined as energy output per unit time, i.e., (E)(Δt). For illustrative purposes, we will assume that the fastball leaves their hand at terminal velocity, ignoring real-life factors such as air drag, dip in linear projectile, and spin. If the pitcher completes the pitch in 0.05 seconds, launching the 0.15-kilogram ball at 25 meters per second, their power is calculated as:

$$P = (KE) \, \Delta t$$
$$= 46.9/0.05 = 938 \text{ watts}$$

These formulations are summarized in table A.1, which also incorporates other definitions that are relevant to the physics of impact.

Table A.1 Physics of force and momentum.

ENTITY	DEFINITION	UNIT
mass (m)	the quantity of matter in a body regardless of its volume or of any forces acting on it	kilogram (kg)
velocity (v)	speed of linear travel of an object	meters/second (m/s)
acceleration (a)	change of velocity over time $= v/t$, or $\Delta v/\Delta t$	meters/second/second (m/s^2)
momentum (p)	also known as quantity of motion $= mv$	kg·m/s
force (F)	also known as rate of change of momentum $F = m \, (\Delta v/\Delta t)$	kg·m/s^2 (newtons)
joule (j)	amount of energy when a force of 1 newton is applied over a distance of 1 meter	kg·m^2/s^2, (n·m)
kinetic energy (KE)	energy stored by an object while in motion $= \frac{1}{2}mv^2$	kg·m^2/s^2, (n·m)

ENTITY	DEFINITION	UNIT
Newton's second law	The rate of change of momentum equals the resultant force in the direction of the resultant force.	
torque (τ; moment of force)	rotational force (τ) that produces a turning effect = (F)(*d*), where F is the magnitude of the force, and *d,* its perpendicular distance to the axis of rotation.	newton-meters (n·m)
watt (w)	A measurement of power (P) = $\Delta E / \Delta t$ 1 joule/second corresponds to the power of an electric circuit with a potential difference of 1 volt and a current of 1 ampere.	joules/second
work, kinetic energy, and force	Work transfers energy from one place to another, or from one form to another. W = (F)(*d*), such as moving an object with a force of 1 newton for a distance *(d)* of 1 meter. By Newton's second law, the work of the resultant force equals the change in stored kinetic energy. W = (F)(*d*) = ΔKE; therefore F = ΔKE/*d*	newton-meters (n·m), joule

Force and Velocity in Performance

An athlete's energy output translates into their capacity to generate power (P), or the amount of transferrable energy per unit time (table A.1). The relationship between power (P) and speed (v) is expressed by the formula:

$$P = (F)(v)$$

where F is the force generated, and v is the velocity of the directional force.

With a constant power output, instantaneous force output (P) is inversely proportional to the speed in applying that force (v), or:

$$F = P/v$$

This formula can be represented as the performance force versus velocity (F-V) curve in athletics (fig. B.1). In practical terms, an athlete with finite power has to slow down movements in order to increase force output. The process involves recruiting muscle motor units to achieve a stronger voluntary muscle response. The recruitment process, manifested by cross-bridge formation between muscle filaments, enables a higher level of force output.[1] The athlete can generate an even larger amount of force by leveraging against the ground reactionary force. They do so by staying in position, as in the case of pushing a car or performing a deadlift, or when carrying a big and heavy box up the stairs one step at a time.

Conversely, a lower force output enhances movement speed. Players from Major League Baseball to Little League prefer a lighter bat for easier control. The lighter bat also generates a faster swing that drives the batted ball farther.[2] Throwing a dart that weighs nothing at all only requires a quick, high-velocity toss, whereas "putting" the considerable heavier shot in track and field demands considerable force generation but produces less velocity.[3]

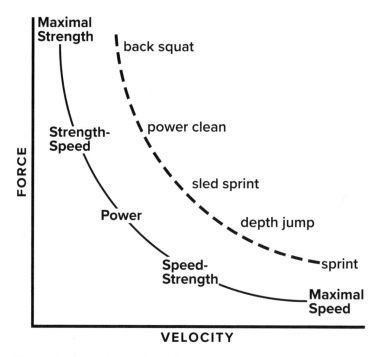

Figure B.1 Inversely correlated force output with speed in athletic performance. Ballistic training (sustained output) favors strength development, while plyometrics (bursts of short-interval maximal output) are conducive for speed. Overall improvements in both power output of force and velocity can be achieved by combined regimens that shift the entire performance curve (solid line) to the right (dashed line).

The balance of strength and speed output is represented by the trough of the F-V performance curve (fig. B.1). This point represents the optimal combination of force development and transition.[4]

Depending on the nature of the sport, an athlete may choose to meet a higher strength demand by increasing their force capacity (toward the top of the curve), or to strive for a more explosive technique delivery by moderating their force output (toward the bottom of curve). Accordingly, the back-fist strike *(uraken uchi)* relies on the speed of delivery to create vibrational shock, whereas a side thrust kick (yoko kekomi geri) is founded on the intensity of impacting force (see chapter 13).

An individual's one-rep max (1-RM) represents the starting point for strength versus speed development (table B.1). In weight lifting, 1-RM is the maximum load that the athlete can lift once before failure sets in. For example, the 1-RM by bench press for an untrained 150-pound male is around 110 pounds, and 70 pounds by an untrained 130-pound female.[5]

Table B.1 Force-velocity performance zones.

ZONE	% ONE-REP MAX (1-RM)	CONDITIONING DRILLS
Maximal Strength	90–100%	back squat
Strength-Speed	80–90%	power clean
Peak Power	30–80%	sled sprint
Speed-Strength	30–60%	depth jump
Maximal Speed	< 30%	sprint

Ballistic training aims to improve force production. These exercises include weight lifting, throwing weights, or jumping with weights. The common practices are the power clean or back squat clean lifts (table B.1). When the athlete works within the strength-speed zone, their power component can be improved by Olympic lifts (snatch, clean, and jerk).

Plyometrics, also called jump training or plyos, are best applied within the speed-strength zone by improving the stretch and contract functions of major muscles.[6] Plyos favors horizontal strength building for enhanced body momentum in running and jumping, as opposed to ballistic training that builds vertical strength as applied to throwing or striking. Plyometrics include jump squats, box jumps, and sprint drills while dragging a light or heavy load.

The majority of martial arts techniques require maximized force output as well as application speed. These performance demands fall within the spectrum of strength-speed to speed-strength zones (table B.1). Practitioners should commit to a balanced conditioning regimen that can increase overall power output. By shifting the entire performance curve to the right (see fig. B.1), one is given the opportunity to improve weaknesses while accentuating the strengths.

GLOSSARY

1-RM, one-rep max: Maximum load that the athlete can lift once before failure.

10,000-hour rule: Theory on the time and effort needed to master a skill set.

abductors: Muscles that move the leg away from the body.

ACTH, adrenocorticotropic hormone: Pituitary gland hormone that stimulates cortisol production.

action potential: Spike of signaling electrical activity by neurons and muscles.

activations, muscle unit: Activation of muscle units, each comprising a motor neuron and all the muscle fibers it stimulates.

adductors: Muscles that move the leg inward toward the median line of the body.

aikidō 合気道: Japanese budō art based on the channeling of internal energy, throws, joint locks, and dynamic body movement.

aikidō-ka: Person who practices aikidō.

amygdala: Focal area within the temporal lobe associated with emotions.

anatta 無我: Mind state of non-self.

animal electricity: Electricity produced in living tissues.

antioxidants: Substances that prevent or slow free radical damage to cells.

Apollonian: Human nature's inclination for order, reason, and discipline.

application kata: Martial arts forms with self-defense movements.

asymmetry, brain: Anatomy and functional differences between left and right brain.

attention functions: Mind's ability to process sensory information and focus on tasks.

aura: Energy field that envelops the body.

aurora australis: Atmospheric light phenomenon in Antarctic.

aurora borealis: Atmospheric light phenomenon in the Arctic.

Australopithecus: Hominin species that existed 4.2–1.9 million years ago.

automation, brain: Subconscious task performance by the brain.

autonomic functions: Vital body functions that operate subconsciously.

AVG, MLB batting average: Statistic to measure a baseball player's success at bat.

baguazhang 八卦掌: Wudang martial art based on the trigrams of the Daoist I Ching.

Bai Hui 百會: Acupuncture point located at the top of the head.

ballistic exercises: Weight-based training.

Beast Barracks (Beast): Seven-week induction summer curriculum for cadets at the U.S. Military Academy at West Point.

bipedalism: Standing, walking, and balancing on two feet.

binary trigger: Decisions based on two choices.

biofield: Informational energy field that surrounds and permeates the body.

biomagnetism: Magnetic fields produced by living organisms.

biomechanical efficiency: Actual versus theoretical maximal output in physical movements.

biomechanics: The mechanics of movement and structure in a living organism.

bō 棒: Wooden staff.

body dynamics: Body movements that enhance biomechanical efficiency.

bokken 木劍: Wooden sword.

boson: A particle that carries a force.

brainstem: Posterior part of the brain that connects the spinal cord.

brain waves, gamma: High-frequency brain waves that reflect intense brain functions.

brain waves, theta: Brain waves associated with learning, memory, and intuition.

breathing, abdominal: Focused diaphragmatic breathing.

breathing, belly: Focused diaphragmatic breathing.

breathing, combat tactical: Practice of slowed breathing to reduce stress.

breathing, Daoist reverse: Breathing that focuses on modulating intra-abdominal pressure.

breathing, diaphragmatic: Conscious or subconscious breathing that utilizes the diaphragm.

breathing, martial arts: See Daoist reverse breathing.

breathing, natural: Gentle and unforced breathing.

breathing, yogic: The three stages of controlled breathing according to pranayama.

Buddhism: Traditions, beliefs, and spiritual practice according to the Buddha.

budō 武道**:** The "martial way" of modern Japanese martial arts.

bujutsu 武術**:** Martial techniques and practical applications.

bunkai 分解**:** Dissect and explain, as applied to kata movements.

bushidō 武士道**:** Way of the warrior class, as defined by a code of ethics.

catecholamines: Hormones made by the adrenal gland.

cerebellum: Hindbrain, located near the brainstem.

cerebrum: Front part of the brain responsible for higher functions.

chadō 茶道**:** "The Way of Tea," tea ceremony.

chaji 茶事**:** Formal, private tea ceremony.

Chán 禪**:** Chinese Buddhist sect that emphasizes enlightenment of one's own mind.

chen qi 沉氣**:** Anchoring of inner energy.

choku-zuki: Straight punch.

chudan: Mid-level.

chunking, data: Grouping individual information pieces into related sets.

clock, internal: Natural, internal system that regulates sleep and wake cycles.

cognition: Mental processes that enable knowledge and comprehension.

cognitive control: Alignment of behavior with preset goals or plans; willpower.

coiled-spring effect: Release of energy stored in compressed joints.

Confucianism: Chinese philosophy and ethical teachings of Confucius.

constructionistic: Human experience founded on self and environment.

consumeristic: Human experience founded on consumption and acquiring material goods.

core: Trunk muscles that stabilize the spine and pelvis.

cortex, anterior cingulate: Region that interphases the prefrontal cortex and the parietal cortex with the limbic system.

cortex, anterior insular: Brain region that orchestrates conscious emotional experiences.

cortex, frontal: Frontal lobes engaged in motor function, problem-solving, memory, and impulse control.

cortex, motor: Brain area for planning, control, and execution of voluntary movements.

cortex, posterior cingulate: Highly active brain area that ties emotions with long-term memory.

cortex, prefrontal: Front part of the frontal lobe involved in planning, decision-making, and social behavior.

cortex, premotor: Cortical region tasked with movement calibration and control.

cortex, primary visual: Cortical region that receives, integrates, and processes visual information from the retina.

corticospinal facilitation: Neural adaptive processes that enhance motor functions.

cortisol: Hormone that regulates body functions in response to stress.

CRH, corticotropin releasing hormone: Hormone involved in stress response; stimulates adrenocorticotropic hormone production.

Crohn's disease: A type of inflammatory bowel disease.

dachi, Hachiji-: "Character for eight" stance; Natural Stance.

dachi, Hangetsu-: Half-Moon Stance.

dachi, Ippon-ashi-: One-Legged Stance.

dachi, Kiba-: Calvary Horse Stance.

dachi, Kōkutsu-: Back Stance.

dachi, Naihanchi-: Inside Straddle Stance.

dachi, Nekoashi-: Cat Foot Stance.

dachi, Sagiashi-: Heron Foot Stance.

dachi, Sanchin-: Three Battles (Hourglass) Stance.

dachi, Shiko-: Square Stance.

dachi, Sōchin- (Fudō-): Diagonal Straddle Leg (Immovable) Stance.

dachi, Zenkutsu-: Front Stance.

dǎ 打: Striking.

daisho 大小: Big and small.

danse macabre: Dance of death.

dantian (tan t'ien) 丹田: Energy center (Daoist concept).

dao (tao) 道: Way, path (Chinese philosophy).

Daoism (Taoism) 道教: Chinese philosophy based on the writings of Laozi.

Dào Jià 道教: Daoist teachings.

Dao, Tian: Way of the Cosmos.

daoyin 導引: Daoist exercises to cultivate jing and to channel and refine qi.

de 德: Moral character.

deashi 出足: Quick start; advancing foot.

default mode network: Large-scale brain network that operates at wakeful rest.

deformable: Change in shape or volume when an external force is applied.

delicious cycle: Brain's feedback loop that awards its own performance by stimulating low-level adrenaline release.

Dentō karate-dō 伝統空手道: Traditional karate-dō practice.

diaphragmatic contraction: Process of diaphragmatic movement to achieve air intake.

Dionysian: Human nature's inclination to sensuality, impulse, and emotions.

dmigs med snying rje: Tibetan concept of compassion that is founded on selflessness.

dō 道: The Way, path (Japanese philosophy).

dōjō kun 道場訓: Training hall rules.

dopamine: A hormone neurotransmitter tied to the brain's reward system.

dorsiflexion: Action of raising the foot upward toward the shin.

EEG (electroencephalography): Test to evaluate electrical activity of the brain.

Eight Elements: Daoist belief in the eight balancing forces of nature.

embryonic tissue organization: Orderly development of the embryo into organs and tissues.

embusen 演武線: Kata starting point and line of movement in Japanese karate.

endorphin: Hormones that reduce pain sensation and boost pleasure.

endurance running theory: Theory that bipedalism was adapted for distance running.

energy, kinetic: Energy acquired from motion.

enteric nervous system: Peripheral nervous system that controls gastrointestinal functions.

epinephrine: See adrenaline.

exhalation, forced: Conscious process of expelling air from the lungs.

fields of elixir: Body's energy centers according to Daoist beliefs; see dantian.

fight-or-flight: Body's reflexive response in the face of danger.

flexors, hip: Muscles that bring either the leg or the trunk forward.

fMRI, functional magnetic resonance imaging: An electromagnetic field-based test that measures brain activity.

folates: Vitamin B9, found in vegetables, fruits, and nuts.

force, electromagnetic: Force produced by moving, charged particles.

force, gravitational: Force that draws together two objects with mass.

force, ground reactionary: Force exerted by the ground when a body pushes against it.

force, strong nuclear: Force that binds protons and neutrons together.

force, weak nuclear: Force emitted over the processes of nuclear decay or fusion.

free radicals: Atoms with unpaired electrons that damage cells.

fudōshin: Unshakeable mind-set; resolve.

gait: Manner of walking.

gamma ray: Penetrating electromagnetic radiation from nuclear decay.

gen 現**:** The present time.

gendai budō 現代武道**:** Modern budō.

geocosmic: Pertaining to astrology.

geri, mae: Front kick.

geri, mikazuki: Sweeping kick with the sole of the foot.

geri, yoko kekomi: Side thrust kick.

gluon: A subatomic particle that transmits force between quarks.

gluteal complex: Backside muscle groups of the pelvis.

go 後**:** The past.

Gōjū-ryū 剛柔流**:** Japanese "hard-soft style" of karate-dō.

Golden Stove 金炉: The lower dantian, where jing processes into qi, emitting heat.

gōngfu 功夫: Chinese martial arts, also called wushu and quánfǎ.

go-no-sen 後の先: After, then before; martial arts timing strategy.

grit: Mental focus, determination, and tenacity.

Grit Scale: Score that measures grittiness based on self-assessment.

g Tum-mo: Tibetan Inner Fire Meditation.

gut-brain axis: Connection of gut health and brain health.

hara 腹: Lower abdomen, considered the wellspring for mastery in an endeavor in Japanese culture.

harai goshi: Sweeping hip throw.

henka 変化: Kata bunkai with minor technique variations.

herd immunity: Reduced likelihood of infection when a vast majority of the population has acquired disease immunity.

hidari ashi kamae: Left-foot-forward opening position.

Hinduism: Indian religion and culture based on Vedic texts.

hokkai-join: Gesture of reality in zazen.

holistic medicine: Health-care and wellness through healing of the mind, body, and soul.

homeostasis: Stable equilibrium of physiologic functions.

Honbu Dōjō 本部道場: Headquarters training hall.

hui 慧: Intelligence.

Hui Yin 會陰: Acupuncture point corresponding to the perineum.

IAP, intrabdominal pressure: Pressurizing the abdominal cavity by core muscle tension.

IBD, inflammatory bowel disease: Chronic inflammation of the digestive tract.

IBMT, integrative body-mind training: Mindfulness meditation that integrates relaxation, breathing, and centering postural exercises.

IBS, irritable bowel syndrome: Chronic disorder of the large intestines.

ibuki 息吹: Breathing, inner strength.

ichi gan, ni soku, san tan, shi riki 一眼二足三胆四力: First, eye; second, feet; third, courage and determination; fourth, strength and effort.

ichi-go ichi-e 一期一会: One lifetime, one encounter.

I Ching 易經: Book of Changes, according to Daoist Philosophy.

ichi-shén ni-waza 一神二技: Spirit first, technique second.

inner alchemy: The Daoist process of cultivating jing, qi, and shén inside the body.

inner crucible: The body's dantians as vessels for performing inner alchemy.

interoceptive sensations: Physical sensations of internal organ function relayed by the emotions.

ippon 一本: Rule of a single determining point in budō competition.

ippon seoi nage: One-arm shoulder throw in judō.

issun no ma'ai 一寸の間合: Distance of one inch.

Itsuku 付居: A fixated mind leading to a total lack of awareness.

Jainism: Ancient Indian religion.

jing 精: Human essence for growth, development, and procreation.

jīngluò 經絡: Meridians; energy flow channels.

jing shén 精神: Stamina and intellect.

jodan: Face-level.

joriki 定力: Stable strength; steadfastness.

judō 柔道: Budō art that uses throw, pin, and joint locks to control an opponent.

judō-ka: One who practices judō.

jujitsu 柔術: Combat-oriented unarmed martial art of grappling, joint locks, pins, holds, strikes, and kicking.

jukendō 銃剣道: Budō art of bayonet fighting.

jutsu 術: Technique; skill; method; conjuring.

kake-no-sen 掛の先: Timing strategy to strike preemptively.

kakushite 隠して: "Hidden" techniques as applied to kata bunkai.

kamae 構え: Opening stance; fighting stance.

kami shi to e 死髪と会: Life or death rests within a hair's breadth.

kanji 漢字: Japanese writing as presented in Chinese characters.

karate-dō 空手道: Unarmed budō art of punching, kicking, sweeping, blocking, and striking.

karate-ka: One who practices karate-dō.

kata 形: Prearranged movement sets for defense and offense.

katana 刀: Curved, single-bladed Japanese sword favored by the samurai.

kendō 剣道: Budō art of swordsmanship.

kendō-ka: One who practices kendō.

kensei 剣聖: "Sword saint," deference to a warrior's legendary swordsmanship.

kenshō 見性: Seeing one's true nature.

ki: See qi.

kiai 気合: Vocalizing a guttural sound from the abdomen.

kihon 基本: Review of fundamental techniques.

kihon kumite 基本組手: Sparring exercises that utilize fundamental techniques.

kime 決め: Whole-body muscle tension during technique execution.

kinematic (kinetic) chain: Sequential joint activation to achieve complex movements.

kinematics, proximal: Movement pattern in the initiating fixed joint, such as the shoulder.

kinesthesia: Awareness of position and movement of body parts.

kobudō 古武道: Ancient budō.

kogo: Competition that alternately assigns attacker or defender roles.

kōhai 後輩: Junior student.

kokoro 心: Heart.

kokyū 呼吸: Awareness of breathing.

koryu bugei 古流武芸: Classical martial arts styles or systems that train the samurai.

kumite 組手: Sparring.

kung fu 功夫: Of craft and effort; now refers to Chinese martial arts.

kyo 居: A preoccupied mind that is vulnerable to attacks.

kyudō 弓道: Budō art of archery.

laterization, brain function: Specialized neural functions on either side of the brain.

learning, adrenalized: Learning processes that overcome loss of mental function due to adrenaline surge.

learning, procedural: Learning of motor skills and habits, and some cognitive skills.

leg, virtual: Biomechanical extrapolation that extends the leg's leverage point.

leukocyte: White blood cells.

limbic system: Brain structures that collectively address emotions and memory.

lobe, frontal: Front brain areas that govern important cognitive functions.

lobe, occipital: The brain's visual processing center.

lobe, parietal: The brain's center for processing touch, taste, and temperature.

lobe, temporal: Forebrain areas responsible for consciousness and long-term memory.

LPHC, lumbopelvic hip complex: Musculoskeletal structures that stabilize the spine and pelvis.

Luohan's Eighteen Hands 羅漢十八手**:** Physical forms brought to the Shaolin Temple by Bodhidharma.

ma'ai 間合**:** Distance between the self and the opponent.

makiwara 卷藁**:** Padded striking post for traditional karate impact training.

meditation, focused attention: Meditative process of sustained attention to an object or event.

Meditation, Inner Fire, g Tum-o: Ancient Tibetan deep meditation practice that increases a person's "inner heat."

meditation, loving kindness: Meditative process to send good wishes, kindness, and comity with silent mantras.

meditation, mindfulness (open monitoring): Meditative process toward total awareness without judgment.

Meiji Restoration: Westernizing reforms and the end of feudalism in Japan from 1868.

membrane potential: Voltage difference across a cell plasma membrane.

memory, procedural: Long-term memory of how to do things; motor skills.

memory registers: Sensory, long-term, and short-term memory storage locations.

memory, working: Short-term memory that enables perception of the present and the linguistic process.

meridian: See jīngluò.

microcosm-macrocosm: Correspondence of pattern, nature, or structure between human beings and the universe according to Greek and Daoist philosophy.

mindfulness: Awareness; coming to terms with feelings, thoughts, and bodily sensations.

mizu no kokoro 水の心: Mind like water; embracing events without being perturbed.

MLB, Major League Baseball: Professional baseball league in the United States.

mobility, hip joint: Freedom of movement in the hips.

mokusō 黙想: Martial arts meditation.

moment arm: Distance between a joint and the line of force acting on the joint.

momentum, angular: Momentum that corresponds to an object's rotation speed.

momentum, linear: Momentum that corresponds to an object's linear traveling speed.

momentum transfer: The amount of momentum that one object gives to another on impact.

motor response, involuntary: Reflexes; automatic muscle responses to a stimulus.

Moving Zen: Also called moving meditation; focusing the mind on task at hand.

muay thai: Thai boxing, based on striking, kicking, and clinching techniques.

mucous membrane: Membrane that lines body cavities and covers the internal organs.

muscle filament: Myosin and actin protein strands that enable muscle contraction.

muscle memory: Procedural memory that enables complex motor tasks without conscious oversight.

muscle sense: Sense of movement.

mushin 無心: "No mind"; an unbiased, unemotional, and freed mind.

mushin no kokoro 無心の心: Mind freed of emotional encumbrances.

mushin no shin 無心の心: See mushin no kokoro.

na 拿: Chinese martial art of grappling and joint locks.

naginata 薙刀: Pole weapon attached with a 12- to 24-inch blade.

Naha-te: Martial arts indigenous to the Naha region of Okinawa.

Naihanchi-dachi: Straddle Stance that is one and a half shoulder widths.

Naihanchi kata: Karate form performed with Naihanchi-dachi.

Nan-Shaolin 南少林: Shaolin quan–influenced Chinese martial arts that originated south of the Huai River.

neiguan 内觀: "Inner vision"; meditative process of inward contemplation.

nèijiāquán 内家拳: Indigenous Wudang martial arts of the south.

nervous system, parasympathetic: Nervous systems associated with maintaining a restive state.

nervous system, sympathetic: Nervous systems associated with ramping up autonomic body functions in fight-or-flight responses.

network, executive and conflict: Network of the attentional system that aligns expectations with external stimuli to come up with an appropriate response.

network, orientation and selection: Network of the attentional system that directs mental focus to the target and filters out irrelevant information.

network, vigilance (alert and arousal): Network of the attentional system that sustains a state of awareness, readying the brain to respond to a stimulus.

neural mapping: Imaging techniques that discriminate brain functional activities.

neuromuscular excitability: Stimulus signal threshold for generating a muscle-fiber response.

neurotransmitter: Signal-relaying chemicals across nerve junctions or synapses.

Newton's laws of motion: Three laws of physics that define the relationship between objects in motion and the outcome of forces acting on them.

nian 念: Presence of mind; wish.

ningen keisei no michi 人間形成の道: Perfection of human character.

nirvana: Transcendent state of enlightenment, peace, and happiness.

niten ichi 二天一: "Two heavens as one"; "two swords as one."

nodachi 野太刀: Field sword that is longer than a katana.

nogare 逃れ: Smooth, silent abdominal breathing.

nondeformable: Sustained shape or volume when an external force is applied.

noradrenaline: Fight-or-flight hormone and neurotransmitter that functions in the brain and body.

norepinephrine: See noradrenaline.

northern lights: Atmospheric light phenomenon in the Arctic.

Okinawa-te: Indigenous martial arts of Okinawa.

ōyō 応用: Technique application as actually performed in a kata.

panic attack: Adrenaline-triggered sudden onset of intense fear and discomfort.

paraspinal complex: Muscle groups that support the back.

pattern recognition: Informational processing based on regularity of data patterns.

Pavlov's response: A conditioned, involuntary response to stimulus.

pelvic floor: Muscles that brace the bottom of the pelvis.

pendulum, inverted: Pendulum with its center of mass above the pivot point.

photon: Massless electromagnetic force carrier emitted as radiation.

Pinan 平安: A set of five basic kata in karate.

plane, vertical: Separates body parts as anterior or posterior.

plyometrics: Jump training to increase speed-strength.

pneuma: Breath, vital spirit.

point of no return: Juncture when intended movement cannot be canceled.

position of engagement: Foot placement to optimize reach for attack or defense.

practice, deliberate: Practice involving focused attention and set goals for improvements.

prana: Breath; life force.

pranayama: Controlled breathing practice to cultivate life force.

probiotics: Live bacteria and yeasts in preserved food that promote gut health.

problem of serial order in behavior: Evidence that behavior planning is based on modifications of prior experience, according to Karl Lashley.

proprioception: Awareness of the position and movement of the body.

proprioceptors: Sensory receptors in muscles and joints that inform on proprioception.

qi: Circulating energy or life force that sustains vitality.

qi, di (earth): Vital energy from the earth that sustains all earthly lives.

qigong 氣功: Craft of energy cultivation through focused breathing and limbering exercises.

qi, ren (human): Essence of a person's being.

qi, tian (cosmos): Vital energy that balances the planets and heavenly events.

qi, wei: Protective qi that shields a person from infection.

qi, ying: Nutritive qi distilled from food.

qi, yuan: Innate qi, congenital essence inherited from parents.

qi, zhen: True qi, from fresh air.

qi, zong: Gathering qi, a blends energy from fresh air and nutrients.

quánfǎ 拳法: See gōngfu.

quantum of execution: Attending to the finest detail in technique execution.

qwīng-gong 輕工: Mystical "light body" martial art to scale walls and skip roofs.

RBT, reality-based training: Training that replicates the actual event's stressors.

rei 靈: Spirit; soul.

resonance signature: Radiological imaging profile of healthy and diseased tissues.

reticular activating system (RAS): Interconnecting neural networks between the spinal cord, sensory pathways, thalamus, and cortex.

RPE, rating of perceived exertion: Measurement of physical exertion based on muscle fatigue, breathing exertion, and heart rate.

ruach: Breath, spirit (Hebrew).

rūḥ: Soul (Arabic).

Ryūkyū Kingdom: Kingdom in the Ryūkyū Islands (Okinawa), 1429–1879.

samu 作務: Physical work done with mindfulness; productive service.

sānbǎo: Three treasures: qi, jing, and shén.

sandan: Third degree.

san jiao 三教: The three teachings of Buddhism, Daoism, and Confucianism.

Sānzhàn (Sanchin) 三戦: Three battles.

second brain: Nervous system of the gut.

seiza 正座: Proper sitting.

sen 先: Before.

senpai: Senior student.

sensei: Teacher.

sensors, multivectoral force: Pressure sensors that measure multidirectional force exertion.

seppuku 腹切り: Ritualized suicide of Japan's ancient samurai warrior class.

shaku 尺: Japanese unit of length, about 12 inches.

shàngling 上靈: Aspirational spirit.

Shaolin quan 少林 拳: Chinese martial arts that originated at the Shaolin Temple.

shén 神: Spirit.

shēn 身: Body.

shiai 試合: Contest, tournament.

shī dì/sidai 師弟: Teacher's younger son.

shifu 師父: Master and father.

shī jiě/sije 師姐: Teacher's older daughter.

shī mèi/simui 師妹: Teacher's younger daughter.

shī xiōng/sihing 師兄: Teacher's older son.

shock waves: The propagation of waves of sharp changes in force or pressure.

shodan: First degree.

shomen: Front of the training hall.

Shōrei-ryū: "The Style of Inspiration," an Okinawan Naha-te style martial art.

shōrinji kempo 少林寺拳法: Japanese martial arts based on Shaolin quan.

Shōrin-ryū: Okinawan martial art based on Shaolin quan.

Shōtōkan 松濤館: Japanese karate-dō style founded by Gichin Funakoshi.

shuai 摔: Chinese martial art of wrestling.

Shuri: Royal capital of the former Ryuku Kingdom, Okinawa.

Shuri-te: Martial arts indigenous to the Shuri region of Okinawa.

sifu: See shifu.

sina'n: Compass.

somatosensory system: Part of the sensory nervous system that monitors changes on the surface or inside the body.

spatiotemporal: Perspective of space and time.

speed, summation of: Increased performance by the cascading contribution of participating segmental joints.

speed, technical: Performance measured by reaction, decision, and execution.

spine, cervical: Top part of the spine connecting cranium and thoracic vertebrae.

spine, lumbar: Vertebral column that makes up the lower back.

spine, thoracic: Vertebral column that runs from base of the neck to the abdomen.

spirituality: The quality of deep feelings and beliefs.

stance, close range: Stance as applied to techniques applied within arm's length.

stance, full range: Stance that requires deep forward bending of the front knee and full reach of the arm and leg.

stance, inside tension: Stances that require inward canting of the knees.

stance, outside tension: Stances that require outward canting of the knees.

stance pressure: Lower body joint compression to achieve ground reactionary force.

stoic: The attitude of focusing on solutions for the present instead of fantasy or idealism; preference of logic over emotion.

structured learning: Process of developing rules for problem-solving.

subame gaeshi: Turning swallow cut.

suki 付**:** "Attached" or "stuck," in the sense that one is preoccupied.

sumō 相撲**:** Competitive full-contact wrestling; Japan's national sport.

synthetic information processing: Cognitive process of instantly comparing real-time observations with prepackaged experience-based knowledge.

taekwondo 跆拳道**:** Korean martial art that favors kicking techniques.

Taiji Pole: Conceptual "energy highway" from the top of the head to the perineum.

taijiquan (tàiji quán, t'ai chi ch'üan) 太極拳**:** Chinese Wudang martial art that practices slowed movements toward self-defense training, health benefits, and meditation.

tai-no-sen 体の先**:** A preemptive counterattack (sen) strategy.

tai sabaki 体捌**:** Defense through body repositioning.

tanden: See dantian and hara.

tàolù 套路**:** Chinese martial arts set routines.

Tekki 鉄騎**:** A series of karate kata performed in Straddle Stance.

temporal binding: Integrative sensory processing to achieve a unified impression of space and time.

tension, internal: To sustain center core integrity by intra-abdominal pressure.

thalamus: Structure above the brain stem that relays motor and sensory signals to the cerebral cortex.

thermospheric: Layer of earth's atmosphere above the mesosphere.

thinking without thinking: Knack to accurately appraise a situation with minimal information.

thin slicing: Learned process of arriving at solutions quickly.

Three Teachings, Harmonious Aggregate of 三教: Coexistence of Buddhism, Daoism, and Confucianism in ancient Chinese culture.

ti 踢: Chinese martial art of kicking.

time, choice reaction: Time taken to decide and then complete a response.

timing: To execute an attack or defense at the right moment.

time, psychological: Mental process of estimating elapsed time.

time, simple reaction: Time needed to complete a physical response to a visual image.

time, total response: Time taken by athlete to react, then complete a movement.

tōde 唐手: Okinawan practice of Chinese martial arts.

todome waza 留め技: Finishing-blow technique.

TOE, theory of everything: A theory that unifies the workings of the four major forces of nature.

Tomari-te: Martial arts indigenous to the Tomari region of Okinawa.

torque: Twisting force.

triad gangs: Chinese organized crime syndicate.

tsuki no kokoro 月の心: Mind like the moon; a state of unperturbed awareness.

V1, visual cortex: Primary visual information processing area in the cerebral cortex.

vagus nerve: Nerve that runs the length of the abdomen and relays sensations of the skin and major organs to the brain.

vestibulum, inner ear: Anatomic structure that detects position and movements of the head.

vicious cycle: Emotionally driven high-adrenaline release that interferes with proper brain function.

vision, peripheral: Side vision that enables viewing of surrounding objects.

visual cues: Visual information that provides context and insight.

visual search: Perceptual task of scanning for a particular object or characteristic.

volume, expiratory reserve: Additional air exchange volume by forced exhalation.

volume, tidal: Volume of air exchange during the normal breathing cycle.

wa 和: Harmony; togetherness.

wàijiāquán 外家拳: Chinse martial arts with foreign influence, such as Shaolin quan.

wakizashi: "Side-inserted"; shorter sword that accompanies the katana.

warm-up decrement: Gradual loss of the effects of warm-up over a period of inactivity.

waza, ashi: Foot (throwing) techniques in judō.

waza, uchi: Striking techniques in Japanese martial arts.

WCS, whole candidate score: U.S. Military Academy at West Point criteria for candidate admission.

wei wu-wei 爲無爲: Observing the Daoist concept of wu-wei.

Wing Chun 詠春: Traditional southern Chinese martial art.

wit and feet: Proper coupling of mental preparedness and body positioning.

Wudang quan 武當拳: Chinese martial arts originating from the Wudang Mountains.

wúji 無極: Limitless; nothingness.

wushu 武術: Chinese martial techniques; see gōngfu.

wushu baoquan li 武術抱拳禮: Wrapped-fist courtesy salutation in Chinese martial arts.

wu-wei 無爲: Daoist attitude of effortless action in alignment with nature.

Wǔ Xíng Quán 五形拳: The Essence of the Five Fists, an ancient Chinese martial arts style.

Wu Zu Quan 五祖拳: Five Ancestors Fist, seventeenth-century southern Chinese style.

yang 陽: Male pole of the dualistic forces of nature in Chinese philosophy.

Yerkes-Dodson law: Dose-dependent polar-opposite effects of adrenaline on cognition.

yi 意: Intent; decision; wish.

yi 義: Justice; righteousness; certitude in morality and deed.

yi yi yin qi 以意引氣: Qi follows yi; physical action is guided by mental intent.

xīn 心: Heart; mind.

xingyiquan 形意拳: One of three major styles of Chinese Wudang quan.

yin 陰: Female pole of the dualistic forces of nature in Chinese philosophy.

yoga: Hindu discipline of breath control and specific body postures.

yūdansha: Practitioners who have achieved black-belt ranks.

Yù xiū qí shēn zhě, xiānzhèng qí xīn 欲修其身者，先正其心**:** To cultivate the body, first right the heart (mind, intent).

zanshin 残心**:** Relaxed alertness; total awareness after completion of action.

zazen 坐禪**:** Sitting in Zen Buddhism meditation to seek one's true nature.

Zen Buddhism: Meditation-based Buddhism to awaken mind and spirit.

Zen drumming: Seeking Zen spirituality through taiko drumming.

ziao zhou tian 小周天**:** Microcosmic orbit, a Daoist meditative practice.

zuki, gyaku-: Counterpunch; straight punch.

zuki, kizami-: Leading-hand straight punch.

zuki, oi-: Stepping punch; lunge punch.

FURTHER READING

On Martial Arts Lineage and Evolution (Chapter 1)

Mark Bishop, *Okinawan Karate: Teachers, Styles and Secret Techniques* (London: A&C Black, 1991).

Itzik Itzhak Cohen, *Karate Uchina-Di: Okinawan Karate: An Exploration of Its Origins and Evolution* (Israel: Itzik Cohen, 2017).

Donn F. Draeger, *Comprehensive Asian Fighting Arts* (New York: Kodansha International, 1981).

Peter A. Lorge, *Chinese Martial Arts from Antiquity to the Twenty-First Century* (Cambridge, UK: Cambridge University Press, 2012).

Patrick McCarthy, *The Bible of Karate: Bubishi* (North Clarendon, VT: Tuttle, 1995).

Jwing-Ming Yang, *The Essence of Shaolin White Crane: Martial Power and Qigong* (Jamaica Plain, MA: YMAA Publication Center, 1996).

Wong Kiew Kit, *The Art of Shaolin Kung Fu* (Rockport, MD: Element Books, 1996).

On Conditioning the Abdominals, Lower Back, and Hip Muscles (Chapter 10)

Jarryd Dent, "Three Major Benefits of Core Training for Sports," CoachUp Nation, November 7, 2018, https://tinyurl.com/ywhzcjuk.

Kellie C. Huxel Bliven and Barton E. Anderson, "Core Stability Training for Injury Prevention," *Sports Health* 5:6 (2013), 514–22, doi:10.1177/1941738113481200.

Elizabeth Kovar, "Beginner Ab and Core Exercises to Increase Stability and Mobility," American Council on Exercise, January 20, 2016, https://tinyurl.com/4hdwva6z.

Tanya Siejhi Gershon, "Hip Abductor Muscles Exercises," LiveStrong.com, June 10, 2019, https://tinyurl.com/ay5ndfj9.

W. Ben Kibler, Joel Press, and Aaron Sciascia, "The Role of Core Stability in Athletic Function," *Sports Medicine* 36:3 (2006), 189–98, doi:10.2165/00007256-200636030-00001.

Marcus Martinez, "All You Need Is a Kettlebell for This Full-Body Workout Program," Yahoo Life, February 4, 2021, https://tinyurl.com/6453cnwe.

Pete McCall, "10 Things to Know About Muscle Fibers," American Council on Exercise, May 7, 2015, https://tinyurl.com/tub57ytk.

On Style-Specific Technique Execution of Budō Arts (Chapter 13)

Iain Abernethy, *Bunkai-Jutsu: The Practical Application of Karate Kata* (Chichester, UK: Summersdale, 2002).

Rupert M. J. Atkinson, *Discovering Aikido: Principles for Practical Learning* (Ramsbury, UK: Crowood Press, 2005).

Wayne Belonoha, *The Wing Chun Compendium, Volumes 1 and 2* (Berkeley, CA: Blue Snake Books, 2005).

Mark Bishop, *Okinawan Karate: Teachers, Styles and Secret Techniques* (London: A&C Black, 1991).

Toshiro Daigo, *Kodokan Judo Throwing Techniques* (Tokyo: Kodansha, 2016).

Donn F. Draeger, *Comprehensive Asian Fighting Arts* (New York: Kodansha International, 1981).

Gichin Funakoshi, *Karate-dō kyōhan: The Master Text* (Tokyo: Kodansha, 2013).

Morio Higaonna, *Traditional Karate-dō: Okinawa Gōjū-ryū, Vol. 1: The Fundamental Techniques* (Tokyo: Sugawara Martial Arts Institute, 1985).

Patrick McCarthy, *The Bible of Karate: Bubishi* (North Clarendon, VT: Tuttle, 1995).

Motobu Choki, *Okinawan Kenpo Karate Jutsu: Kumite,* trans. Eric Shahan (Japan: Eric Michael Shahan, 2018).

Masatoshi Nakayama, *Dynamic Karate* (Palo Alto, CA: Kodansha, 1966).

Hidetaka Nishiyama and Richard Carl Brown, *Karate: The Art of "Empty Hand" Fighting* (Rutland, VT: Tuttle, 1960).

Wong Kiew Kit, *The Art of Shaolin Kung Fu* (Rockport, MD: Element Books, 1996).

Wong Kiew Kit, *The Complete Book of Tai Chi Chuan: A Comprehensive Guide to the Principles and Practice,* revised edition (Boston: Tuttle, 2016).

Gogen Yamaguchi, *Gōjū-ryū: Karate Dō Kyohan* (Hamilton, Canada: Masters Publication, 2006).

Jwing-Ming Yang, *The Root of Chinese Chi Kung: The Secrets of Chi Kung Training* (Jamaica Plain, MA: YMAA Publication Center, 1989).

Jwing-Ming Yang, *The Essence of Shaolin White Crane: Martial Power and Qigong* (Jamaica Plain, MA: YMAA Publication Center, 1996).

NOTES

Preface

1 Wikipedia, s.v. "Karate in the United States," 2018.

2 Japan Karate Association, "Supreme Master Funakoshi Gichin: The Father of Modern Karate," n.d., https://tinyurl.com/5a7ks9fh.

3 Uplifter, "Top 5 Most Popular Martial Arts in the World," April 5, 2019, https://tinyurl.com /mr2r398z; IBISWorld, "Martial Arts Studios Industry in the US," January 22, 2021, https://tinyurl .com/vhy6v93m.

4 Patrick Blennerhassett, "It's Time to Admit Most Traditional Martial Arts Are Fake and Don't Actually Teach You How to Fight," Sensei Says, *South China Morning Post,* October 24, 2019, https:// tinyurl.com/t75am9xc.

5 Ibid.; Charlie Campbell, "Meet the Chinese MMA Fighter Taking on Grandmasters of Kung Fu," Time.com, November 8, 2018, https://tinyurl.com/3wxzp5u9.

6 Will Henshaw, "The Future of Martial Arts: What Martial Arts Will Look Like in 100 Years," WOMA TV, December 3, 2015, https://tinyurl.com/y7erchmz.

7 Vicki Zakrzewski, "How Humility Will Make You the Greatest Person Ever," *Greater Good Magazine,* January 12, 2016, https://tinyurl.com/2ktkhr28.

8 Wikipedia, s.v. "Chinese Martial Arts," 2018.

9 Yuval Noah Harari, *Homo Deus: A Brief History of Tomorrow* (New York: Harper Perennial, 2018).

10 Dalai Lama, "On Buddha Nature, "The Buddha" blog, PBS, March 9, 2010, https://tinyurl.com /t97ha7f2; A. Lambert, "Zen Buddhism = Humanism?" *Houston Chronicle,* 2011.

11 Kris Wilder, "Sanchin Kata—Ancient Wisdom," YMAA, March 8, 2010, https://tinyurl.com /rjcj36eb.

12 Wikipedia, s.v. "Sanchin," 2018.

Chapter 1

1 Statista, "Number of Participants in Martial Arts in the United States from 2006–2017 (in Millions)," Statista.com, 2019.

2 H. G. R. Robinson, "United States Military and Martial Arts: A History of the Strategic Air Command (SAC) and Its Combative Measures Program," *International Armed Services Judo and Jujitsu Academy,* 2018, https://asjja.com.

3 Gichin Funakoshi, *Karate-dō: My Way of Life* (New York: Kodansha International, 1975).

4 Wikipedia, s.v. "Curtis LeMay," 2018.

5 P. Velez, "Our History," JKA San Antonio, 2010.

6 Wikipedia, s.v. "Judō in the United States," 2020.

7 Funakoshi, *Karate-dō: My Way of Life*.

8 Wikipedia, s.v. "Karate in the United States," 2018.

9 Ibid.

10 Ibid.

11 Wikipedia, s.v. "Okinawan kobudō," 2018; Hidetaka Nishiyama and Richard Carl Brown, *Karate: The Art of "Empty Hand" Fighting* (Rutland, VT: Tuttle, 1960); Masatoshi Nakayama, *Best Karate, Volume 2: Fundamentals* (New York: Kodansha International, 1978).

12 Wikipedia, s.v. "Martial Arts," 2018.

13 John Clements, "A Short Introduction to Historical European Martial Arts," *Meibukan* 1 (2006), 2–4.

14 Mark Bishop, *Okinawan Karate: Teachers, Styles and Secret Techniques* (London: A&C Black, 1991); Itzik Itzhak Cohen, *Karate Uchina-Di: Okinawan Karate: An Exploration of Its Origins and Evolution* (Israel: Itzik Cohen, 2017); Donn F. Draeger, *Comprehensive Asian Fighting Arts* (New York: Kodansha International, 1981); Peter A. Lorge, *Chinese Martial Arts from Antiquity to the Twenty-First Century* (Cambridge, UK: Cambridge University Press, 2012); Patrick McCarthy, *The Bible of Karate: Bubishi* (North Clarendon, VT: Tuttle, 1995); Jwing-Ming Yang, *The Essence of Shaolin White Crane: Martial Power and Qigong* (Jamaica Plain, MA: YMAA Publication Center, 1996); Wong Kiew Kit, *The Art of Shaolin Kung Fu* (Rockport, MD: Element Books, 1996).

15 Abdughani Mamadazimov, "Horse in North and Ship in South," SEnECA blog, 2018, https://tinyurl.com/6j2e6fj8.

16 Wikipedia, s.v. "Chinese Martial Arts," 2018.

17 Ibid.; Draeger, *Comprehensive Asian Fighting Arts;* Lorge, *Chinese Martial Arts;* Yang, *Essence of Shaolin White Crane;* S. Henning, "Ignorance, Legend, and Taijiquan," *Journal of the Chen-Style Taijiquan Research Association of Hawaii* 2:3 (1994), 1–7; Matthew Schafer, "Shaolin vs. Wudang vs. History," Schafer's Self-Defense Corner blog, December 24, 2008, https://tinyurl.com/mzbnf77j.

18 Wikipedia, s.v. "Luohan Quan," 2018.

19 Yang, *Essence of Shaolin White Crane*.

20 Wikipedia, s.v. "Southern Shaolin Monastery," 2018.

21 Ibid.

22 Wong, *The Art of Shaolin Kung Fu*.

23 Jwing-Ming Yang, *The Root of Chinese Chi Kung: The Secrets of Chi Kung Training* (Jamaica Plain, MA: YMAA Publication Center, 1989).

24 Wong Kiew Kit, *The Complete Book of Tai Chi Chuan: A Comprehensive Guide to the Principles and Practice,* revised edition (Boston: Tuttle, 2016).

25 Wong, *The Art of Shaolin Kung Fu*.

26 Ibid.

27 David Chow and Richard C. Spangler, *Kung Fu: History, Philosophy and Technique* (Burbank, CA: Action Pursuit Group, 1980).

28 Susan Debra Blum and Lionel M. Jensen, eds., *China Off Center: Mapping the Margins of the Middle Kingdom* (Honolulu: University of Hawaii Press, 2002).

29 Bishop, *Okinawan Karate.*

30 Andreas Quast, "The Weapons Ban Theories," Ryūkyū Bugei blog, April 27, 2017, https://ryukyu-bugei .com/?p=7281.

31 Wikipedia, s.v. "Okinawan kobudō," 2018.

32 Ibid.

33 McCarthy, *Bible of Karate.*

34 Wikipedia, s.v. "Kūsankū," 2017.

35 Wikipedia, s.v. "Sakugawa Kanga," 2018.

36 Wikipedia, s.v. "Kūsankū (kata)," 2018.

37 Ibid.

38 Cohen, *Karate Uchina-Di.*

39 Bishop, *Okinawan Karate.*

40 Yang, *Essence of Shaolin White Crane.*

41 *Encyclopaedia Britannica,* s.v. "Samurai—Japanese Warrior," 2019.

42 Erard, "Real Fighting."

43 Gichin Funakoshi, *Karate-dō kyōhan: The Master Text* (Tokyo: Kodansha, 2013).

44 Funakoshi, *Karate-dō: My Way of Life.*

45 Ibid.

46 Ibid.

47 Reidar P. Lystad, Kobi Gregory, and Juno Wilson, "The Epidemiology of Injuries in Mixed Martial Arts: A Systemic Review and Meta-Analysis," *Orthopaedic Journal of Sports Medicine* 2:1 (January 2014), doi:10.1177/2325967113518492.

48 Henshaw, "Future of Martial Arts."

Chapter 2

1 Chad Hansen and Edward N. Zalta, eds., "Daoism," *The Stanford Encyclopedia of Philosophy,* 2017, https://plato.stanford.edu.

2 Wikipedia, s.v. "Taoism," 2019; Ronnie Littlejohn, "Daoist Philosophy," Internet Encyclopedia of Philosophy, n.d., https://iep.utm.edu/daoism.

3 Elizabeth Emrich, "How to Live Forever: Daoism in the Ming and Qing Dynasties," Johnson Museum of Art, Cornell University, October 14, 2006, https://tinyurl.com/6ztx2apd.

4 Wikipedia, s.v. "Taoism," 2019.

5 Wikipedia, s.v. "Three Teachings," 2020.

6 Littlejohn, "Daoist Philosophy."

7 Hansen and Zalta, "Daoism."

8 Wikipedia, s.v. "Taoism," 2019; Littlejohn, "Daoist Philosophy."

9 Alexandra David-Néel, *Magic and Mystery in Tibet* (New York: Claude Kendall, 1932).

10 M. A. Wenger and B. K. Bagchi, "Studies of Autonomic Functions in Practitioners of Yoga in India," *Behavioral Science* (October 1961), 312–23, doi:10.1002/bs.3830060407.

11 M. Laurino, D. Menicucci, F. Mastorci, P. Allegrini, A. Piarulli, E. P. Scilingo, R. Bedini, et al., "Mind-Body Relationships in Elite Apnea Divers During Breath Holding: A Study of Autonomic Responses to Acute Hypoxemia," *Frontiers in Neuroengineering* 5:4 (2012), 1–10, doi:10.3389/fneng.2012.00004.

12 Paul C. Castle, Neil Maxwell, Alan Allchorn, Alexis Mauger, and Danny K. White, "Deception of Ambient and Body Core Temperature Improves Self-Paced Cycling in Hot, Humid Conditions," *European Journal of Applied Physiology* 12:1 (2012), 377–85, doi:10.1007/s00421-011-1988-y.

13 S. Grant, "10 Amazing Examples of Mind Over Matter," Listverse, May 21, 2013, https://tinyurl.com/y9enj5sb.

14 Norman Doidge, *The Brain That Changes Itself* (New York: Penguin, 2007).

15 H. Wahbeh, A. Sagher, W. Back, P. Pundhir, and F. Travis, "A Systematic Review of Transcendent States Across Medication and Contemplative Traditions," *Explore (NY)* 14:1 (2014), 19–35.

16 Marcus E. Raichle, Ann Mary MacLeod, Abraham Z. Snyder, William J. Powers, Debra A. Gusnard, and Gordon L. Shulman, "A Default Mode of Brain Function," *Proceedings of the National Academy of Sciences USA* 98:2 (January 16, 2001), 676–82, doi:10.1073/pnas.98.2.676.

17 Alice G. Walton, "7 Ways Meditation Can Actually Change the Brain," *Forbes*, February 9, 2015, https://tinyurl.com/5kzzz35n.

18 Funakoshi, *Karate-dō kyōhan.*

19 Jim Taylor, "What Mental Training for Sports Is Really All About," *Psychology Today*, November 12, 2018, https://tinyurl.com/pfpw4hxr.

20 Funakoshi, *Karate-dō kyōhan;* Robin L. Rielly, *Karate Training: The Samurai Legacy and Modern Practice* (Tokyo: Tuttle, 1985); Vincent Paul Cooper, "Mizu no Kokoro—A Mind Like Water," Trans4mind, n.d., https://tinyurl.com/53men64u.

21 Wendell E. Wilson, "Mushin and Zanshin," Essays on the Martial Arts, 2010, https://tinyurl.com/yt52teav.

22 Michael Spivey, *The Continuity of Mind* (New York: Oxford University Press, 2007).

23 Zach Schonbrun, *The Performance Cortex: How Neuroscience Is Redefining Athletic Genius* (New York: Penguin, 2018).

24 Rielly, *Karate Training.*

25 Cooper, "Mizu no Kokoro"; C. Carlson and M. Charlie, "Biological Baseball," San Francisco Exploratorium, 1999, https://tinyurl.com/4bb6s2px.

26 Levi Gadye, "The Tools That Let Neuroscientists Study (and Even Repair) Brain Circuits," BrainFacts.org, February 13, 2018, https://tinyurl.com/yzsfj57h.

Chapter 3

1 Wikipedia, s.v. "Japanese Sword," 2019.

2 National Academy of Sports Medicine, *Essentials of Sports Performance Training*, 2nd ed. (Burlington, MA: Jones and Bartlett, 2018).

3 Ibid.

4   ISSA, "Exercises to Improve Quickness and Why Clients Need Them," ISSA blog, n.d., https://tinyurl.com/jn7469n4.

5   Yi-Yuan Tang and Michael I. Posner, "Attention Training and Attention State Training," *Trends in Cognitive Sciences* 13:5 (May 2009), 222–27, doi:10.1016/j.tics.2009.01.009.

6   Schonbrun, *Performance Cortex.*

7   Malcolm Gladwell, *Outliers: The Story of Success* (Boston: Little, Brown and Co., 2008).

8   David Epstein, *The Sports Gene: Inside the Science of Extraordinary Athletic Performance* (New York: Penguin, 2013).

9   Carlson and Charlie, "Biological Baseball."

10   Schonbrun, *Performance Cortex.*

11   Shuji Mori, Yoshio Ohtani, and Kuniyasu Imanaka, "Reaction Times and Anticipatory Skills of Karate Athletes," *Human Movement Science* 21:2 (July 2002), 213–30, doi:10.1016/s0167-9457(02)00103-3.

12   Schonbrun, *Performance Cortex.*

13   K. Anders Ericsson, "Deliberate Practice and Acquisition of Expert Performance: A General Overview," *Academic Emergency Medicine* 15:11 (2008), 988–94, doi:10.1111/j.1553-2712.2008.00227.x; Nathan Colin Wong, "The 10,000 Hour Rule," *Journal of the Canadian Urological Association* 9:9–10 (Sept.–Oct. 2015), 299, doi:10.5489/cuaj.3267.

14   Mori et al., "Reaction Times."

15   Ibid.

16   Ibid.

17   Schonbrun, *Performance Cortex.*

18   A. Mark Williams and David Elliott, "Anxiety, Expertise and Visual Search Strategy in Karate," *Journal of Sport and Exercise Psychology* 21 (1999), 362–75, doi:10.1123/jsep.21.4.362.

19   F. B. Theiss, "Karate: A Scientific View," *Samurai,* summer 1971, 42–47.

20   Jonathan St. B. T. Evans and Keith E. Stanovich, "Dual-Process Theories of Higher Cognition: Advancing the Debate," *Perspectives on Psychological Science* 8:3 (2014), 223–41, doi:10.1177/1745691612460685.

21   Daniel Kahneman, *Thinking, Fast and Slow* (New York: Farrar, Straus and Giroux, 2011).

22   Seymour Epstein, "Integration of the Cognitive and the Psychodynamic Unconscious," *American Psychologist* 49:8 (1994), 709–24, doi:10.1037/0003-066X.49.8.709.

23   Christof Koch, "Intuition May Reveal Where Expertise Resides in the Brain," *Scientific American,* May 1, 2015, https://tinyurl.com/2kc5ncsj.

24   Evans and Stanovich, "Dual-Process Theories."

25   Kahneman, *Thinking.*

26   Aidan Moran, "Thinking in Action: Some Insights from Cognitive Sport Psychology," *Thinking Skills and Creativity* 7:2 (2012), doi:10.1016/j.tsc.2012.03.005.

27   Kielan Yarrow, Peter Brown, and John W Krakauer, "Inside the Brain of an Elite Athlete: The Neural Processes That Support High Achievement in Sports," *Nature Review Neuroscience* 10 (2009), 585–96, doi:10.1038/nrn2672; Philip D. Harvey, "Domains of Cognition and Their Assessment," *Dialogues in Clinical Neuroscience* 21:3 (2019), 227–37, doi:10.31887/DCNS.2019.21.3/pharvey.

28 Mario Jovanovic, Goran Sporis, Darija Omrcen, and Fredi Fiorentini, "Effects of Speed, Agility, Quickness Training Method on Power Performance in Elite Soccer Players," *Journal of Applied Sport Science Research* 25:5 (May 2011), 1285–92, doi:10.1519/JSC.0b013e3181d67c65; Kelly Baggett, "Quickness and Absolute Speed vs Sports Speed and Explosiveness," Higher-Faster-Sports.com, n.d., https://tinyurl.com/ahr8yuwh; K. Devine, "The Deception: Speed vs Quickness (Hint: They Are Not the Same)," blog, 2012.

29 Kate Allgood, "Improving Performance with Better Attention," NeuroTracker, March 29, 2018, https://tinyurl.com/trr9eark.

30 Yarrow et al., "Inside the Brain of an Elite Athlete."

31 Kate Moran, "How Chunking Helps Content Processing," Nielsen Norman Group, March 10, 2016, https://tinyurl.com/b65ukyny.

32 Wikipedia, s.v. "Chunking (psychology)," 2019.

33 Yarrow et al., "Inside the Brain of an Elite Athlete."

34 Bruce Abernathy, Jörg Schorer, Robin C. Jackson, and Norbert Hagemann, "Perceptual Training Methods Compared: The Relative Efficacy of Different Approaches to Enhancing Sport-Specific Anticipation," *Journal of Experimental Psychology Applied* 18:2 (May 2012), 143–53, doi:10.1037/a0028452.

35 Ibid.; Moşoi Adrian Alexandru and Balint Lorand, "Motor Behavior and Anticipation: A Pilot Study of Junior Tennis Players," *Procedua: Social and Behavioral Sciences* 187 (2015), 448–53, doi:10.1016/j.sbspro.2015.03.084.

36 P. M. Fitts, "The Information Capacity of the Human Motor System in Controlling the Amplitude of Movement," *Journal of Experimental Psychology* 47:6 (1954), 381–91, doi:10.1037/h0055392.

37 Tor D. Wager and Edward E. Smith, "Neuroimaging Studies of Working Memory: A Meta-Analysis," *Cognitive, Affective and Behavioral Neuroscience* 3:4 (December 2003), 256–74, doi:10.3758/cabn.3.4.255.

38 Michael Kent, "Fitts and Posner's Stages of Learning," *The Oxford Dictionary of Sports Science & Medicine,* 3rd ed. (Oxford, UK: Oxford University Press, 2007).

39 Richard M. Shiffrin and Walter Schneider, "Controlled and Automatic Human Information Processing: II. Perceptual Learning, Automatic Attending, and a General Theory," *Psychological Review* 84 (1977), 127–90, doi:10.1037/0033-295X.84.2.127.

40 Janelle Weaver, "Motor Learning Unfolds Over Different Timescales in Distinct Neural Systems," *PLoS Biology* 13:12 (December 8, 2015), e1002313, doi:10.1371/journal.pbio.1002313; Hwal Lee and Lisa M. Miller, "Brain Training: Three Psychological Skills to Cope with Performance Stress and Anxiety," *Training & Conditioning,* May 12, 2017, https://tinyurl.com/snykz5rp.

41 Jürgen Beckmann, Peter Gröpel, and Felix Ehrlenspiel, "Preventing Motor Skill Failure through Hemisphere-Specific Priming: Cases from Choking under Pressure," *Journal of Experimental Psychology: General* 142:3 (2013), 679–91, doi:10.1037/a0029852.

42 Fitts, "Information Capacity."

43 Robin C. Jackson and Damian Farrow, "Implicit Perceptual Training: How, When, and Why?" *Human Movement Science* 24:3 (July 2005), 308–25, doi:10.1016/j.humov.2005.06.003.

44 Yarrow et al., "Inside the Brain of an Elite Athlete."

45 Jovanovic et al., "Effects of Speed"; Devine, "The Deception"; Baggett, "Quickness and Absolute Speed."

Chapter 4

1 Wikipedia, s.v. "Starting Blocks," 2018.

2 Anita D. Barber, Priti Srinivasan, Suresh E. Joel, Brian S. Caffo, James J. Pekar, and Stewart H. Mostofsky, "Motor 'Dexterity'?: Evidence That Left Hemisphere Lateralization of Motor Circuit Connectivity Is Associated with Better Motor Performance in Children," *Cerebral Cortex* 22:1 (2012), 51–59, doi:10.1093/cercor/bhr062; Adam Eikenberry, Jim McAuliffe, Timothy N. Welsh, Carlos Zerpa, Moira McPherson, and Ian J. Newhouse, "Starting with the 'Right' Foot Minimizes Sprint Start Time," *Acta Psychologica* 127:2 (March 2008), 495–500, doi:10.1016 /j.actpsy.2007.09.002; Dominique Stasulli, "Reaction Time in Track and Field Athletes," Freelap, January 13, 2015, https://tinyurl.com/phcy6bx3; C. Collet, "Strategic Aspects of Reaction Time in World-Class Sprinters," *Perceptual and Motor Skills* 88:1 (February 1999), 65–75, doi:10.2466 /pms.1999.88.1.65; Aditi S. Majumdar and Robert A. Robergs, "The Science of Speed: Determinants of Performance in the 100 m Sprint," *International Journal of Sports Science & Coaching* 6:3 (2011), 479–93, doi:10.1260/1747-9541.6.3.479.

3 Eikenberry et al., "Starting with the 'Right' Foot."

4 Ibid.

5 Mori et al., "Reaction Times"; Abernathy, "Perceptual Training Methods Compared"; David Epstein, "Are Athletes Really Getting Faster, Better, Stronger?" TED Talks, video, March 2014, https://tinyurl.com/2a652r49; Carlson and Charlie, "Biological Baseball."

6 Eikenberry et al., "Starting with the 'Right' Foot."

7 C. Collet, "Strategic Aspects of Reaction Time."

8 Stasulli, "Reaction Time in Track and Field Athletes."

9 Ibid.

10 Andrew E. Papale and Bryan M. Hooks, "Circuit Changes in Motor Cortex During Motor Skill Learning," *Neuroscience* 368 (2018), 283–97, doi:10.1016/j.neuroscience.2017.09.010.

11 Shailesh S. Kantak, James W. Stinear, Ethan R. Buch, and Leonardo G. Cohen, "Rewiring the Brain: Potential Role of the Premotor Cortex in Motor Control, Learning, and Recovery of Function Following Brain Injury," *Neurorehabilitation and Neural Repair* 26:3 (March–April 2016), 282–92, doi:10.1177/1545968311420845.

12 Dale Purves, George J. Augustine, David Fitzpatrick, Lawrence C. Katz, Anthony-Samuel LaMantia, James O. McNamara, and S. Mark Williams, eds., *Neuroscience,* 2nd. ed. (Sunderland, MA: Sinauer Associates, 2001).

13 Wikipedia, s.v. "Motor Control," 2020.

14 David P. Broadbent, Joe Causer, A. Mark Williams, and Paul R. Ford, "Perceptual-Cognitive Skill Training and Its Transfer to Expert Performance in the Field: Future Research Directions," *European Journal of Sport Science* 15:4 (2015), 322–31, doi:10.1080/17461391.2014.957727.

15 Midori Kodama, Takashi Ono, Fumio Yamashita, Hiroki Ebata, Meigen Liu, Shoko Kasuga, and Junichi Ushiba, "Structural Gray Matter Changes in the Hippocampus and the Primary Motor Cortex on An-Hour-to-One-Day Scale Can Predict Arm-Reaching Performance Improvement," *Frontiers in Human Neuroscience* 12:209 (2018), doi:10.3389/fnhum.2018.00209.

16 Papale and Hooks, "Circuit Changes in Motor Cortex."

17 Yarrow et al., "Inside the Brain of an Elite Athlete."

18 Weaver, "Motor Learning Unfolds Over Different Timescales."

19 A. Adams and M. Uyehara, "20 Top Fighters in Japan," *Black Belt 1971 Yearbook* (Los Angeles: Black Belt Inc., 1971).

20 Ibid.

21 M. Sekine, "Takeshi Oishi, All Japan Champion," *Samurai* 1 (1971), 37.

22 Ibid.

23 Fiorenzo Moscatelli, Giovanni Messina, Anna Valenzano, Vincenzo Monda, Andrea Viggiano, Antonietta Messina, Annamaria Petito, et al., "Functional Assessment of Corticospinal System Excitability in Karate Athletes," *PLoS ONE* 11:5 (May 24, 2016), e015998, doi:10.1371/journal.pone.0155998.

24 Yarrow et al., "Inside the Brain of an Elite Athlete."

25 Annie Bosler and Don Greene, "How to Practice Effectively … for Just About Anything," AnnieBosler.com, video, https://youtu.be/f2O6mQkFiiw.

26 R. Edward Roberts, Elaine J. Anderson, and Masud Husain, "White Matter Microstructure and Cognitive Function," *Neuroscientist* 19:1 (2013), 8–15, doi:10.1177/1073858411421218.

27 Bosler and Greene, "How to Practice Effectively."

28 Moscatelli et al., "Functional Assessment of Corticospinal System Excitability."

29 Anna Burdukiewicz, Jadwiga Pietraszewska, Justyna Andrzejewska, Krystyna Chromik, and Aleksandra Stachoń, "Asymmetry of Musculature and Hand Grip Strength in Bodybuilders and Martial Artists," *International Journal of Environmental Research and Public Health* 17:13 (2020), 4695, doi:10.3390/ijerph17134695.

30 Giorgio Vallortigara and Lesley J. Rogers, "Survival with an Asymmetrical Brain: Advantages and Disadvantages of Cerebral Lateralization," *Behavioral and Brain Sciences* 28:4 (2005), 575–89, doi:10.1017/S0140525X05000105.

31 John I. Todor, "Sequential Motor Ability of Left-Handed Inverted and Non-Inverted Writers," *Acta Psychologica* 44 (1980), 165–73, doi:10.1016/0001-6918(80)90065-7; Gavin Buckingham, Julie C. Main, and David P. Carey, "Asymmetries in Motor Attention During a Cued Bimanual Reaching Task: Left and Right Handers Compared," *Cortex* 47:4 (2011), 432–40, doi:10.1016/j.cortex.2009.11.003.

32 Barber et al., "Motor 'Dexterity'?"

33 Sean J. Maloney, "The Relationship Between Asymmetry and Athletic Performance: A Critical Review," *Journal of Strength and Conditioning Research* 33:9 (2019), 2579–93, doi:10.1519/JSC.0000000000002608.

34 Eikenberry et al., "Starting with the 'Right' Foot"; Majumdar and Roberts, "The Science of Speed."

35 Eikenberry et al., "Starting with the 'Right' Foot."

36 Maloney, "Relationship Between Asymmetry and Athletic Performance."

37 Maxim Mikheev, Christine Mohr, Sergei Afanasiev, Theodor Landis, and Gregor Thut, "Motor Control and Cerebral Hemispheric Specialization in Highly Qualified Judo Wrestlers," *Neuropsychologia* 40:8 (2002), 1209–19, doi:10.1016/S0028-3932(01)00227-5.

38 Burdukiewicz et al., "Asymmetry of Musculature."

39 Mikheev et al., "Motor Control."

40 Jia Han, Gordon Waddington, Roger Adams, Judith Anson, and Yu Liu, "Assessing Proprioception: A Critical Review of Methods," *Journal of Sport and Health Science* 5:1 (2016), 80–90, doi:10.1016/j.jshs.2014.10.004.

41 Zeynep Çelik, "Kinaesthesia," in *Sensorium: Embodied Experience, Technology, and Contemporary Art,* edited by Caroline A. Jones (Cambridge, MA: MIT Press, 2006).

42 Han et al., "Assessing Proprioception."

43 Mario Prsa, Karin Morandell, Angie Geraldine Cuenu, and Daniel Huber, "Feature-Selective Encoding of Substrate Vibrations in the Forelimb Somatosensory Cortex," *Nature Review Neuroscience* 567:7748 (2019), 384, doi:10.1038/s41586-019-1015-8.

44 Joshua E. Aman, Naveen Elangovan, I-Ling Yeh, and Jürgen Konczak, "The Effectiveness of Proprioceptive Training for Improving Motor Function: A Systematic Review," *Frontiers in Human Neuroscience* 8 (2014), 1075, doi:10.3389/fnhum.2014.01075.

45 Wikipedia, s.v. "Basal Ganglia," 2019.

Chapter 5

1 Shahram Heshmat, "Can You Be Addicted to Adrenaline?" *Psychology Today,* August 8, 2015, https://tinyurl.com/cudfcju7.

2 Vicky Lebeau, "The Strange Case of Dr. Jekyll and Mr. Hyde," *Britannica Online Encyclopedia,* 2019, https://tinyurl.com/3mxcm57d.

3 Lisa Sanders, "Mysterious Psychosis," *New York Times Magazine,* March 10, 2009, https://tinyurl.com/55wdfw4t.

4 Mitsuo Ishida, *Hormone Hunters: The Discovery of Adrenaline* (Kyoto, Japan: Kyoto University Press, 2018).

5 Elizabeth A. Shirtcliff, Michael J. Vitacco, Alexander R. Graf, Andrew J. Gostisha, Jenna L. Merz, and Carolyn Zahn-Waxler, "Neurobiology of Empathy and Callousness: Implications for the Development of Antisocial Behavior," *Behavioral Sciences and the Law* 27:2 (March–April 2009), 137–71, doi:10.1002/bsl.862.

6 Mark A. Staal, "Stress, Cognition, and Human Performance: A Literature Review and Conceptual Framework," NASA Ames Research Center, August 2004, https://tinyurl.com/nufrpvkx.

7 Amy F. T. Arnsten, "Stress Signalling Pathways That Impair Prefrontal Cortex Structure and Function," *Nature Reviews: Neuroscience* 10:6 (2009), 410–22, doi:10.1038/nrn2648; D. E. Broadbent, *Decision and Stress* (London: Academic Press, 1971); Karl Schlimm, "Channeling Adrenaline: Upset Prevention and Recovery Training," *Skies,* 2019, https://tinyurl.com/2c5v7cdd.

8 Wikipedia, s.v. "Panic Attack," 2019.

9 Arnsten, "Stress Signalling Pathways."

10 Ibid.

11 Epstein, "Are Athletes Really Getting Faster."

12 Arnsten, "Stress Signalling Pathways"; Broadbent, *Decision and Stress.*

13 Laura Niedziocha, "Does Exercise Cause an Adrenaline Rush?" Healthfully.com, November 28, 2018, https://tinyurl.com/tmz53vmv.

14 James R. Hansen, *First Man: The Life of Neil A. Armstrong* (New York: Simon and Schuster, 2018).

15 NASA, *Apollo 11 Mission Report,* November 1969 (Houston: NASA, 1969), https://tinyurl.com/wrtb3uja.

16 Hans Selye, "Stress Without Distress," in *Psychopathology of Human Adaptation,* edited by George Serban, 137–46 (Philadelphia: J. B. Lippincott, 1974).

17 Wikipedia, s.v. "Aviation Safety," 2020.

18 Robert L. Helmreich, "On Error Management: Lessons from Aviation," *BMJ* 320 (2000), 781–85, doi:10.1136/bmj.320.7237.781.

19 Wikipedia, s.v. "Aviation Safety," 2020.

20 Helmreich, "On Error Management."

21 Kathleen Vonk, "Police Performance Under Stress," *Law and Order* 56:10 (October 2008), 86–90, https://tinyurl.com/t77wrm26.

22 Howard Kunreuther and Robert Meyer, *The Ostrich Paradox: Why We Underprepare for Disasters* (Upper Saddle River, NJ: Wharton School Publishing, 2017).

23 Schlimm, "Channeling Adrenaline."

24 Vonk, "Police Performance."

25 Eric Ravenscraft, "Use the Combat Breathing Technique to Help Control Nervous Shaking," *Lifehacker*, September 16, 2015, https://tinyurl.com/y5rmhj8c.

26 Steven E. Petersen and Michael I. Posner, "The Attention System of the Human Brain," *Annual Review of Neuroscience* 13 (1990), 25–42, doi:10.1146/annurev.ne.13.030190.000325; Jin Fan, Bruce D. McCandliss, John Fossella, Jonathan I. Flombaum, and Michael I. Posner, "The Activation of Attentional Networks," *Neuroimage* 26:2 (2005), 471–79, doi:10.1016/j.neuroimage.2005.02.004.

27 Steven E. Petersen and Michael I. Posner, "The Attention System of the Human Brain: 20 Years After," *Annual Review of Neuroscience* 35 (2012), 73–89, doi:10.1146/annurev-neuro-062111-150525; Alfredo Spagna, Melissa-Ann Mackie, and Jin Fan, "Supramodal Executive Control of Attention," *Frontiers in Psychology*, February 24, 2015, doi:10.3389/fpsyg.2015.00065.

28 Petersen and Posner, "Attention System of the Human Brain: 20 Years After."

29 Michael I. Posner and Stanislas Dehaene, "Attentional Networks," *Trends in Neurosciences* 17:2 (1994) 75–79, doi:10.1016/0166-2236(94)90078-7; Petersen and Posner, "Attention System of the Human Brain: 20 Years After."

30 Ashleigh Johnstone and Paloma Marí-Beffa, "The Effects of Martial Arts Training on Attentional Networks in Typical Adults," *Frontiers in Psychology* 9:80 (2018), 1–9, doi:10.3389/fpsyg.2018.00080; Mori et al., "Reaction Times."

31 K. Daniels and E. Thornton, "Length of Training, Hostility, and the Martial Arts: A Comparison with Other Sporting Groups," *British Journal of Sports Medicine* 26:3 (1992), 118–20, doi:10.1136/bjsm.26.3.118.

32 Marianha Alesi, Antonino Bianco, Johnny Padulo, Francesco Paolo Vella, Marco Petrucci, Antonio Paoli, Antonio Palma, and Annamaria Pepi, "Motor and Cognitive Development: The Role of Karate," *Muscles, Ligaments and Tendons Journal* 4:2 (2014), 114–20, https://tinyurl.com/wszy2ua6.

33 Tang and Posner, "Attention Training"; Johnstone and Marí-Beffa, "Effects of Martial Arts Training"; Petersen and Posner, "Attention System of the Human Brain: 20 Years After."

34 Tang and Posner, "Attention Training"; Petersen and Posner, "Attention System of the Human Brain: 20 Years After."

35 Petersen and Posner, "Attention System of the Human Brain: 20 Years After."

36 Tang and Posner, "Attention Training."

37 B. A. Vogt and O. Devinsky, "Topography and Relationship of Mind and Brain," *Progress in Brain Research* 122 (2000), 11–22, doi:10.1016/s0079-6123(08)62127-5.

Chapter 6

1 B. Alvelve, "Biopsychology," *What Is Time?* Alvelve.com, 2019.

2 Catalin V. Buhusi and Warren H. Meck, "What Makes Us Tick? Functional and Neural Mechanisms of Interval Timing," *Nature Reviews Neuroscience* 6:10 (2005), doi:10.1038/nrn1764.

3 James W. Bisley and Michael E. Goldberg, "Neuronal Activity in the Lateral Intraparietal Area and Spatial Attention," *Science* 299:5603 (2003), 81–86, doi:10.1126/science.1077395.

4 B. Alvelve, "Time Perception," *What Is Time?* Alvelve.com, 2019.

5 Mihaly Csikszentmihalyi, *Flow: The Psychology of Optimal Experience* (New York: Harper Collins, 2008).

6 Alvelve, "Biopsychology."

7 Christopher Chabris and Daniel Simons, *The Invisible Gorilla: How Our Intuitions Deceive Us* (New York: Crown, 2011).

8 Daniel Simons, "Selective Attention Test," YouTube, video, March 10, 2010, https://youtu.be/vJG698U2Mvo.

9 Kahneman, *Thinking, Fast and Slow.*

10 Jinichi Tokeshi, *Kendo: Elements, Rules, and Philosophy* (Honolulu: University of Hawaii Press, 2003).

11 Tomas Sluyter, "Waza Explained," Renshinjuku Kendo, October 4, 2013, https://tinyurl.com/2r2we9pr.

12 Hidetaka Nishiyama, *The Traditional Karate Coach's Manual* (Los Angeles: International Traditional Karate Federation, 1989).

Chapter 7

1 K. Anders Ericsson, Ralf T. Krampe, and Clemens Tesch-Römer, "The Role of Deliberate Practice in the Acquisition of Expert Performance," *Psychological Review* 100:3 (1993), 363–406, doi:10.1037/0033-295X.100.3.363.

2 Gladwell, *Outliers;* Drake Baer, "Malcolm Gladwell Explains What Everyone Gets Wrong about His Famous '10,000 Hour Rule,'" *Business Insider India,* June 2, 2014, https://tinyurl.com/eyc3xmy3.

3 Malcolm Gladwell, *Blink: The Power of Thinking Without Thinking* (New York: Back Bay Books, 2005).

4 Leslie A. Zebrowitz, "First Impressions from Faces," *Current Directions in Psychological Science* 26:3 (2018), 237–42, doi:10.1177/0963721416683996; Nalini Ambady, "The Perils of Pondering: Intuition and Thin Slice Judgments," *Psychological Inquiry* 21 (2010), 271–78, doi:10.1080/1047840X.2010.524882.

5 Art Markman, "Creativity Is Memory," *Psychology Today,* October 6, 2015, https://tinyurl.com/axka4a96.

6 Ambady, "Perils of Pondering."

7 Schonbrun, *Performance Cortex.*

8 Bruce R. Etnyre and Hally B. W. Poindexter, "Characteristics of Motor Performance, Learning, Warm-Up Decrement, and Reminiscence During a Balance Task," *Perceptual and Motor Skills* 89:3 (1995), 1027–30, doi:10.2466/pms.1995.80.3.1027.

9   Chris Lonsdale and Jimmy T M Tam, "On the Temporal and Behavioural Consistency of Pre-performance Routines: An Intra-individual Analysis of Elite Basketball Players' Free Throw Shooting Accuracy," *Journal of Sports Sciences* 26:3 (2008), 259–66, doi:10.1080/02640410701473962.

10   Nishiyama, *Traditional Karate Coach's Manual.*

11   Kyjean Tomboc, "What Happens When You Neglect Your Lower Body Composition," InBodyUSA .com, November 10, 2016, https://tinyurl.com/zr2szped.

12   Gladwell, *Outliers;* Ericsson et al., "Role of Deliberate Practice."

13   Michael Miller, "The Great Practice Myth: Debunking the 10,000 Hour Rule and What It Actually Takes to Get to the Mountaintop," 6Seconds.org, n.d. https://tinyurl.com/kh3usrbs.

14   Carol Dweck, *Mindset: The New Psychology of Success* (New York: Ballantine, 2016).

15   Epstein, *Sports Gene.*

16   Ericsson, "Deliberate Practice and Acquisition of Expert Performance"; Wong, "10,000 Hour Rule."

17   K. Anders Ericsson and Robert Pool, *Peak: Secrets from the New Science of Expertise* (New York: Eamon Doplan/Houghton Mifflin Harcourt, 2016).

18   Mori et al., "Reaction Times"; Bruce Abernethy, "Visual Search Strategies and Decision-Making Sport," *International Journal of Sport Psychology* 22 (1991), 189–210; Nicolas Milazzo, Damian Farrow, and Jean F Fournier, "Effect of Implicit Perceptual-Motor Training on Decision-Making Skills and Underpinning Gaze Behavior in Combat Athletes," *Perceptual and Motor Skills* 123:1 (2016), 300–23, doi:10.1177/0031512516656816.

19   Nicole W. Forrester, "How Olympians Train Their Brains to Become Mentally Tough," TheConversation.com, February 21, 2018, https://tinyurl.com/y94uv66w.

20   Jackson Yee, "Improving Your Self-Discipline with Your Training," EliteFTS.com, August 29, 2011, https://tinyurl.com/8aw34vn8.

21   Taylor, "What Mental Training for Sports Is Really All About."

22   Laura C. Healy, Nikos Ntoumanis, and Joan L. Duda, "Goal Motives and Multiple-Goal Striving in Sport and Academia: A Person-Centered Investigation of Goal Motives and Inter-Goal Relations," *Journal of Science and Medicine in Sport* 19:12 (2016), 1010–14, doi:10.1016/j.jsams.2016.03.001.

23   James J. Gibson, "A Theory of Direct Visual Perception," in *Vision and Mind: Selected Readings in the Philosophy of Perception,* edited by Alva Noë and Evan Thompson, 77–91 (Cambridge, MA: MIT Press, 2002).

24   Jackson and Farrow, "Implicit Perceptual Training."

25   Julia Uddén, Vasiliki Folia, and Karl Magnus Petersson, "The Neuropharmacology of Implicit Learning," *Current Neuropharmacology* 8:4 (2010), 367–81, doi:10.2174/157015910793358178.

26   Alexandru and Lorand, "Motor Behavior and Anticipation."

27   Beckmann et al., "Preventing Motor Skill Failure."

28   Lee and Miller, "Brain Training."

29   Shojiro Sugiyama, "Sport Karate vs. Budō Karate," 1994.

Chapter 8

1   Kwang Hyun Ko, "Origins of Bipedalism," *Brazilian Archives of Biology and Technology* 58:6 (2015), doi:10.1590/S1516-89132015060399.

2 Erin Wayman, "Becoming Human: The Evolution of Walking Upright," *Smithsonian Magazine,* August 6, 2012, https://tinyurl.com/4p4t3svd; James Shreeve, "New Skeleton Gives Path from Trees to Ground an Odd Turn," *Science* 272:5262 (1996), 654, doi:10.1126/science.272.5262.654.

3 Dennis M. Bramble and Daniel E. Lieberman, "Endurance Running and the Evolution of *Homo,*" *Nature* 432:7015 (2004), 345–52, doi:10.1038/nature03052.

4 Kevin D. Hunt, "The Evolution of Human Bipedality: Ecology and Functional Morphology," *Journal of Human Evolution* 26:3 (1994), 183–202, doi:10.1006/jhev.1994.1011; C. Owen Lovejoy, "The Origin of Man," *Science* 211:4480 (1981), 341–50, doi:10.1126/science.211.4480.341; Ko, "Origins of Bipedalism"; Wikipedia, s.v. "Bipedalism," 2019.

5 Bramble and Lieberman, "Endurance Running."

6 Ibid.; Nathan Bernardo, "How Human Evolution Allowed for the Development of Martial Arts," HowTheyPlay.com, August 6, 2020, https://tinyurl.com/5yzdmstk; David R. Carrier, "The Advantage of Standing Up to Fight and the Evolution of Habitual Bipedalism in Hominins," *PLoS ONE* 6:5 (May 18, 2011), e19630, doi:10.1371/journal.pone.0019630; David R. Carrier and Michael H. Morgan, "Protective Buttressing of the Hominin Face," *Biological Reviews* 90:1 (2015), 330–46, doi:10.1111/brv.12112.

7 Carrier and Morgan, "Protective Buttressing."

8 Richard W. Young, "Evolution of the Human Hand: The Role of Throwing and Clubbing," *Journal of Anatomy* 202:1 (2003), 165–74, doi:10.1046/j.1469-7580.2003.00144.x.

9 Michael H. Morgan and David R. Carrier, "Protective Buttressing of the Human Fist and the Evolution of Hominin Hands," *Journal of Experimental Biology* 216:Part 2 (2013), 236–44, doi:10.1242/jeb.075713.

10 Carrier, "Advantage of Standing Up."

11 C. W. Chan and A. Rudins, "Foot Biomechanics During Walking and Running," *Mayo Clinic Proceedings* 69:5 (1994), 448–61, doi:10.1016/s0025-6196(12)61642-5.

12 Bramble and Lieberman, "Endurance Running."

13 Elaine E. Kozma, Nicole M. Webb, William E. H. Harcourt-Smith, David A. Raichlen, Kristiaan D'Août, Mary H. Brown, Emma M. Finestone, et al., "Hip Extensor Mechanics and the Evolution of Walking and Climbing Capabilities in Humans, Apes, and Fossil Hominins," *Proceedings of the National Academy of Sciences USA* 115:16 (April 17, 2018), 4134–39, doi:10.1073/pnas.1715120115.

14 Carrier, "Advantage of Standing Up."

15 Thomas J. Roberts and Nicolai Konow, "How Tendons Buffer Energy Dissipation by Muscles," *Exercise and Sport Sciences Reviews* 41:4 (2013), 186–93, doi:10.1097/JES.0b013e3182a4e6d5.

16 William Irvin Sellers, Todd C. Pataky, Paolo Caravaggi, and Robin Huw Crompton, "Evolutionary Robotic Approaches in Primate Gait Analysis," *International Journal of Primatology* 31 (2010), 321–38, doi:10.1007/s10764-010-9396-4.

17 Bramble and Lieberman, "Endurance Running."

18 University of Helsinki, "What Is Time?" *Science Daily,* April 15, 2005, https://tinyurl.com/dvxh6kz8.

19 Brian Greene, *The Elegant Universe: Superstrings, Hidden Dimensions, and the Quest for the Ultimate Theory* (New York: W. W. Norton, 2003); Annenberg Learner, "Journey into the Fourth Dimension," *Mathematics Illuminated,* website, 2020, https://tinyurl.com/y4yewwpv.

Chapter 9

1 Aikinomichi Wien, "Aikido: Yoko Okamoto Sensei Berlin 2017," YouTube, video, February 5, 2018, https://youtu.be/KrpFhe6jFyI.

2 Corpore, "21 Posture Quotes That Will Immediately 'Sit You Straight,'" CoporeWear.com, blog, 2019, https://tinyurl.com/brjdz3dv.

3 Wikipedia, s.v. "Musashi Miyamoto," 2020.

4 Richard E. Rowell, "One-Inch Distance: Life and Death in the Thickness of Paper," *Budo Theory: Exploring Martial Arts Principles* (Nanton, Canada: Richard E. Rowell, 2011).

5 Musashi Miyamoto, *A Book of Five Rings. A Guide to Strategy,* trans. Victor Harris (Woodstock, NY: Overlook Press, 1974); "Traditional Sword Fighting Style for a Samurai," Kasoku Sekai, blog, August 7, 2014, https://tinyurl.com/4ejxsjv9.

6 Laurence R. Harris and Charles Mander, "Perceived Distance Depends on the Orientation of Both the Body and the Visual Environment," *Journal of Vision* 14:12 (2014), 1–8, doi:10.1167/14.12.17.

7 Ottica Santona, "Posture and Vision," n.d., Santona.it, https://tinyurl.com/wvtz8287.

8 Mark Filllipi, "Vision and Posture," NaturalEyeCare.com, n.d., https://tinyurl.com/z74suk82.

9 Kathrine Jáuregui-Renaud, "Postural Balance and Peripheral Neuropathy," in Peripheral Neuropathy: *A New Insight into the Mechanism, Evaluation and Management of a Complex Disorder,* ed. Nizar Souayah, Intechopen.com, March 27, 2013, doi:10.5772/55344.

10 T. Wise, "I Thought It'd Be Easy to Fix My Bad Posture—I Was Wrong," Yahoo Life, August 7, 2019, https://tinyurl.com/yhjd9hjf.

11 Wikipedia, s.v. "Joseph Pilates," 2019.

12 Corpore, "21 Posture Quotes."

13 https://idealalignment.com.

14 "Proper Posture for Higher Engagement and Cognitive Performance," American Posture Institute, blog, September 19, 2017, https://tinyurl.com/3pufsnt6; Markus Muehlhan, Michael Marxen, Julia Landsiedel, Hagen Malberg, and Sebastian Zaunseder, "The Effect of Body Posture on Cognitive Performance: A Question of Sleep Quality," *Frontiers in Human Neuroscience* 8:171 (2014), doi:10.3389/fnhum.2014.00171.

15 Jáuregui-Renaud, "Postural Balance."

16 Edgar Garcia-Rill, Yutaka Homma, and Robert D Skinner, "Arousal Mechanisms Related to Posture and Locomotion. 2. Ascending Modulation," *Progress in Brain Research* 143 (2004), 291–98, https://tinyurl.com/4dxrjhz7.

17 Muehlhan et al., "Effect of Body Posture."

18 Garcia-Rill et al., "Arousal Mechanisms."

19 Andrea Berencsi, Masami Ishihara, and Kuniyasu Imanaka, "The Functional Role of Central and Peripheral Vision in the Control of Posture," *Human Movement Science* 24:5–6 (2005), 687–709, doi:10.1016/j.humov.2005.10.014.

20 Jáuregui-Renaud, "Postural Balance."

21 Ibid.

22 Wikipedia, s.v. "Stance (Martial Arts)," 2018.

23 Gichin Funakoshi, *To-Te Jitsu* (Hamilton, Canada: Masters, 1997).

24 Funakoshi, *Karate-dō kyōhan.*

25 Marc Keys, "Weight Training Guide for Developing Speed and Power," Origin Fitness, January 16, 2016, https://tinyurl.com/knz5dwf.

26 Antonio Srado, "The Biomechanics of a Boxer's Cross Punch," Antonio Srado, blog, October 21, 2013, https://tinyurl.com/425tf6a2.

27 Rick Hamilton, "Are Hard Punches Based off of Triceps Muscles or by Practicing Your Punches Over and Over?" *Quora Digest,* 2018.

28 Randall G. Hassell and Dale F. Poertner, "Scientific Karate," *Samurai* 1 (1978), 6–9.

29 Ibid.

30 Ibid.; Masatoshi Nakayama, *Dynamic Karate* (Palo Alto, CA: Kodansha, 1966); Hamilton, "Are Hard Punches Based off of Triceps Muscles."

31 Michael P. Garofalo, "Yang Style Taijiquan. Quotations, Sayings, Wisdom, Poems, Aphorisms, Classics, Principles, Guides, Concepts, Terms, Miscellaneous," Cloud Hands, blog, 2018, https://tinyurl.com/ysy22yz4; William Gleason, *Aikido and Words of Power: The Sacred Sounds of Koto-tama* (Rochester, VT: Destiny Books, 2009).

32 David A. Winter, *Biomechanics and Motor Control of Human Movement,* 4th ed. (Hoboken, NJ: Wiley & Sons, 2009).

33 Carrier, "Advantage of Standing Up."

34 Nakayama, *Dynamic Karate.*

35 L. G. Brown, "The World Tournament," *Samurai,* summer ed. 1971, Los Angeles.

36 James T. Webber and David A. Raichlen, "The Role of Plantigrady and Heel-Strike in the Mechanics and Energetics of Human Walking with Implications for the Evolution of the Human Foot," *Journal of Experimental Biology* 219:23 (December 2016), 3729–37, doi:10.1242/jeb.138610.

37 Kathryn Knight, "Humans Walk on Virtual Length Legs," *Journal of Experimental Biology* 219:23 (December 2016), 3671–72, doi:10.1242/jeb.153080.

38 Hassell and Poertner, "Scientific Karate"; Nakayama, *Dynamic Karate.*

39 Theiss, "Karate: A Scientific View."

40 Ibid.

41 Bosler and Greene, "How to Practice Effectively."

Chapter 10

1 Tarina van der Stockt, Evan Thomas, Kim Jackson, Olajumoke Ogunleye, and Rewan Elsayed Elkanafany, "Kinetic Chain," Physiopedia.com, 2012, https://tinyurl.com/aussdts.

2 Ibid.

3 Hassell and Poertner, "Scientific Karate."

4 Bart H. F. J. M. Koopman, "Dynamics of Human Movement," *Technology and Health Care* 18:4–5 (2010), 371–85, doi:10.3233/THC-2010-0599.

5 Van der Stockt et al., "Kinetic Chain."

6 S. Robinson, "The Seven Biomechanical Principles," Exercise Science Portfolio, 2019.

7 ISSA, "Your Guide to the Kinetic Chain," ISSA blog, n.d., https://tinyurl.com/r9h9pfhj.

8 Mike Reinold, "The Problem with the Kinetic Chain Concept," MikeReinold.com, November 29, 2011, https://tinyurl.com/3usj97zc.

9 Van der Stockt et al., "Kinetic Chain."

10 Samuel K. Chu, Prakash Jayabalan, W. Ben Kibler, and Joel Press, "The Kinetic Chain Revisited: New Concepts on Throwing Mechanics and Injury," *PM&R Journal of Injury, Function, and Rehabilitation* 8:3 (2016), 569–77, doi:10.1016/j.pmrj.2015.11.015; Shane T. Seroyer, Shane J. Nho, Bernard R. Bach, Charles A. Bush-Joseph, Gregory P. Nicholson, and Anthony A. Romeo, "The Kinetic Chain in Overhand Pitching: Its Potential Role for Performance Enhancement and Injury Prevention," *Sports Health* 2:2 (2010), 135–46, doi:10.1177/1941738110362656.

11 Seroyer et al., "Kinetic Chain in Overhand Pitching."

12 Corenna R. Dolce, Pitching in Baseball, n.d., https://tinyurl.com/3sy2mdvm.

13 Corenna R. Dolce, "Biomechanics of Baseball Pitch," Pitching in Baseball, n.d., https://tinyurl.com/4vecdknu.

14 Winter, *Biomechanics and Motor Control.*

15 Reinold, "The Problem with the Kinetic Chain Concept"; Hayden Giuliani, "The Rotational Athlete and the Importance of the Glutes," Athletic Lab, November 9, 2017, https://tinyurl.com/3x7pfa2t.

16 Theiss, "Karate: A Scientific View"; Nakayama, *Dynamic Karate.*

17 Hassell and Poertner, "Scientific Karate."

18 Nakayama, *Dynamic Karate.*

19 Chris Sharrock, Jarrod Cropper, Joel Mostad, Matt Johnson, and Terry Malone, "A Pilot Study of Core Stability and Athletic Performance: Is There a Relationship?" *International Journal of Sports Physical Therapy* 6:2 (2011), 63–74.

20 Erin Ritterbusch, "The Importance of the Lumbopelvic Hip Complex," Athletic Lab, April 12, 2017, https://tinyurl.com/y6654h2b.

21 W. Ben Kibler, Joel Press, and Aaron Sciascia, "The Role of Core Stability in Athletic Function," *Sports Medicine* 36:3 (2006), 189–98, doi:10.2165/00007256-200636030-00001.

22 Hamilton, "Are Hard Punches Based off of Triceps Muscles"; Gretchen D. Oliver, Priscilla M. Dwelly, Nicholas D. Sarantis, Rachael A. Helmer, and Jeffery A. Bonacci, "Muscle Activation of Different Core Exercises," *National Strength and Conditioning Association Journal* 24:11 (2010), 3069–74, doi:10.1519/JSC.0b013e3181d321da.

23 Hamilton, "Are Hard Punches Based off of Triceps Muscles."

24 Gary B. Wilkerson, Jessica L. Giles, Dustin K. Seibel, "Prediction of Core and Lower Extremity Strains and Sprains in Collegiate Football Players: A Preliminary Study," *Journal of Athletic Training* 47:3 (2012), 264–72, doi:10.4085/1062-6050-47.3.17.

25 C. W. Nicol, *Moving Zen: Karate as a Way to Gentleness* (London: Paul H. Crompton, 1989).

26 Paul Lam, "The Combined 42 Forms," Tai Chi for Health Institute, 2007, https://tinyurl.com/srunjtp7.

27 Wikipedia, s.v. "24-Form Tai Chi Chuan," 2019.

28 Lam, "Combined 42 Forms."

29 Nakayama, *Best Karate: Fundamentals.*

30 Todd Bumgardner, "Core Strength: Your Ultimate Guide to Core Training," BodyBuilding.com, July 27, 2018, https://tinyurl.com/478rf9n2; Jen Glantz, "I Took a Sword Fight Class: It's Not the Workout You Think It Is," NBCNews.com, September 22, 2019, https://tinyurl.com/wuyfyfvm.

31 Aaron Swanson, "Basic Biomechanics: Moment Arm and Torque," AaronSwansonPT.com, July 3, 2011, https://tinyurl.com/26jt9772.

32 Matt Phillips, "Introduction to Running Biomechanics," Runners Connect, n.d., https://tinyurl .com/dc8zwc89; Neil Edward Bezodis, Steffen Willwacher, and Aki Ilkka Tapio Salo, "The Biomechanics of the Track and Field Sprint Start: A Narrative Review," *Sports Medicine* 49 (2019), 1345–64, doi:10.1007/s40279-019-01138-1.

33 Donald A. Neumann, "Kinesiology of the Hip: A Focus on Muscular Actions," *Journal of Orthopedic and Sports Physical Therapy* 40:2 (2010), 82–94, doi:10.2519/jospt.2010.3025.

34 Tanya Siejhi Gershon, "Hip Abductor Muscles Exercises," LiveStrong.com, June 10, 2019, https://tinyurl.com/ay5ndfj9.

35 R. D. McLeish and J. Charnley, "Abduction Forces in the One-legged Stance," *Journal of Biomechanics* 3:2 (March 1970), 191–94, doi:10.1016/0021-9290(70)90006-0.

36 Neumann, "Kinesiology of the Hip."

37 Ohio State Wexner Medical Center, "Dormant Butt Syndrome May Be to Blame for Knee, Hip, and Back Pain," Ohio State University, May 23, 2016, https://tinyurl.com/38fmx7dt; Eun-Kyung Kim, "The Effect of Gluteus Medius Strengthening on the Knee Joint Function Score and Pain in Meniscal Surgery Patients," *Journal of Physical Therapy Science* 28:10 (2016), 2751–53, doi:10.1589 /jpts.28.2751.

38 Kay M. Crossley, Wan-Jing Zhang, Anthony G. Schache, Adam Bryant, and Sallie M. Cowan, "Performance on the Single-Leg Squat Task Indicates Hip Abductor Muscle Function," *American Journal of Sports Medicine* 39:4 (April 2011), 866–73, doi:10.1177/0363546510395456.

39 Cara L. Lewis, Eric Foch, Marc M. Luko, Kari L. Loverro, and Anne Khuu, "Differences in Lower Extremity and Trunk Kinematics between Single Leg Squat and Step Down Tasks," *PLoS ONE* 10:5 (2015), e0126258, doi:10.1371/journal.pone.0126258.

40 Neumann, "Kinesiology of the Hip."

41 Nick Ng, "Do You Use Your Hip Abductors and Adductors in Running?" The Nest, n.d., https://tinyurl.com/mxmau2sk.

42 Neumann, "Kinesiology of the Hip."

Chapter 11

1 Robert S. Fitzgerald and Neil S. Cherniack, "Historical Perspectives on the Control of Breathing," *Comprehensive Physiology* 2:2 (April 2012), 915–32, doi:10.1002/cphy.c100007.

2 Georg Feuerstein, *Yoga: The Technology of Ecstasy,* 1st ed. (New York: TarcherPerigee, 1989).

3 J.-W. Fitting, "From Breathing to Respiration," *Respiration* 89:1 (January 2015), 82–87, doi:10.1159 /000369474.

4 Jane Barthelemy, "3 Breathing Techniques for Qigong Practitioners," Five Seasons, blog, September 23, 2016, https://tinyurl.com/3vnm3tmw.

5 James Mallinson and Mark Singleton, *Roots of Yoga* (New York: Penguin Classics, 2016).

6   Sarah Novotny and Len Kravitz, "The Science of Breathing," University of New Mexico, n.d., https://tinyurl.com/2nrjau2x.

7   Ishwar V. Basavaraddi, "Yoga: Its Origin, History and Development," Government of India Ministry of External Affairs, April 23, 2015, https://tinyurl.com/yk38b5n4.

8   Wikipedia, s.v. "Qi," 2020.

9   Malini Nair, "Beyond Hinduism, Yoga Also Has Roots in Buddhist, Jain, and Sufi Traditions," Quartz India, January 29, 2017, https://tinyurl.com/trpvrwfj.

10 Xiaorong Chen, Jiabao Cui, Ru Li, Richard Norton, Joel Park, Jian Kong, and Albert Yeung, "Dao Yin (a.k.a. Qigong): Origin, Development, Potential Mechanisms, and Clinical Applications," *Evidence-Based Complementary and Alternative Medicine* 2019:3705120, doi:10.1155/2019/3705120.

11 Wikipedia, s.v. "Qigong," 2019.

12 Ibid.

13 Roger Jahnke, Linda Larkey, Carol Rogers, Jennifer Etnier, and Fang Lin, "A Comprehensive Review of Health Benefits of Qigong and Tai Chi," *American Journal of Health Promotion* 24:6 (2010), e1–e25, doi:10.4278/ajhp.081013-LIT-248.

14 Barthelemy, "3 Breathing Techniques."

15 Ibid.

16 *New World Encyclopedia*, s.v. "Microcosm and Macrocosm," 2018.

17 Wikipedia, s.v. "Microcosmic Orbit," 2019.

18 Xuan-Yun Zhou, "Daoist Breathing Techniques," YMAA, May 20, 2009, https://tinyurl.com/2s4xfj5z.

19 Smita Pandit, "Diaphragmatic Breathing Explained," VisualStories.com, n.d., https://tinyurl.com/yyjyxbvx.

20 G. K. Pal, S. Velkumary, and Madanmohan, "Effect of Short-Term Practice of Breathing Exercises on Autonomic Functions in Normal Human Volunteers," *Indian Journal of Medical Research* 120:2 (August 2004), 115–21.

21 Michael Christopher Melnychuk, Paul M. Dockree, Redmond G. O'Connell, Peter R. Murphy, Joshua H. Balsters, and Ian H. Robertson, "Coupling of Respiration and Attention via the Locus Coeruleus: Effects of Meditation and Pranayama," *Psychophysiology* April 22, 2018, doi:10.1111/psyp.13091.

22 Wendy Bumgardner, "How to Reach the Anaerobic Zone During Exercise," Verywell Fit, February 27, 2020, https://tinyurl.com/5cj543nw.

23 Bruno Bordoni and Emiliano Zanier, "Anatomic Connections of the Diaphragm: Influence of Respiration on the Body System," *Journal of Multidisciplinary Healthcare* 6 (2013), 281–91, doi:10.2147/JMDH.S45443.

24 Wikipedia, s.v. "Rectus abdominis," 2020.

25 Doug Keller, *Refining the Breath: The Yogic Practice of Pranayama* (South Riding, VA: Do Yoga Productions, 2003).

26 "Diaphragm," *Interactive Respiratory Physiology: Encyclopedia* (Baltimore: Johns Hopkins School of Medicine, 1995).

27 Christopher Watson and Roy Suenaka, *Complete Aikido: Aikido Kyohan: The Definitive Guide to the Way of Harmony* (North Clarendon, VT: Tuttle, 1997); Scott Heaney, "What Is Ibuki and Nogare," The Martial Way, n.d., https://tinyurl.com/rtk6wfxr.

28 Heaney, "What Is Ibuki and Nogare."

29 Watson and Suenaka, *Complete Aikido.*

30 Minhhuy Ho, "The Concept of Ki in Aikido, A Literature Survey," The Aikido FAQ, n.d., www .aikidofaq.com/philosophy.

Chapter 12

1 Wikipedia, s.v. "David Carradine," 2020.

 2 Wikipedia, s.v. "Bruce Lee," 2019.

 3 Ibid.

 4 Wikipedia, s.v. "Kung fu (term)," 2018.

 5 Bruce Lee, *The Art of Expressing the Human Body,* trans. John R. Little (Tokyo: Tuttle, 1998).

 6 Wikipedia, s.v. "Bruce Lee," 2019.

 7 Funakoshi, *Karate-dō: My Way of Life.*

 8 Jennifer Warner, "Catcher's Mitts Strike Out at Hand Protection," WebMD, July 1, 2005, https:// tinyurl.com/4289c46d.

 9 Cecil Adams, "The True Force of a Boxer's Punch," *Connect Savannah,* July 20, 2010, https://tinyurl .com/2h2t4bxr.

10 Ibid.

11 Nakayama, *Dynamic Karate.*

12 Ibid.

13 Ibid.; Jearl D. Walker, "Karate Strikes," *American Journal of Physics* 43:10 (October 1, 1975), 845, doi:10.1119/1.9966.

14 "Michael S. Feld, Physics Professor, Dies at Age 69," MIT News, April 11, 2010, https://tinyurl.com /224kzfmu.

15 Curtis Rist, "The Physics of ... Karate," *Discover,* January 19, 2000, https://tinyurl.com/crzhk5un.

16 Michael S. Feld, Ronald E. McNair, and Stephen R. Wilk, "The Physics of Karate," *Scientific American* 240 (April 1979), 150–58, https://tinyurl.com/f3dzsek.

17 Ibid.

18 Ibid.

19 Nishiyama, *Traditional Karate Coach's Manual.*

20 Oliver et al., "Muscle Activation."

21 Nishiyama, *Traditional Karate Coach's Manual;* Hassell and Poertner, "Scientific Karate."

22 Aman et al., "Effectiveness of Proprioceptive Training."

23 Koopman, "Dynamics of Human Movement."

24 Giuliani, "Rotational Athlete."

25 Leslie V. Simon, Richard A. Lopez, and Kevin C. King, "Blunt Force Trauma," *StatPearls,* January 2021, https://tinyurl.com/3jkh6csp.

Chapter 13

1 Nakayama, *Dynamic Karate;* Walker, "Karate Strikes"; J. Mack, S. Stojsih, D. Sherman, N. Dau, and C. Bir, "Amateur Boxer Biomechanics and Punch Force," 28th International Conference on Biomechanics in Sports, 2010, Marquette, MI, https://tinyurl.com/22sdyvu4.

Chapter 14

1 Kristina Kaine, "Is There a Difference Between the Spirit and the Soul?" HuffPost, January 11, 2016, https://tinyurl.com/25znc5hw.

2 Online Etymology Dictionary, s.v. "Spirit," 2017.

3 Wikipedia, s.v. "Spirit," 2020.

4 Jim Myers, "Ruach: Spirit or Wind or ?" Biblical Heritage Center, n.d., https://tinyurl.com/2tfk5adv.

5 Wikipedia, s.v. "Spirit," 2020; Catherine Tims, "Body, Soul, and Spirit," Autumn Damask, September 7, 2019, https://tinyurl.com/cm4rn76s.

6 Kees Waaijman, *Spirituality: Forms, Foundations, Methods* (Leuven, Belgium: Peeters, 2003).

7 Wikipedia, s.v. "Spirituality," 2020.

8 Kaine, "Is There a Difference."

9 Wikipedia, s.v. "Spirituality," 2020.

10 Wikipedia, s.v. "Chinese Martial Arts," 2018.

11 Matthew Lee, "Analyzing Wude: The Martial Ethics behind Wushu," Jiaoo Wushu, January 6, 2016, https://tinyurl.com/7t8uxayv.

12 Sor-Hoon Tan, "The Concept of Yi (义) in the Mencius and the Problems of Distributive Justice," *Australasian Journal of Philosophy* 92:3 (2014), 489–505, doi:10.1080/00048402.2014.882961.

13 Wikipedia, s.v. "Guan Yu," 2020.

14 Lee, "Analyzing Wude."

15 Tim Clark, "The Bushidō Code: The Eight Virtues of the Samurai," in *A Man's Life, Featured, Martial Arts, on Virtue,* ed. Brett McKay and Kate McKay, September 14, 2008, ArtOfManliness.com, https://tinyurl.com/bpvmmx6u.

16 Oleg Benesch, *Inventing the Way of the Samurai. Nationalism, Internationalism, and Bushidō in Modern Japan* (Oxford, UK: Oxford University Press, 2014).

17 Inazō Nitobe, *Bushidō: The Soul of Japan* (1899; repr. Tokyo: Kodansha International, 2012).

18 Kiyoshi Matsumoto, "Japan's Hidden Moral Code," TalkAboutJapan.com, July 18, 2018, https://tinyurl.com/y9rd4nj6; Oleg Benesch, "The Samurai Next Door: Chinese Examinations of the Japanese Martial Spirit," *Extrême-Orient Extrême-Occident* 38 (2014), 129–68, doi:10.4000/extremeorient.376.

19 Wikipedia, s.v. *"Bushidō: The Soul of Japan,"* 2019.

20 Benesch, *Inventing the Way of the Samurai.*

21 Benesch, "The Samurai Next Door."

22 Matsumoto, "Japan's Hidden Moral Code."

23 Tim Hanlon, "My Black Belt and Karate-dō," *Shōtōkan Karate Magazine,* June 2020.

24 Nicol, *Moving Zen.*

25 Heinrich Dumoulin, *Zen Buddhism: A History—Japan, Volume 2* (New York: Macmillan, 2005).

26 Wikipedia, s.v. "Dhyāna in Buddhism," 2020.

27 Dumoulin, *Zen Buddhism, vol. 2.*

28 Philip Smith, "Moving Zen: Seido Karate in Pictures and Words," *Tricycle,* Spring 2000, https://tinyurl.com/386jbw7a.

29 Hansen and Zalta, "Daoism."

30 Wikipedia, s.v. "Rinzai School," 2020.

31 Daisetz Teitaro Suzuki, *Manual of Zen Buddhism* (1935; repr. Baghdad, Iraq: M. G. Sheet, 2005), https://tinyurl.com/z32hhf2c.

32 Wikipedia, s.v. "Mushin (mental state)," 2018.

33 Wikipedia, s.v. "Bushidō: The Soul of Japan," 2019.

34 K. Kato, "The Way of the Samurai," in *Samurai* (Brooklyn, NY: Complete Sports Publications, 1974).

35 William Scott Wilson, *Ideals of the Samurai: Writings of Japanese Warriors* (1982; repr. Valencia CA: Black Belt Books, 2014).

36 Oleg Benesch, "Imperial Japan Saw Itself As a 'Warrior Nation'—and the Idea Lingers Today," Yahoo Style, December 22, 2017, https://tinyurl.com/5a29uv8m.

37 *Britannica Online Encyclopedia,* s.v. "Mishima Yukio," 2007.

38 Evan Andrews, "What Is Seppuku?" History.com, August 22, 2018, https://tinyurl.com/aeayr23x.

39 *Britannica Online Encyclopedia,* s.v. "Mishima Yukio," 2007.

40 Kallie Szczepanski, "About Seppuku (or Harakiri)," ThoughtCo, March 1, 2019, https://tinyurl.com/chv5dwh7.

41 Andrews, "What Is Seppuku?"

42 Szczepanski, "About Seppuku."

43 *Britannica Online Encyclopedia,* s.v. "Mishima Yukio," 2007.

44 Kato, "Way of the Samurai."

45 Szczepanski, "About Seppuku."

Chapter 15

1 Mike Randall, "Class of 2021 West Point Cadets Celebrate 12-Mile 'March Back,'" [Middletown, NY] *Times Herald-Record,* August 14, 2017, https://tinyurl.com/7knzf65k.

2 Wikipedia, s.v. "United States Military Academy," 2020.

3 "The Life and Times of a West Point Cadet," WestPointCadet, blog, June 27, 2013, https://tinyurl.com/hv8y2e84.

4 Angela L. Duckworth, Christopher Peterson, Michael D. Matthews, Dennis R. Kelly, "Grit: Perseverance and Passion for Long-Term Goals," *Journal of Personality and Social Psychology* 92:6 (2007), 1087–1101, doi:10.1037/0022-3514.92.6.1087.

5 Ibid.

6 Ibid.

7 Ibid.

8 Lea Winerman, "What Sets High Achievers Apart?" *Monitor on Psychology* 44:11 (2013), 28, https://tinyurl.com/f9w33dae.

9 Catharine M. Cox, *Genetic Studies of Genius. II. The Early Mental Traits of Three Hundred Geniuses* (Stanford, CA: Stanford University Press, 1926).

10 Duckworth et al., "Grit."

11 Margaret M. Perlis, "5 Characteristics of Grit—How Many Do You Have?" *Forbes,* October 29, 2013, https://tinyurl.com/cktbfaea.

12 Duckworth et al., "Grit."

13 Ibid.; Perlis, "5 Characteristics."

14 Lex Borghans, Huub Meijers, and Bas ter Weel, "The Role of Noncognitive Skills in Explaining Cognitive Test Scores," *Economic Inquiry* 46:1 (2008), 2–12, doi:10.1111/j.1465-7295.2007.00073.x, https://tinyurl.com/vwtp7wnx.

15 Urvashi Dutta and Anita Puri Singh, "Studying Spirituality in the Context of Grit and Resilience of College-Going Young Adults," *International Journal for Innovative Research in Multidisciplinary Field* 3:9 (2017), 50–55, https://tinyurl.com/2hdzka4p.

16 Barb Leonard and Mary Jo Kreitzer, "What Is Life Purpose?" University of Minnesota, n.d., https://tinyurl.com/24ub4aut.

17 Gregg Prescott, "What Is My Role or Purpose in This Spiritual Awakening?" In5D, February 12, 2018, https://tinyurl.com/57fxk7k3.

18 Funakoshi, *Karate-dō: My Way of Life.*

19 Perlis, "5 Characteristics of Grit."

20 Mel Schwartz, *The Possibility Principle: How Quantum Physics Can Improve the Way You Think, Live, and Love* (Boulder, CO: Sounds True, 2017).

21 Perlis, "5 Characteristics of Grit."

22 Ibid.

23 Massimo Pigliucci and Gregory Lopez, *A Handbook for New Stoics: How to Thrive in a World Out of Your Control—52 Week-by-Week Lessons* (New York: Experiment, 2019).

24 Oxford Lexico, s.v. "Community," 2019.

25 Cooper, "Mizu no Kokoro"; Carlson and Charlie, "Biological Baseball."

26 Wikipedia, s.v. "14th Dalai Lama," 2020.

27 356 Matthieu Ricard, Antoine Lutz, and Richard J. Davidson, "Mind of the Meditator," *Scientific American* 311:5 (November 2014), 38–45, doi:10.1038/scientificamerican1114-38.

28 John Geirland, "Buddha on the Brain," *Wired,* January 2, 2006, https://tinyurl.com/tjvhkw.

29 Marc Kaufman, "Dalai Lama Talks to Scientists," *Washington Post,* November 13, 2005.

30 Dalai Lama, *The Universe in a Single Atom: The Convergence of Science and Spirituality* (New York: Morgan Road, 2005); Ricard et al., "Mind of the Meditator."

31 Kathy Gilsinan, "The Buddhist and the Neuroscientist: What Compassion Does to the Brain," Atlantic.com, July 4, 2015, https://tinyurl.com/p436an3h.

32 Antoine Lutz, Lawrence L. Greischar, Nancy B. Rawlings, Matthieu Ricard, and Richard J. Davidson, "Long-Term Meditators Self-Induce High-Amplitude Gamma Synchrony During Mental Practice," *Proceedings of the National Academy of Sciences USA* 101:46 (2004), 16369–73, doi:10.1073/pnas.0407401101.

33 Giovanni, "Types of Meditation—An Overview of 23 Meditation Techniques," Awaken, blog, November 19, 2020, https://tinyurl.com/x7n7xsbs.

34 Dominique P. Lippelt, Bernhard Hommel, and Lorenza S. Colzato, "Focused Attention, Open Monitoring, and Loving Kindness Meditation: Effects on Attention, Conflict Monitoring, and Creativity—A Review," *Frontiers in Psychology* 5 (September 2014), 1083, doi:10.3389/fpsyg.2014.01083; Ricard et al., "Mind of the Meditator."

35 Shohaku Okumura, *The Mountains and Waters Sutra: A Practitioner's Guide to Dogen's "Sansuikyo"* (Somerville, MA: Wisdom Publications, 2018).

36 Fynn-Mathis Trautwein, Philipp Kanske, Anne Böckler, and Tania Singer, "Differential Benefits of Mental Training Types for Attention, Compassion, and Theory of Mind," *Cognition* 194 (January 2020), 104039, doi:10.1016/j.cognition.2019.104039.

37 Ricard et al., "Mind of the Meditator."

38 Lutz et al., "Long-Term Meditators."

39 Yi-Yuan Tang, Yinghua Ma, Junhong Wang, Yaxin Fan, Shigang Feng, Qilin Lu, Qingbao Yu, Danni Sui, et al., "Short-Term Meditation Training Improves Attention and Self-Regulation," *Proceedings of the National Academy of Sciences USA* 104 (2007), 17152–56, doi:10.1073/pnas.0707678104; "Body-Mind Training Is the Best Type of Meditation, Says Science," AFP, July 25, 2016, https://tinyurl.com/3r7eacjm.

40 Tang et al., "Short-Term Meditation Training."

41 Lutz et al., "Long-Term Meditators."

42 Ibid.

43 Yung-Jong Shiah, "From Self to Nonself: The Nonself Theory," *Frontiers in Psychology* 4:7 (February 2016), 124, doi:10.3389/fpsyg.2016.00124.

44 Ibid.; Rune E. A. Johansson, *The Psychology of Nirvana* (London: Allen & Unwin, 1969).

45 Matt Danzico, "Brains of Buddhist Monks Scanned in Meditation Study," BBC News, April 24, 2011, https://tinyurl.com/da76d4ab.

46 Nicholas F. Gier, "The Virtues of Asian Humanism," *Journal of Oriental Studies* 12 (October 2002), 14–28.

47 Sue McGreevey, "Eight Weeks to a Better Brain," *Harvard Gazette,* January 21, 2011, https://tinyurl.com/39p9wwut.

48 Ricard et al., "Mind of the Meditator"; Christopher G. Davey and Ben J. Harrison, "The Brain's Center of Gravity: How the Default Mode Network Helps Us to Understand the Self," *World Psychiatry* 17:3 (2018), 278–79, doi:10.1002/wps.20553.

49 McGreevey, "Eight Weeks to a Better Brain."

50 Ricard et al., "Mind of the Meditator."

51 Masahiro Fujino, Yoshiyuki Ueda, Hiroaki Mizuhara, Jun Saiki, and Michio Nomura, "Open Monitoring Meditation Reduces the Involvement of Brain Regions Related to Memory Function," *Scientific Reports* 8:9968 (2018), doi:10.1038/s41598-018-28274-4.

52 Littlejohn, "Daoist Philosophy."

53 Ricard et al., "Mind of the Meditator"; Fujino et al., "Open Monitoring Meditation."

54 Yamamoto Tsunetomo, *Hagakure: The Book of the Samurai,* trans. William Scott Wilson (Boston: Shambhala, 2012).

Chapter 16

1 Jerry Alan Johnson, *The Secret Teachings of Chinese Energetic Medicine: Energetic Anatomy and Physiology* (Pacific Grove, CA: International Institute of Medical Qigong, 2014).

2 Jwing-Ming Yang, "What Is Qi and What Is Qigong?" YMAA, March 14, 2016, https://tinyurl.com/hyvy8crz.

3 Ibid.

4 Marty Eisen, "Qi in Traditional Chinese Medicine," Qi Encyclopedia, February 16, 2016, https://tinyurl.com/axm9r2hz.

5 Jerry Alan Johnson, *The Secret Teachings of Chinese Energetic Medicine, Volume 2: Energetic Alchemy, Dao Yin Therapy, Healing Qi Deviations, and Spirit Pathology* (Pacific Grove, CA: International Institute of Medical Qigong, 2014).

6 Eisen, "Qi in Traditional Chinese Medicine."

7 Johnson, *Secret Teachings Vol. 2.*

8 Ibid.

9 Johnson, *Secret Teachings: Energetic Anatomy.*

10 Ibid.

11 Johnson, *Secret Teachings Vol. 2.*

12 Daisetz Teitaro Suzuki, *The Zen Doctrine of No Mind* (York Beach, ME: Weiser Books, 1991).

13 Johnson, *Secret Teachings Vol. 2.*

14 Doug French, "Trump's 'Gut Feeling,'" Mises Institute, January 12, 2019, https://tinyurl.com/js8yb4rb.

15 Ian Miller, "The Gut-Brain Axis: Historical Reflections," *Microbial Ecology in Health and Disease* 29:2 (2018), 1542921, doi:10.1080/16512235.2018.1542921.

16 "Facts About IBS," IFFGD, n.d., https://tinyurl.com/67tudxet.

17 David J. Gracie, Elspeth A. Guthrie, P. John Hamlin, and Alexander C Ford, "Bi-Directionality of Brain-Gut Interactions in Patients with Inflammatory Bowel Disease," *Gastroenterology* 154:6 (May 2018), 1635–46, doi:10.1053/j.gastro.2018.01.027.

18 Ibid.

19 John B. Furness, "Enteric Nervous System," Scholarpedia, 2007, doi:10.4249/scholarpedia.4064.

20 Michael D. Gershon, *The Second Brain: A Groundbreaking New Understanding of Nervous Disorders of the Stomach and Intestine* (New York: HarperCollins, 1998).

21 John Coates, *The Hour Between Dog and Wolf: How Risk Taking Transforms Us, Body and Mind* (New York: Penguin, 2012).

22 Erik Ceunen, Johan W. S. Vlaeyen, and Ilse Van Diest, "On the Origin of Interoception," *Frontiers in Psychology,* May 23, 2016, doi:10.3389/fpsyg.2016.00743.

23 Ibid.

24 Sean P. Durham, "Why Your Gut Feeling Is More Powerful Than Logic," Medium.com, September 1, 2019, https://tinyurl.com/v2b7xe4x.

25 Gregory Scott Brown, "Food and the Pursuit of Happiness," *Psychology Today,* July 8, 2018, https://tinyurl.com/yxrtr9tm.

26 Eisen, "Qi in Traditional Chinese Medicine."

27 James Lake, "The Impact of Air Pollution on Mental Health," *Psychology Today,* January 8, 2020, https://tinyurl.com/h8b366u9.

28 Jim E. Banta, Gina Segovia-Siapco, Christine Betty Crocker, Danielle Montoya, and Noara Alhusseini, "Mental Health Status and Dietary Intake Among California Adults: A Population-Based Survey," *Food Sciences and Nutrition* 70:6 (September 2019), 1, doi:10.1080/09637486.2019.1570085.

29 "Proteins, Carbs, and Fats: How They Affect Your Performance," Composition ID, January 18, 2018, https://tinyurl.com/42sfsw5y.

30 Alice Gomstyn, "Food for Your Mood: How What You Eat Affects Your Mental Health," Aetna, n.d., https://tinyurl.com/2ub4kfj3.

31 Wikipedia, s.v. "Aurora," 2020.

32 Allison Reimer, "My Magical Northern Lights, Have You Seen Them?" September 21, 2017, https://tinyurl.com/vhyj9pwx.

33 Wikipedia, s.v. "Aurora," 2020.

34 Wikipedia, s.v. "Nuclear Fusion," 2020.

35 Deepti Mathur and Samir D. Mathur, "Stacking Waves: Bosons and Fermions," *Quantum Mechanics: A Mini Course,* Ohio State University, November 27, 2017, https://tinyurl.com/3x2pwcrb.

36 Wikipedia, s.v. "Boson," 2020.

37 Jaakko Malmivuo and Robert Plonsey, *Bioelectromagnetism: Principles and Applications of Bioelectric and Biomagnetic Fields* (New York: Oxford University Press, 1995).

38 Sebastian Anthony, "Will Your Body Be the Battery of the Future?" ExtremeTech, September 5, 2012, https://tinyurl.com/jr32zzu4.

39 Paul A. Swinton, Ray Lloyd, Justin W. L. Keogh, Ioannis Agouris, and Arthur D. Stewart, "A Biomechanical Comparison of the Traditional Squat, Powerlifting Squat, and Box Squat," *Journal of Strength and Conditioning Research* 26:7 (July 2012), 1805–16, doi:10.1519/JSC.0b013e3182577067.

40 Anthony, "Will Your Body Be the Battery of the Future?"

41 Christina L. Ross, "Energy Medicine: Current Status and Future Perspectives," *Global Advances in Health and Medicine* 8 (2019), 1–10, doi:10.1177/2164956119831221.

42 T. M. Srinivasan, "Biophotons as Subtle Energy Carriers," *International Journal of Yoga* 10:2 (May–August 2017), 57–58, doi:10.4103/ijoy.IJOY_18_17.

43 Felix Scholkmann, Daniel Fels, and Michal Cifra, "Non-Chemical and Non-Contact Cell-to-Cell Communication: A Short Review," *American journal of translational research* 5:6 (2013), 586–93.

44 Fritz-Albert Popp, Ulrich Warnke, Herbert L. Konig, and Walter Peschka, eds., *Electromagnetic Bio-Information* (Munich: Urban & Schwarzenberg, 1989); Beverly Rubik, David Muehsam, Richard Hammerschlag, and Shamini Jain, "Biofield Science and Healing: History, Terminology, and Concepts," *Global Advances in Health and Medicine* 4 (2015), 8–14, doi:10.7453/gahmj.2015.038.suppl.

45 Rubik et al., "Biofield Science and Healing."

46 Ibid.

47 Johnson, *Secret Teachings: Energetic Anatomy.*

48 Popp et al., *Electromagnetic Bio-Information.*

Chapter 17

1 Friedrich Nietzsche, *The Birth of Tragedy: Out of the Spirit of Music,* ed. Michael Tanner, trans. Shaun Whiteside, (London: Penguin Random House, 1994).

2 Allison Stieger, "Living with Apollo and Dionysus," Mythic Stories, blog, May 2, 2013, https://tinyurl.com/adbv72ub.

3 Nietzsche, *Birth of Tragedy.*

4 "COVID-19 in the USA," Johns Hopkins University of Medicine Coronavirus Resource Center, 2020, https://coronavirus.jhu.edu.

5 John Bacon and Greg Hilburn, "Gulf Coast Braces for 'Unprecedented' Challenge as Tropical Storms Marco and Laura Surge Toward Landfall," *USA Today,* August 23, 2020, https://tinyurl.com/tjvybnn9; Amir Vera and Jamiel Lynch, "Deadly California Wildfires Scorch More Than 1 Million Acres with No End in Sight," CNN, August 23, 2020, https://tinyurl.com/3pcv9d74.

6 U.S. Centers for Disease Control and Prevention, "1918 Pandemic (H1N1 Virus)," CDC National Center for Immunization and Respiratory Diseases, March 20, 2019, https://tinyurl.com/4spdhwvn.

7 Paul Burnett, "The Red Scare," n.d., https://tinyurl.com/2h7m2h9t.

8 Wikipedia, s.v. "Luohan Quan," 2020.

9 Yang, *Essence of Shaolin White Crane.*

10 Kadien S. Hill, "From Fact to Fiction: The Wuxia Experience and the Wushu Practice," bachelor's thesis, Georgia Southern University, 2018, https://tinyurl.com/4vb8uc8p.

11 S. Pangambam, "5 Hindrances to Self-Mastery: Shi Heng YI (Transcript)," The Singju Post, April 24, 2020, https://tinyurl.com/22nwt7kv.

12 Ibid.

13 Funakoshi, *Karate-dō: My Way of Life.*

14 Robert M. Gagné, *The Conditions of Learning and Theory of Instruction* (1965; repr. Belmont, CA: Wadsworth Publishing, 1985).

15 Shojiro Sugiyama, *25 Shoto-kan Kata* (Chicago: Shojiro Sugiyama, 1984).

16 International Traditional Karate Federation, *Traditional Karate Competition Rules* (Los Angeles: ITKF, 2009).

17 Bishop, *Okinawan Karate;* McCarthy, *Bible of Karate.*

18 Hirokazu Kanazawa, *Shōtōkan Karate International Kata, Volume 2* (Tokyo: Ikeda Shoten, 1982).

19 Gregory Allen, "Kata Chintō—Fighting to the East," White Crane Education, n.d., https://tinyurl.com/9tbuazfz.

20 Ibid.

21 Wikipedia, s.v. "Wanshū," 2019.

22 Andrew Griffiths, "The History of Nijūshiho," The History of Fighting, blog, February 22, 2013, https://tinyurl.com/3tw8rf38.

23 Tom Braithwaite, "Southern Dragon Style Kung Fu," Dragon Martial Arts Association, n.d., https://tinyurl.com/ym748npe.

24 Wikipedia, s.v. "Gojūshiho," 2019.

25 Hirokazu Kanazawa, *Shōtōkan Karate International Kata, Volume 1* (Tokyo: Ikeda Shoten, 1982).

26 Wikipedia, s.v. "Seisan," 2019.

27 Jesse Enkamp, "Mysticism in Karate (Pt.1)," Karate by Jesse, n.d., https://tinyurl.com/3rxe3fym.

28 Stephen St. Laurent, "Torajiro Mori: The Essence of Kata," *Samurai* 1:3 (March 1975).

29 Daxue, "The Great Learning," trans. James Legge, Chinese Classics and Translations, November 1, 2005, https://tinyurl.com/63484erb.

30 Shiffrin and Schneider, "Controlled and Automatic Human Information Processing."

31 Kris Eiring, "Imagery, Visualization, and Performance," Sport Psychology, n.d., https://tinyurl.com/2m2jst5r.

32 Lawrence A. Kane and Kris Wilder, *The Way of Kata: A Comprehensive Guide to Deciphering Martial Applications* (Wolfeboro, NH: YMAA, 2005).

33 Sugiyama, "Sport Karate."

34 ITKF, *Traditional Karate Competition Rules.*

35 Ibid.

36 Leigh Ann Schildmeier, "What Is Kata?" Park Avenue Solutions, n.d., https://tinyurl.com/thezrtx7.

Postscript

1 Kenneth Bi, dir., *The Drummer* (2007; Hong Kong: Emperor Motion Pictures).

2 Chen Qing Shan, *A Scholar's Path: An Anthology of Classical Chinese Poems and Prose of Chen Qing Shan,* trans. Peter Chen and Michael Chan (Hackensack, NJ: World Scientific, 2010), doi:10.1142/7840.

3 Harari, *Homo Deus.*

Appendix A

1 Hassell and Poertner, "Scientific Karate."

Appendix B

1 Keys, "Weight Training Guide"; Detric Smith, "Heavier Isn't Always Better: How the Force-Velocity Curve Impacts Your Training," Stack, July 8, 2019, https://tinyurl.com/4ecj97xh; Emma F. Hodson-Tole and James M. Wakeling, "Motor Unit Recruitment for Dynamic Tasks: Current Understanding and Future Directions," *Journal of Comparative Physiology* 179:1 (January 2009), 57–66, doi:10.1007/s00360-008-0289-1; Charles Molnar and Jane Gair, "Muscle Contraction and Locomotion," *Concepts of Biology, 1st Canadian Edition,* BCcampus, May 14, 2015, https://opentextbc.ca/biology.

2 Daniel A. Russell, "Bat Weight, Swing Speed and Ball Velocity," Physics and Acoustics of Baseball and Softball Bats, March 27, 2008, https://tinyurl.com/y6en38bb.

3 Smith, "Heavier Isn't Always Better."

4 Ibid.

5 J. Griffing, "Bench Press Standards," ExRx.net, n.d.

6 Wikipedia, s.v. "Plyometrics," 2020.

BIBLIOGRAPHY

Abernathy, Bruce, Jörg Schorer, Robin C. Jackson, and Norbert Hagemann. "Perceptual Training Methods Compared: The Relative Efficacy of Different Approaches to Enhancing Sport-Specific Anticipation." *Journal of Experimental Psychology Applied* 18:2 (May 2012), 143–53. doi:10.1037/a0028452.

Abernethy, Bruce. "Visual Search Strategies and Decision-Making Sport." *International Journal of Sport Psychology* 22 (1991), 189–210.

Abernethy, Iain. *Bunkai-Jutsu: The Practical Application of Karate Kata.* Chichester, UK: Summersdale, 2002.

Adams, A., and M. Uyehara. "20 Top Fighters in Japan." *Black Belt 1971 Yearbook* Los Angeles: Black Belt Inc., 1971.

Adams, Cecil. "The True Force of a Boxer's Punch." *Connect Savannah.* July 20, 2010. https://tinyurl.com/2h2t4bxr.

Alesi, Marianha, Antonino Bianco, Johnny Padulo, Francesco Paolo Vella, Marco Petrucci, Antonio Paoli, Antonio Palma, and Annamaria Pepi. "Motor and Cognitive Development: The Role of Karate." *Muscles, Ligaments and Tendons Journal* 4:2 (2014), 114–20. https://tinyurl.com/wszy2ua6.

Alexandru, Moşoi Adrian, and Balint Lorand. "Motor Behavior and Anticipation: A Pilot Study of Junior Tennis Players." *Procedia: Social and Behavioral Sciences* 187 (2015), 448–53. doi:10.1016/j.sbspro.2015.03.084.

Allen, Gregory. "Kata Chintō—Fighting to the East." White Crane Education. n.d. https://tinyurl.com/9tbuazfz.

Allgood, Kate. "Improving Performance with Better Attention." NeuroTracker. March 29, 2018. https://tinyurl.com/trr9eark.

Alvelve, B. "Biopsychology." *What Is Time?* Alvelve.com. 2019.

Alvelve, B. "Time Perception." *What Is Time?* Alvelve.com. 2019.

Aman, Joshua E., Naveen Elangovan, I-Ling Yeh, Jürgen Konczak. "The Effectiveness of Proprioceptive Training for Improving Motor Function: A Systematic Review." *Frontiers in Human Neuroscience* 8 (2014), 1075. doi:10.3389/fnhum.2014.01075.

Ambady, Nalini. "The Perils of Pondering: Intuition and Thin Slice Judgments." *Psychological Inquiry* 21 (2010), 271–78. doi:10.1080/1047840X.2010.524882.

American Posture Institute. "Proper Posture for Higher Engagement and Cognitive Performance." Blog. September 19, 2017. https://tinyurl.com/3pufsnt6.

Andrews, Evan. "What Is Seppuku?" History.com. August 22, 2018. https://tinyurl.com/aeayr23x.

Annenberg Learner. "Journey into the Fourth Dimension." *Mathematics Illuminated.* Website. 2020. https://tinyurl.com/y4yewwpv.

Anthony, Sebastian. "Will Your Body Be the Battery of the Future?" ExtremeTech. September 5, 2012. https://tinyurl.com/jr32zzu4.

Arnsten, Amy F. T. "Stress Signaling Pathways That Impair Prefrontal Cortex Structure and Function." *Nature Reviews: Neuroscience* 10:6 (2009), 410–22. doi:10.1038/nrn2648.

Atkinson, Rupert M. J. *Discovering Aikido: Principles for Practical Learning.* Ramsbury, UK: Crowood Press, 2005.

Bacon, John, and Greg Hilburn. "Gulf Coast Braces for 'Unprecedented' Challenge as Tropical Storms Marco and Laura Surge Toward Landfall." *USA Today.* August 23, 2020. https://tinyurl.com/tjvybnn9.

Baer, Drake. "Malcolm Gladwell Explains What Everyone Gets Wrong about His Famous '10,000 Hour Rule.'" *Business Insider India,* June 2, 2014. https://tinyurl.com/eyc3xmy3.

Baggett, Kelly. "Quickness and Absolute Speed vs Sports Speed and Explosiveness." Higher-Faster-Sports.com. n.d. https://tinyurl.com/ahr8yuwh.

Banta, Jim E., Gina Segovia-Siapco, Christine Betty Crocker, Danielle Montoya, and Noara Alhusseini. "Mental Health Status and Dietary Intake Among California Adults: A Population-Based Survey." *Food Sciences and Nutrition* 70:6 (September 2019), 1. doi:10.1080/09637486.2019.1570085.

Barber, Anita D., Priti Srinivasan, Suresh E. Joel, Brian S. Caffo, James J. Pekar, and Stewart H. Mostofsky. "Motor 'Dexterity'?: Evidence That Left Hemisphere Lateralization of Motor Circuit Connectivity Is Associated with Better Motor Performance in Children." *Cerebral Cortex* 22:1 (2012), 51–59. doi:10.1093/cercor/bhr062.

Barthelemy, Jane. "3 Breathing Techniques for Qigong Practitioners." Five Seasons. Blog. September 23, 2016. https://tinyurl.com/3vnm3tmw.

Basavaraddi, Ishwar V. "Yoga: Its Origin, History and Development." Government of India Ministry of External Affairs. April 23, 2015. https://tinyurl.com/yk38b5n4.

Beckmann, Jürgen, Peter Gröpel, and Felix Ehrlenspiel. "Preventing Motor Skill Failure through Hemisphere-Specific Priming: Cases from Choking under Pressure." *Journal of Experimental Psychology: General* 142:3 (2013), 679–91. doi:10.1037/a0029852.

Belonoha, Wayne. *The Wing Chun Compendium, Volumes 1 and 2.* Berkeley, CA: Blue Snake Books, 2005.

Benesch, Oleg. "Imperial Japan Saw Itself as a 'Warrior Nation'—and the Idea Lingers Today." Yahoo Style. December 22, 2017. https://tinyurl.com/5a29uv8m.

Benesch, Oleg. *Inventing the Way of the Samurai. Nationalism, Internationalism, and Bushidō in Modern Japan.* Oxford, UK: Oxford University Press, 2014.

Benesch, Oleg. "The Samurai Next Door: Chinese Examinations of the Japanese Martial Spirit." *Extrême-Orient Extrême-Occident* 38 (2014), 129–68. doi:10.4000/extremeorient.376.

Berencsi, Andrea, Masami Ishihara, and Kuniyasu Imanaka. "The Functional Role of Central and Peripheral Vision in the Control of Posture." *Human Movement Science* 24:5–6 (2005), 687–709. doi:10.1016/j.humov.2005.10.014.

Bernardo, Nathan. "How Human Evolution Allowed for the Development of Martial Arts." HowTheyPlay.com, August 6, 2020. https://tinyurl.com/5yzdmstk.

Bezodis, Neil Edward, Steffen Willwacher, and Aki Ilkka Tapio Salo. "The Biomechanics of the Track and Field Sprint Start: A Narrative Review." *Sports Medicine* 49 (2019), 1345–64. doi:10.1007/s40279-019-01138-1.

Bi, Kenneth. Director. *The Drummer.* 2007. Hong Kong: Emperor Motion Pictures.

Bishop, Mark. *Okinawan Karate: Teachers, Styles and Secret Techniques.* London: A&C Black, 1991.

Bisley, James W., and Michael E. Goldberg. "Neuronal Activity in the Lateral Intraparietal Area and Spatial Attention." *Science* 299:5603 (2003), 81–86. doi:10.1126/science.1077395.

Blennerhassett, Patrick. "It's Time to Admit Most Traditional Martial Arts Are Fake and Don't Actually Teach You How to Fight." Sensei Says. *South China Morning Post.* October 24, 2019. https://tinyurl.com/t75am9xc.

Blum, Susan Debra, and Lionel M. Jensen, editors. *China Off Center: Mapping the Margins of the Middle Kingdom.* Honolulu: University of Hawaii Press, 2002.

Bordoni, Bruno, and Emiliano Zanier. "Anatomic Connections of the Diaphragm: Influence of Respiration on the Body System." *Journal of Multidisciplinary Healthcare* 6 (2013), 281–91. doi:10.2147/JMDH.S45443.

Borghans, Lex, Huub Meijers, and Bas ter Weel. "The Role of Noncognitive Skills in Explaining Cognitive Test Scores." *Economic Inquiry* 46:1 (2008), 2–12. doi:10.1111/j.1465-7295.2007.00073.x, https://tinyurl.com/vwtp7wnx.

Bosler, Annie, and Don Greene. "How to Practice Effectively ... for Just About Anything." AnnieBosler.com. Video. https://youtu.be/f2O6mQkFiiw.

Braithwaite, Tom. "Southern Dragon Style Kung Fu." Dragon Martial Arts Association. n.d. https://tinyurl.com/ym748npe.

Bramble, Dennis M., and Daniel E. Lieberman. "Endurance Running and the Evolution of *Homo*." *Nature* 432:7015 (2004), 345–52. doi:10.1038/nature03052.

Broadbent, D. E. *Decision and Stress*. London: Academic Press, 1971.

Broadbent, David P., Joe Causer, A. Mark Williams, and Paul R. Ford. "Perceptual-Cognitive Skill Training and Its Transfer to Expert Performance in the Field: Future Research Directions." *European Journal of Sport Science* 15:4 (2015), 322–31. doi:10.1080/17461391.2014.957727.

Brown, Gregory Scott. "Food and the Pursuit of Happiness." *Psychology Today*. July 8, 2018. https://tinyurl.com/yxrtr9tm.

Brown, L. G. "The World Tournament." *Samurai*. Summer edition 1971. Los Angeles.

Buckingham, Gavin, Julie C. Main, and David P. Carey. "Asymmetries in Motor Attention During a Cued Bimanual Reaching Task: Left and Right Handers Compared." *Cortex* 47:4 (2011), 432–40. doi:10.1016/j.cortex.2009.11.003.

Buhusi, Catalin V., and Warren H. Meck. "What Makes Us Tick? Functional and Neural Mechanisms of Interval Timing." *Nature Reviews Neuroscience* 6:10 (2005). doi:10.1038/nrn1764.

Bumgardner, Todd. "Core Strength: Your Ultimate Guide to Core Training." BodyBuilding.com. July 27, 2018. https://tinyurl.com/478rf9n2.

Bumgardner, Wendy. "How to Reach the Anaerobic Zone During Exercise." Verywell Fit. February 27, 2020. https://tinyurl.com/5cj543nw.

Burdukiewicz, Anna, Jadwiga Pietraszewska, Justyna Andrzejewska, Krystyna Chromik, and Aleksandra Stachoń. "Asymmetry of Musculature and Hand Grip Strength in Bodybuilders and Martial Artists." *International Journal of Environmental Research and Public Health* 17:13 (2020), 4695. doi:10.3390/ijerph17134695.

Burnett, Paul. "The Red Scare." n.d. https://tinyurl.com/2h7m2h9t.

Campbell, Charlie. "Meet the Chinese MMA Fighter Taking on Grandmasters of Kung Fu." Time.com. November 8, 2018. https://tinyurl.com/3wxzp5u9.

Carlson, C., and M. Charlie. "Biological Baseball." San Francisco Exploratorium. 1999. https://tinyurl.com/4bb6s2px.

Carrier, David R. "The Advantage of Standing Up to Fight and the Evolution of Habitual Bipedalism in Hominins." *PLoS ONE* 6:5 (May 18, 2011), e19630. doi:10.1371/journal.pone.0019630.

Carrier, David R., and Michael H. Morgan. "Protective Buttressing of the Hominin Face." *Biological Reviews* 90:1 (2015), 330–46. doi:10.1111/brv.12112.

Castle, Paul C., Neil Maxwell, Alan Allchorn, Alexis Mauger, and Danny K. White. "Deception of Ambient and Body Core Temperature Improves Self-Paced Cycling in Hot, Humid Conditions." *European Journal of Applied Physiology* 12:1 (2012), 377–85. doi:10.1007/s00421-011-1988-y.

Çelik, Zeynep. "Kinaesthesia." in *Sensorium: Embodied Experience, Technology, and Contemporary Art*. Edited by Caroline A. Jones. Cambridge, MA: MIT Press, 2006.

Ceunen, Erik, Johan W. S. Vlaeyen, and Ilse Van Diest. "On the Origin of Interoception." *Frontiers in Psychology*. May 23, 2016. doi:10.3389/fpsyg.2016.00743.

Chabris, Christopher, and Daniel Simons. *The Invisible Gorilla: How Our Intuitions Deceive Us*. New York: Crown, 2011.

Chan, C. W., and A. Rudins. "Foot Biomechanics During Walking and Running." *Mayo Clinic Proceedings* 69:5 (1994), 448–61. doi:10.1016/s0025-6196(12)61642-5.

Chen, Xiaorong, Jiabao Cui, Ru Li, Richard Norton, Joel Park, Jian Kong, and Albert Yeung, "Dao Yin (a.k.a. Qigong): Origin, Development, Potential Mechanisms, and Clinical Applications." *Evidence-Based Complementary and Alternative Medicine* 2019:3705120. doi:10.1155/2019/3705120.

Choki, Motobu. *Okinawan Kenpo Karate Jutsu: Kumite.* Translated by Eric Shahan. Japan: Eric Michael Shahan, 2018.

Chow, David, and Richard C. Spangler. *Kung Fu: History, Philosophy and Technique.* Burbank, CA: Action Pursuit Group, 1980.

Chu, Samuel K., Prakash Jayabalan, W. Ben Kibler, and Joel Press. "The Kinetic Chain Revisited: New Concepts on Throwing Mechanics and Injury." *PM&R Journal of Injury, Function, and Rehabilitation* 8:3 (2016), 569–77. doi:10.1016/j.pmrj.2015.11.015.

Clark, Tim. "The Bushidō Code: The Eight Virtues of the Samurai." In *A Man's Life, Featured, Martial Arts, on Virtue.* Edited by Brett McKay and Kate McKay. September 14, 2008. ArtOfManliness.com. https://tinyurl.com/bpvmmx6u.

Clements, John. "A Short Introduction to Historical European Martial Arts." *Meibukan* 1 (2006), 2–4.

Coates, John. *The Hour Between Dog and Wolf: How Risk Taking Transforms Us, Body and Mind.* New York: Penguin, 2012.

Cohen, Itzik Itzhak. *Karate Uchina-Di: Okinawan Karate: An Exploration of Its Origins and Evolution.* Israel: Itzik Cohen, 2017.

Collet, C. "Strategic Aspects of Reaction Time in World-Class Sprinters." *Perceptual and Motor Skills* 88:1 (February 1999), 65–75. doi:10.2466/pms.1999.88.1.65.

Cooper, Vincent Paul. "Mizu no Kokoro—A Mind Like Water." Trans4mind. n.d. https://tinyurl.com/53men64u.

Cox, Catharine M. *Genetic Studies of Genius. II. The Early Mental Traits of Three Hundred Geniuses.* Stanford, CA: Stanford University Press, 1926.

Crossley, Kay M., Wan-Jing Zhang, Anthony G. Schache, Adam Bryant, and Sallie M. Cowan. "Performance on the Single-Leg Squat Task Indicates Hip Abductor Muscle Function." *American Journal of Sports Medicine* 39:4 (April 2011), 866–73. doi:10.1177/0363546510395456.

Csikszentmihalyi, Mihaly. *Flow: The Psychology of Optimal Experience.* New York: Harper Collins, 2008.

Daigo, Toshiro. *Kodokan Judo Throwing Techniques.* Tokyo: Kodansha, 2016.

Dalai Lama. "On Buddha Nature. "The Buddha." Blog. PBS. March 9, 2010. https://tinyurl.com/t97ha7f2.

Dalai Lama. *The Universe in a Single Atom: The Convergence of Science and Spirituality.* New York: Morgan Road, 2005.

Daniels, K., and E. Thornton. "Length of Training, Hostility, and the Martial Arts: A Comparison with Other Sporting Groups." *British Journal of Sports Medicine* 26:3 (1992), 118–20. doi:10.1136/bjsm.26.3.118.

Danzico, Matt. "Brains of Buddhist Monks Scanned in Meditation Study." BBC News. April 24, 2011. https://tinyurl.com/da76d4ab.

Davey, Christopher G., and Ben J. Harrison. "The Brain's Center of Gravity: How the Default Mode Network Helps Us to Understand the Self." *World Psychiatry* 17:3 (2018), 278–79. doi:10.1002/wps.20553.

David-Néel, Alexandra. *Magic and Mystery in Tibet.* New York: Claude Kendall, 1932. Translated from *Mystiques et Magiciens du Thibet,* Paris: Pion, 1929.

Daxue. "The Great Learning." Translated by James Legge. Chinese Classics and Translations. November 1, 2005. https://tinyurl.com/63484erb.

Dent, Jarryd. "Three Major Benefits of Core Training for Sports." CoachUp Nation. November 7, 2018. https://tinyurl.com/ywhzcjuk.

Devine, K. "The Deception: Speed vs Quickness (Hint: They Are Not the Same)." Blog. 2012.

Doidge, Norman. *The Brain That Changes Itself.* New York: Penguin, 2007.

Dolce, Corenna R. Pitching in Baseball. n.d. https://tinyurl.com/3sy2mdvm.

Draeger, Donn F. *Comprehensive Asian Fighting Arts.* New York: Kodansha International, 1981.

Duckworth, Angela L., Christopher Peterson, Michael D. Matthews, Dennis R. Kelly. "Grit: Perseverance and Passion for Long-Term Goals." *Journal of Personality and Social Psychology* 92:6 (2007), 1087–1101. doi:10.1037/0022-3514.92.6.1087.

Dumoulin, Heinrich. *Zen Buddhism: A History—Japan, Volume 2.* New York: Macmillan, 2005.

Durham, Sean P. "Why Your Gut Feeling Is More Powerful Than Logic." Medium.com. September 1, 2019. https://tinyurl.com/v2b7xe4x.

Dutta, Urvashi, and Anita Puri Singh. "Studying Spirituality in the Context of Grit and Resilience of College-Going Young Adults." *International Journal for Innovative Research in Multidisciplinary Field* 3:9 (2017), 50–55. https://tinyurl.com/2hdzka4p.

Dweck, Carol. *Mindset: The New Psychology of Success.* New York: Ballantine, 2016.

Eikenberry, Adam, Jim McAuliffe, Timothy N. Welsh, Carlos Zerpa, Moira McPherson, and Ian J. Newhouse. "Starting with the 'Right' Foot Minimizes Sprint Start Time." *Acta Psychologica* 127:2 (March 2008), 495–500. doi:10.1016/j.actpsy.2007.09.002.

Eiring, Kris. "Imagery, Visualization, and Performance." *Sport Psychology.* n.d. https://tinyurl.com/2m2jst5r.

Eisen, Marty. "Qi in Traditional Chinese Medicine." *Qi Encyclopedia.* February 16, 2016. https://tinyurl.com/axm9r2hz.

Emrich, Elizabeth. "How to Live Forever: Daoism in the Ming and Qing Dynasties." Johnson Museum of Art, Cornell University. October 14, 2006. https://tinyurl.com/6ztx2apd.

Enkamp, Jesse. "Mysticism in Karate (Pt.1)." Karate by Jesse. n.d. https://tinyurl.com/3rxe3fym.

Epstein, David. "Are Athletes Really Getting Faster, Better, Stronger?" TED Talks. Video. March 2014. https://tinyurl.com/2a652r49.

Epstein, David. *The Sports Gene: Inside the Science of Extraordinary Athletic Performance.* New York: Penguin, 2013.

Epstein, Seymour. "Integration of the Cognitive and the Psychodynamic Unconscious." *American Psychologist* 49:8 (1994), 709–24. doi:10.1037/0003-066X.49.8.709.

Erard, Guillaume. "Real Fighting Is Not the Primary Purpose of Budō." GuillaumeErard.com. April 13, 2018. https://tinyurl.com/2wz3ub5w.

Ericsson, K. Anders. "Deliberate Practice and Acquisition of Expert Performance: A General Overview." *Academic Emergency Medicine* 15:11 (2008), 988–94. doi:10.1111/j.1553-2712.2008.00227.x.

Ericsson, K. Anders, Ralf T. Krampe, and Clemens Tesch-Römer. "The Role of Deliberate Practice in the Acquisition of Expert Performance." *Psychological Review* 100:3 (1993), 363–406. doi:10.1037/0033-295X.100.3.363.

Ericsson, K. Anders, and Robert Pool. *Peak: Secrets from the New Science of Expertise.* New York: Eamon Doplan/Houghton Mifflin Harcourt, 2016.

Etnyre, Bruce R., and Hally B. W. Poindexter. "Characteristics of Motor Performance, Learning, Warm-Up Decrement, and Reminiscence During a Balance Task." *Perceptual and Motor Skills* 89:3 (1995), 1027–30. doi:10.2466/pms.1995.80.3.1027.

Evans, Jonathan St. B. T., and Keith E. Stanovich. "Dual-Process Theories of Higher Cognition: Advancing the Debate." *Perspectives on Psychological Science* 8:3 (2014), 223–41. doi:10.1177/1745691612460685.

Fan, Jin, Bruce D. McCandliss, John Fossella, Jonathan I. Flombaum, and Michael I. Posner. "The Activation of Attentional Networks." *Neuroimage* 26:2 (2005), 471–79. doi:10.1016/j.neuroimage.2005.02.004.

Feld, Michael S., Ronald E. McNair, and Stephen R. Wilk. "The Physics of Karate." *Scientific American* 240 (April 1979), 150–58. https://tinyurl.com/f3dzsek.

Feuerstein, Georg. *Yoga: The Technology of Ecstasy.* 1st ed. New York: TarcherPerigee, 1989.

Filllipi, Mark. "Vision and Posture." NaturalEyeCare.com. n.d. https://tinyurl.com/z74suk82.

Fitting, J.-W. "From Breathing to Respiration." *Respiration* 89:1 (January 2015), 82–87. doi:10.1159/000369474.

Fitts, P. M. "The Information Capacity of the Human Motor System in Controlling the Amplitude of Movement." *Journal of Experimental Psychology* 47:6 (1954), 381–91. doi:10.1037/h0055392.

Fitzgerald, Robert S., and Neil S. Cherniack. "Historical Perspectives on the Control of Breathing." *Comprehensive Physiology* 2:2 (April 2012), 915–32. doi:10.1002/cphy.c100007.

Forrester, Nicole W. "How Olympians Train Their Brains to Become Mentally Tough." TheConversation.com. February 21, 2018. https://tinyurl.com/y94uv66w.

French, Doug. "Trump's 'Gut Feeling.'" Mises Institute. January 12, 2019. https://tinyurl.com/js8yb4rb.

Fujino, Masahiro, Yoshiyuki Ueda, Hiroaki Mizuhara, Jun Saiki, and Michio Nomura. "Open Monitoring Meditation Reduces the Involvement of Brain Regions Related to Memory Function." *Scientific Reports* 8:9968 (2018). doi:10.1038/s41598-018-28274-4.

Funakoshi, Gichin. *Karate-dō kyōhan: The Master Text*. Tokyo: Kodansha, 2013.

Funakoshi, Gichin. *Karate-dō: My Way of Life*. New York: Kodansha International, 1975.

Funakoshi, Gichin. *To-Te Jitsu*. Hamilton, Canada: Masters, 1997.

Furness, John B. "Enteric Nervous System." Scholarpedia. 2007. doi:10.4249/scholarpedia.4064.

Gadye, Levi. "The Tools That Let Neuroscientists Study (and Even Repair) Brain Circuits." BrainFacts.org. February 13, 2018. https://tinyurl.com/yzsfj57h.

Gagné, Robert M. *The Conditions of Learning and Theory of Instruction*. 1965. Reprinted Belmont, CA: Wadsworth Publishing, 1985.

Garcia-Rill, Edgar, Yutaka Homma, and Robert D Skinner. "Arousal Mechanisms Related to Posture and Locomotion. 2. Ascending Modulation." *Progress in Brain Research* 143 (2004), 291–98. https://tinyurl.com/4dxrjhz7.

Garofalo, Michael P. "Yang Style Taijiquan. Quotations, Sayings, Wisdom, Poems, Aphorisms, Classics, Principles, Guides, Concepts, Terms, Miscellaneous." Cloud Hands. Blog. 2018. https://tinyurl.com/ysy22yz4.

Geirland, John. "Buddha on the Brain." *Wired*. January 2, 2006. https://tinyurl.com/tjvhkw.

Gershon, Michael D. *The Second Brain: A Groundbreaking New Understanding of Nervous Disorders of the Stomach and Intestine*. New York: HarperCollins, 1998.

Gershon, Tanya Siejhi. "Hip Abductor Muscles Exercises." LiveStrong.com. June 10, 2019. https://tinyurl.com/ay5ndfj9.

Gibson, James J. "A Theory of Direct Visual Perception." In *Vision and Mind: Selected Readings in the Philosophy of Perception*. Edited by Alva Noë and Evan Thompson. 77–91. Cambridge, MA: MIT Press, 2002.

Gier, Nicholas F. "The Virtues of Asian Humanism." *Journal of Oriental Studies* 12 (October 2002), 14–28.

Gilsinan, Kathy. "The Buddhist and the Neuroscientist: What Compassion Does to the Brain." Atlantic.com. July 4, 2015. https://tinyurl.com/p436an3h.

Giuliani, Hayden. "The Rotational Athlete and the Importance of the Glutes." Athletic Lab. November 9, 2017. https://tinyurl.com/3x7pfa2t.

Gladwell, Malcolm. *Blink: The Power of Thinking Without Thinking*. New York: Back Bay Books, 2005.

Gladwell, Malcolm. *Outliers: The Story of Success*, Boston: Little, Brown and Co., 2008.

Glantz, Jen. "I Took a Sword Fight Class: It's Not the Workout You Think It Is." NBCNews.com. September 22, 2019. https://tinyurl.com/wuyfyfvm.

Gleason, William. *Aikido and Words of Power: The Sacred Sounds of Kototama*. Rochester, VT: Destiny Books, 2009.

Gomstyn, Alice. "Food for Your Mood: How What You Eat Affects Your Mental Health." Aetna. n.d. https://tinyurl.com/2ub4kfj3.

Gracie, David J., Elspeth A. Guthrie, P. John Hamlin, and Alexander C Ford. "Bi-Directionality of Brain-Gut Interactions in Patients with Inflammatory Bowel Disease." *Gastroenterology* 154:6 (May 2018), 1635–46. doi:10.1053/j.gastro.2018.01.027.

Grant, S. "10 Amazing Examples of Mind Over Matter." Listverse. May 21, 2013. https://tinyurl.com /y9enj5sb.

Greene, Brian. *The Elegant Universe: Superstrings, Hidden Dimensions, and the Quest for the Ultimate Theory.* New York: W. W. Norton, 2003.

Griffiths, Andrew. "The History of Nijūshiho." The History of Fighting. Blog. February 22, 2013. https:// tinyurl.com/3tw8rf38.

Hamilton, Rick. "Are Hard Punches Based off of Triceps Muscles or by Practicing Your Punches Over and Over?" *Quora Digest.* 2018.

Han, Jia, Gordon Waddington, Roger Adams, Judith Anson, and Yu Liu. "Assessing Proprioception: A Critical Review of Methods." *Journal of Sport and Health Science* 5:1 (2016), 80–90. doi:10.1016 /j.jshs.2014.10.004.

Hanlon, Tim. "My Black Belt and Karate-dō." *Shōtōkan Karate Magazine.* June 2020.

Hansen, Chad, and Edward N. Zalta, editors. "Daoism." *The Stanford Encyclopedia of Philosophy.* 2017. https://plato.stanford.edu.

Hansen, James R. *First Man: The Life of Neil A. Armstrong.* New York: Simon and Schuster, 2018.

Harari, Yuval Noah. *Homo Deus: A Brief History of Tomorrow.* New York: Harper Perennial, 2018.

Harris, Laurence R., and Charles Mander. "Perceived Distance Depends on the Orientation of Both the Body and the Visual Environment." *Journal of Vision* 14:12 (2014), 1–8. doi:10.1167/14.12.17.

Harvey, Philip D. "Domains of Cognition and Their Assessment." *Dialogues in Clinical Neuroscience* 21:3 (2019), 227–37. doi:10.31887/DCNS.2019.21.3/pharvey.

Hassell, Randall G., and Dale F. Poertner. "Scientific Karate." *Samurai* 1 (1978), 6–9.

Healy, Laura C., Nikos Ntoumanis, and Joan L. Duda. "Goal Motives and Multiple-Goal Striving in Sport and Academia: A Person-Centered Investigation of Goal Motives and Inter-Goal Relations." *Journal of Science and Medicine in Sport* 19:12 (2016), 1010–14. doi:10.1016/j.jsams.2016.03.001.

Heaney, Scott. "What Is Ibuki and Nogare." The Martial Way. n.d. https://tinyurl.com/rtk6wfxr.

Helmreich, Robert L. "On Error Management: Lessons from Aviation." *BMJ* 320 (2000), 781–85. doi:10.1136/bmj.320.7237.781.

Henning, S. "Ignorance, Legend, and Taijiquan." *Journal of the Chen-Style Taijiquan Research Association of Hawaii* 2:3 (1994), 1–7.

Henshaw, Will. "The Future of Martial Arts: What Martial Arts Will Look Like in 100 Years." WOMA TV. December 3, 2015. https://tinyurl.com/y7erchmz.

Heshmat, Shahram. "Can You Be Addicted to Adrenaline?" *Psychology Today.* August 8, 2015. https:// tinyurl.com/cudfcju7.

Higaonna, Morio. *Traditional Karate-dō: Okinawa Gōjū-ryū, Vol. 1: The Fundamental Techniques.* Tokyo: Sugawara Martial Arts Institute, 1985.

Hill, Kadien S. "From Fact to Fiction: The Wuxia Experience and the Wushu Practice." Bachelor's thesis. Georgia Southern University. 2018. https://tinyurl.com/4vb8uc8p.

Ho, Minhhuy. "The Concept of Ki in Aikido, A Literature Survey." The Aikido FAQ. n.d. www.aikidofaq .com/philosophy.

Hodson-Tole, Emma F., and James M. Wakeling. "Motor Unit Recruitment for Dynamic Tasks: Current Understanding and Future Directions." *Journal of Comparative Physiology* 179:1 (January 2009), 57–66. doi:10.1007/s00360-008-0289-1.

Hunt, Kevin D. "The Evolution of Human Bipedality: Ecology and Functional Morphology." *Journal of Human Evolution* 26:3 (1994), 183–202. doi:10.1006/jhev.1994.1011.

Huxel Bliven, Kellie C., and Barton E. Anderson. "Core Stability Training for Injury Prevention." *Sports Health* 5:6 (2013), 514–22. doi:10.1177/1941738113481200.

IBISWorld. "Martial Arts Studios Industry in the US." January 22, 2021. https://tinyurl.com/vhy6v93m.

International Traditional Karate Federation. *Traditional Karate Competition Rules.* Los Angeles: ITKF, 2009.

Ishida, Mitsuo. *Hormone Hunters: The Discovery of Adrenaline.* Kyoto, Japan: Kyoto University Press, 2018.

ISSA. "Exercises to Improve Quickness and Why Clients Need Them." ISSA blog. n.d. https://tinyurl .com/jn7469n4.

ISSA. "Your Guide to the Kinetic Chain." ISSA blog. n.d. https://tinyurl.com/r9h9pfhj.

Jackson, Robin C., and Damian Farrow. "Implicit Perceptual Training: How, When, and Why?" *Human Movement Science* 24:3 (July 2005), 308–25. doi:10.1016/j.humov.2005.06.003.

Jahnke, Roger, Linda Larkey, Carol Rogers, Jennifer Etnier, and Fang Lin. "A Comprehensive Review of Health Benefits of Qigong and Tai Chi." *American Journal of Health Promotion* 24:6 (2010), e1–e25. doi:10.4278/ajhp.081013-LIT-248.

Japan Karate Association. "Supreme Master Funakoshi Gichin: The Father of Modern Karate." n.d. https://tinyurl.com/5a7ks9fh.

Jáuregui-Renaud, Kathrine. "Postural Balance and Peripheral Neuropathy." In *Peripheral Neuropathy: A New Insight into the Mechanism, Evaluation and Management of a Complex Disorder.* Edited by Nizar Souayah. Intechopen.com. March 27, 2013. doi:10.5772/55344.

Johansson, Rune E. A. *The Psychology of Nirvana.* London: Allen & Unwin, 1969.

Johnson, Jerry Alan. *The Secret Teachings of Chinese Energetic Medicine: Energetic Anatomy and Physiology.* Pacific Grove, CA: International Institute of Medical Qigong, 2014.

Johnson, Jerry Alan. *The Secret Teachings of Chinese Energetic Medicine, Volume 2: Energetic Alchemy, Dao Yin Therapy, Healing Qi Deviations, and Spirit Pathology.* Pacific Grove, CA: International Institute of Medical Qigong, 2014.

Johnstone, Ashleigh, and Paloma Marí-Beffa. "The Effects of Martial Arts Training on Attentional Networks in Typical Adults." *Frontiers in Psychology* 9:80 (2018), 1–9. doi:10.3389/fpsyg.2018.00080.

Jovanovic, Mario, Goran Sporis, Darija Omrcen, and Fredi Fiorentini. "Effects of Speed, Agility, Quickness Training Method on Power Performance in Elite Soccer Players." *Journal of Applied Sport Science Research* 25:5 (May 2011), 1285–92. doi:10.1519/JSC.0b013e3181d67c65.

Kahneman, Daniel. *Thinking, Fast and Slow.* New York: Farrar, Straus and Giroux, 2011.

Kaine, Kristina. "Is There a Difference Between the Spirit and the Soul?" HuffPost. January 11, 2016. https://tinyurl.com/25znc5hw.

Kanazawa, Hirokazu. *Shōtōkan Karate International Kata, Volume 1.* Tokyo: Ikeda Shoten, 1982.

Kanazawa, Hirokazu. *Shōtōkan Karate International Kata, Volume 2.* Tokyo: Ikeda Shoten, 1982.

Kane, Lawrence A., and Kris Wilder. *The Way of Kata: A Comprehensive Guide to Deciphering Martial Applications.* Wolfeboro, NH: YMAA, 2005.

Kantak, Shailesh S., James W. Stinear, Ethan R. Buch, and Leonardo G. Cohen. "Rewiring the Brain: Potential Role of the Premotor Cortex in Motor Control, Learning, and Recovery of Function Following Brain Injury." *Neurorehabilitation and Neural Repair* 26:3 (March–April 2016), 282–92. doi:10.1177/1545968311420845.

Kato, K. "The Way of the Samurai." In *Samurai.* Brooklyn, NY: Complete Sports Publications, 1974.

Kaufman, Marc. "Dalai Lama Talks to Scientists." *Washington Post.* November 13, 2005.

Keller, Doug. *Refining the Breath: The Yogic Practice of Pranayama* (South Riding, VA: Do Yoga Productions, 2003).

Kent, Michael. "Fitts and Posner's Stages of Learning." *The Oxford Dictionary of Sports Science & Medicine,* 3rd ed. Oxford, UK: Oxford University Press, 2007.

Keys, Marc. "Weight Training Guide for Developing Speed and Power." *Origin Fitness.* January 16, 2016. https://tinyurl.com/knz5dwf.

Kibler, W. Ben, Joel Press, and Aaron Sciascia. "The Role of Core Stability in Athletic Function." *Sports Medicine* 36:3 (2006), 189–98. doi:10.2165/00007256-200636030-00001.

Kim, Eun-Kyung. "The Effect of Gluteus Medius Strengthening on the Knee Joint Function Score and Pain in Meniscal Surgery Patients." *Journal of Physical Therapy Science* 28:10 (2016), 2751–53. doi:10.1589/jpts.28.2751.

Knight, Kathryn. "Humans Walk on Virtual Length Legs." *Journal of Experimental Biology* 219:23 (December 2016), 3671–72. doi:10.1242/jeb.153080.

Ko, Kwang Hyun. "Origins of Bipedalism." *Brazilian Archives of Biology and Technology* 58:6 (2015). doi:10.1590/S1516-89132015060399.

Koch, Christof. "Intuition May Reveal Where Expertise Resides in the Brain." *Scientific American.* May 1, 2015. https://tinyurl.com/2kc5ncsj.

Kodama, Midori, Takashi Ono, Fumio Yamashita, Hiroki Ebata, Meigen Liu, Shoko Kasuga, and Junichi Ushiba. "Structural Gray Matter Changes in the Hippocampus and the Primary Motor Cortex on An-Hour-to-One-Day Scale Can Predict Arm-Reaching Performance Improvement." *Frontiers in Human Neuroscience* 12:209 (2018). doi:10.3389/fnhum.2018.00209.

Koopman, Bart H. F. J. M. "Dynamics of Human Movement." *Technology and Health Care* 18:4–5 (2010), 371–85. doi:10.3233/THC-2010-0599.

Kovar, Elizabeth. "Beginner Ab and Core Exercises to Increase Stability and Mobility." American Council on Exercise. January 20, 2016. https://tinyurl.com/4hdwva6z.

Kozma, Elaine E., Nicole M. Webb, William E. H. Harcourt-Smith, David A. Raichlen, Kristiaan D'Août, Mary H. Brown, Emma M. Finestone, Stephen R. Ross, Peter Aerts, and Herman Pontzer. "Hip Extensor Mechanics and the Evolution of Walking and Climbing Capabilities in Humans, Apes, and Fossil Hominins." *Proceedings of the National Academy of Sciences USA* 115:16 (April 17, 2018), 4134–39. doi:10.1073/pnas.1715120115.

Kunreuther, Howard, and Robert Meyer. *The Ostrich Paradox: Why We Underprepare for Disasters.* Upper Saddle River, NJ: Wharton School Publishing, 2017.

Lake, James. "The Impact of Air Pollution on Mental Health." *Psychology Today.* January 8, 2020. https://tinyurl.com/h8b366u9.

Lam, Paul. "The Combined 42 Forms." Tai Chi for Health Institute. 2007. https://tinyurl.com/srunjtp7.

Lambert, A. "Zen Buddhism = Humanism?" *Houston Chronicle.* 2011.

Lashley, Karl S. "The Problem of Serial Order in Behavior." In *Cerebral Mechanisms in Behavior: The Hixon Symposium,* edited by Lloyd A. Jeffress, 112–46. New York: Wiley, 1951.

Laurino, M., D. Menicucci, F. Mastorci, P. Allegrini, A. Piarulli, E. P. Scilingo, R. Bedini, et al. "Mind-Body Relationships in Elite Apnea Divers During Breath Holding: A Study of Autonomic Responses to Acute Hypoxemia." *Frontiers in Neuroengineering* 5:4 (2012), 1–10. doi:10.3389/fneng.2012.00004.

Lebeau, Vicky. "The Strange Case of Dr. Jekyll and Mr. Hyde." *Britannica Online Encyclopedia.* 2019. https://tinyurl.com/3mxcm57d.

Lee, Bruce. *The Art of Expressing the Human Body.* Translated by John R. Little. Tokyo: Tuttle, 1998.

Lee, Hwal, and Lisa M. Miller. "Brain Training: Three Psychological Skills to Cope with Performance Stress and Anxiety." *Training & Conditioning.* May 12, 2017. https://tinyurl.com/snykz5rp.

Lee, Matthew. "Analyzing Wude: The Martial Ethics behind Wushu." Jiayoo Wushu. January 6, 2016. https://tinyurl.com/7t8uxayv.

Leonard, Barb, and Mary Jo Kreitzer. "What Is Life Purpose?" University of Minnesota. n.d. https://tinyurl.com/24ub4aut.

Lewis, Cara L., Eric Foch, Marc M. Luko, Kari L. Loverro, and Anne Khuu. "Differences in Lower Extremity and Trunk Kinematics between Single Leg Squat and Step Down Tasks." *PLoS ONE* 10:5 (2015), e0126258. doi:10.1371/journal.pone.0126258.

Lippelt, Dominique P., Bernhard Hommel, and Lorenza S. Colzato. "Focused Attention, Open Monitoring, and Loving Kindness Meditation: Effects on Attention, Conflict Monitoring, and Creativity—A Review." *Frontiers in Psychology* 5 (September 2014), 1083. doi:10.3389/fpsyg.2014.01083.

Littlejohn, Ronnie. "Daoist Philosophy." *Internet Encyclopedia of Philosophy.* n.d. https://iep.utm.edu/daoism.

Lonsdale, Chris, and Jimmy T M Tam. "On the Temporal and Behavioural Consistency of Pre-performance Routines: An Intra-individual Analysis of Elite Basketball Players' Free Throw Shooting Accuracy." *Journal of Sports Sciences* 26:3 (2008), 259–66. doi:10.1080/02640410701473962.

Lorge, Peter A. *Chinese Martial Arts from Antiquity to the Twenty-First Century*. Cambridge, UK: Cambridge University Press, 2012.

Lovejoy, C. Owen. "The Origin of Man." *Science* 211:4480 (1981), 341–50. doi:10.1126/science.211.4480.341.

Lutz, Antoine, Lawrence L. Greischar, Nancy B. Rawlings, Matthieu Ricard, and Richard J. Davidson. "Long-Term Meditators Self-Induce High-Amplitude Gamma Synchrony During Mental Practice." *Proceedings of the National Academy of Sciences USA* 101:46 (2004), 16369–73. doi:10.1073/pnas.0407401101.

Lystad, Reidar P., Kobi Gregory, and Juno Wilson. "The Epidemiology of Injuries in Mixed Martial Arts: A Systemic Review and Meta-Analysis." *Orthopaedic Journal of Sports Medicine* 2:1 (January 2014). doi:10.1177/2325967113518492.

Mack, J., S. Stojsih, D. Sherman, N. Dau, and C. Bir. "Amateur Boxer Biomechanics and Punch Force." 28th International Conference on Biomechanics in Sports, 2010, Marquette, MI. https://tinyurl.com/22sdyvu4.

Majumdar, Aditi S., and Robert A. Robergs. "The Science of Speed: Determinants of Performance in the 100 m Sprint." *International Journal of Sports Science & Coaching* 6:3 (2011), 479–93. doi:10.1260/1747-9541.6.3.479.

Mallinson, James, and Mark Singleton. *Roots of Yoga*. New York: Penguin Classics, 2016.

Malmivuo, Jaakko, and Robert Plonsey. *Bioelectromagnetism: Principles and Applications of Bioelectric and Biomagnetic Fields*. New York: Oxford University Press, 1995.

Maloney, Sean J. "The Relationship Between Asymmetry and Athletic Performance: A Critical Review." *Journal of Strength and Conditioning Research* 33:9 (2019), 2579–93. doi:10.1519/JSC.0000000000002608.

Mamadazimov, Abdughani. "Horse in North and Ship in South." SEnECA blog. 2018. https://tinyurl.com/6j2e6fj8.

Markman, Art. "Creativity Is Memory." *Psychology Today*. October 6, 2015. https://tinyurl.com/axka4a96.

Martinez, Marcus. "All You Need Is a Kettlebell for This Full-Body Workout Program." Yahoo Life. February 4, 2021. https://tinyurl.com/6453cnwe.

Mathur, Deepti, and Samir D. Mathur. "Stacking Waves: Bosons and Fermions." *Quantum Mechanics: A Mini Course*. Ohio State University. November 27, 2017. https://tinyurl.com/3x2pwcrb.

Matsumoto, Kiyoshi. "Japan's Hidden Moral Code." TalkAboutJapan.com. July 18, 2018. https://tinyurl.com/y9rd4nj6.

McCall, Pete. "10 Things to Know About Muscle Fibers." American Council on Exercise. May 7, 2015. https://tinyurl.com/tub57ytk.

McCarthy, Patrick. *The Bible of Karate: Bubishi*. North Clarendon, VT: Tuttle, 1995.

McGreevey, Sue. "Eight Weeks to a Better Brain." *Harvard Gazette*. January 21, 2011. https://tinyurl.com/39p9wwut.

McLeish, R. D., and J. Charnley. "Abduction Forces in the One-Legged Stance." *Journal of Biomechanics* 3:2 (March 1970), 191–94. doi:10.1016/0021-9290(70)90006-0.

Melnychuk, Michael Christopher, Paul M. Dockree, Redmond G. O'Connell, Peter R. Murphy, Joshua H. Balsters, and Ian H. Robertson. "Coupling of Respiration and Attention via the Locus Coeruleus: Effects of Meditation and Pranayama." *Psychophysiology* April 22, 2018. doi:10.1111/psyp.13091.

Mikheev, Maxim, Christine Mohr, Sergei Afanasiev, Theodor Landis, and Gregor Thut. "Motor Control and Cerebral Hemispheric Specialization in Highly Qualified Judo Wrestlers." *Neuropsychologia* 40:8 (2002), 1209–19. doi:10.1016/S0028-3932(01)00227-5.

Milazzo, Nicolas, Damian Farrow, and Jean F Fournier. "Effect of Implicit Perceptual-Motor Training on Decision-Making Skills and Underpinning Gaze Behavior in Combat Athletes." *Perceptual and Motor Skills* 123:1 (2016), 300–23. doi:10.1177/0031512516656816.

Miller, Ian. "The Gut-Brain Axis: Historical Reflections." *Microbial Ecology in Health and Disease* 29:2 (2018), 1542921. doi:10.1080/16512235.2018.1542921.

Miller, Michael. "The Great Practice Myth: Debunking the 10,000 Hour Rule and What It Actually Takes to Get to the Mountaintop." 6Seconds.org, n.d. https://tinyurl.com/kh3usrbs.

Miyamoto, Musashi. *A Book of Five Rings. A Guide to Strategy.* Translated by Victor Harris. Woodstock, NY: Overlook Press, 1974.

Molnar, Charles, and Jane Gair. "Muscle Contraction and Locomotion." *Concepts of Biology, 1st Canadian Edition.* BCcampus, May 14, 2015. https://opentextbc.ca/biology.

Moran, Aidan. "Thinking in Action: Some Insights from Cognitive Sport Psychology." *Thinking Skills and Creativity* 7:2 (2012). doi:10.1016/j.tsc.2012.03.005.

Moran, Kate. "How Chunking Helps Content Processing." Nielsen Norman Group. March 10, 2016. https://tinyurl.com/b65ukyny.

Morgan, Michael H., and David R. Carrier. "Protective Buttressing of the Human Fist and the Evolution of Hominin Hands." *Journal of Experimental Biology* 216:Part 2 (2013), 236–44. doi:10.1242/jeb.075713.

Mori, Shuji, Yoshio Ohtani, and Kuniyasu Imanaka. "Reaction Times and Anticipatory Skills of Karate Athletes." *Human Movement Science* 21:2 (July 2002), 213–30. doi:10.1016/s0167-9457(02)00103-3.

Moscatelli, Fiorenzo, Giovanni Messina, Anna Valenzano, Vincenzo Monda, Andrea Viggiano, Antonietta Messina, Annamaria Petito, Antonio Ivano Triggiani, Michela Anna Pia Ciliberti, Marcellino Monda, et al. "Functional Assessment of Corticospinal System Excitability in Karate Athletes." *PLoS ONE* 11:5 (May 24, 2016), e015998. doi:10.1371/journal.pone.0155998.

Muehlhan, Markus, Michael Marxen, Julia Landsiedel, Hagen Malberg, and Sebastian Zaunseder. "The Effect of Body Posture on Cognitive Performance: A Question of Sleep Quality." *Frontiers in Human Neuroscience* 8:171 (2014). doi:10.3389/fnhum.2014.00171.

Myers, Jim. "Ruach: Spirit or Wind or ?" Biblical Heritage Center. n.d. https://tinyurl.com/2tfk5adv.

Nair, Malini. "Beyond Hinduism, Yoga Also Has Roots in Buddhist, Jain, and Sufi Traditions." Quartz India. January 29, 2017. https://tinyurl.com/trpvrwfj.

Nakayama, Masatoshi. *Best Karate, Volume 2: Fundamentals.* New York: Kodansha International, 1978.

Nakayama, Masatoshi. *Dynamic Karate.* Palo Alto, CA: Kodansha, 1966.

NASA. *Apollo 11 Mission Report.* November 1969. Houston: NASA, 1969. https://tinyurl.com/wrtb3uja.

National Academy of Sports Medicine. *Essentials of Sports Performance Training,* 2nd edition. Burlington, MA: Jones and Bartlett, 2018.

Neumann, Donald A. "Kinesiology of the Hip: A Focus on Muscular Actions." *Journal of Orthopedic and Sports Physical Therapy* 40:2 (2010), 82–94. doi:10.2519/jospt.2010.3025.

Ng, Nick. "Do You Use Your Hip Abductors and Adductors in Running?" The Nest. n.d. https://tinyurl.com/mxmau2sk.

Nicol, C. W. *Moving Zen: Karate as a Way to Gentleness.* London: Paul H. Crompton, 1989.

Niedziocha, Laura. "Does Exercise Cause an Adrenaline Rush?" Healthfully.com. November 28, 2018. https://tinyurl.com/tmz53vmv.

Nietzsche, Friedrich. *The Birth of Tragedy: Out of the Spirit of Music.* Edited by Michael Tanner. Translated by Shaun Whiteside. London: Penguin Random House, 1994.

Nishiyama, Hidetaka. *The Traditional Karate Coach's Manual.* Los Angeles: International Traditional Karate Federation, 1989.

Nishiyama, Hidetaka, and Richard Carl Brown. *Karate: The Art of "Empty Hand" Fighting.* Rutland, VT: Tuttle, 1960.

Nitobe, Inazō. *Bushidō: The Soul of Japan.* 1899. Reprinted Tokyo: Kodansha International, 2012.

Novotny, Sarah, and Len Kravitz. "The Science of Breathing." University of New Mexico. n.d. https://tinyurl.com/2nrjau2x.

Ohio State Wexner Medical Center. "Dormant Butt Syndrome May Be to Blame for Knee, Hip, and Back Pain." Ohio State University. May 23, 2016. https://tinyurl.com/38fmx7dt.

Okumura, Shohaku. *The Mountains and Waters Sutra: A Practitioner's Guide to Dogen's "Sansuikyo."* Somerville, MA: Wisdom Publications, 2018.

Oliver, Gretchen D., Priscilla M. Dwelly, Nicholas D. Sarantis, Rachael A. Helmer, and Jeffery A. Bonacci. "Muscle Activation of Different Core Exercises." *National Strength and Conditioning Association Journal* 24:11 (2010), 3069–74. doi:10.1519/JSC.0b013e3181d321da.

Pal, G. K., S. Velkumary, and Madanmohan. "Effect of Short-Term Practice of Breathing Exercises on Autonomic Functions in Normal Human Volunteers." *Indian Journal of Medical Research* 120:2 (August 2004), 115–21.

Pandit, Smita. "Diaphragmatic Breathing Explained." VisualStories.com. n.d. https://tinyurl.com/yyjyxbvx.

Pangambam, S. "5 Hindrances to Self-Mastery: Shi Heng YI (Transcript)." *The Singju Post.* April 24, 2020. https://tinyurl.com/22nwt7kv.

Papale, Andrew E., and Bryan M. Hooks. "Circuit Changes in Motor Cortex During Motor Skill Learning." *Neuroscience* 368 (2018), 283–97. doi:10.1016/j.neuroscience.2017.09.010.

Perlis, Margaret M. "5 Characteristics of Grit—How Many Do You Have?" *Forbes.* October 29, 2013. https://tinyurl.com/cktbfaea.

Petersen, Steven E., and Michael I. Posner. "The Attention System of the Human Brain." *Annual Review of Neuroscience* 13 (1990), 25–42. doi:10.1146/annurev.ne.13.030190.000325.

Petersen, Steven E., and Michael I. Posner. "The Attention System of the Human Brain: 20 Years After." *Annual Review of Neuroscience* 35 (2012), 73–89. doi:10.1146/annurev-neuro-062111-150525.

Phillips, Matt. "Introduction to Running Biomechanics." Runners Connect. n.d. https://tinyurl.com/dc8zwc89.

Pigliucci, Massimo, and Gregory Lopez. *A Handbook for New Stoics: How to Thrive in a World Out of Your Control—52 Week-by-Week Lessons.* New York: Experiment, 2019.

Popp, Fritz-Albert, Ulrich Warnke, Herbert L. Konig, and Walter Peschka, editors. *Electromagnetic Bio-Information.* Munich: Urban & Schwarzenberg, 1989.

Posner, Michael I., and Stanislas Dehaene. "Attentional Networks." *Trends in Neurosciences* 17:2 (1994) 75–79. doi:10.1016/0166-2236(94)90078-7.

Prescott, Gregg. "What Is My Role or Purpose in This Spiritual Awakening?" In5D. February 12, 2018. https://tinyurl.com/57fxk7k3.

Prsa, Mario. Karin Morandell, Angie Geraldine Cuenu, and Daniel Huber. "Feature-Selective Encoding of Substrate Vibrations in the Forelimb Somatosensory Cortex." *Nature Review Neuroscience* 567:7748 (2019), 384. doi:10.1038/s41586-019-1015-8.

Purves, Dale, George J. Augustine, David Fitzpatrick, Lawrence C. Katz, Anthony-Samuel LaMantia, James O. McNamara, and S. Mark Williams, editors. *Neuroscience,* 2nd. ed. Sunderland, MA: Sinauer Associates, 2001.

Quast, Andreas. "The Weapons Ban Theories." Ryūkyū Bugei blog. April 27, 2017. https://ryukyu-bugei.com/?p=7281.

Raichle, Marcus E., Ann Mary MacLeod, Abraham Z. Snyder, William J. Powers, Debra A. Gusnard, and Gordon L. Shulman. "A Default Mode of Brain Function." *Proceedings of the National Academy of Sciences USA* 98:2 (January 16, 2001), 676–82. doi:10.1073/pnas.98.2.676.

Randall, Mike. "Class of 2021 West Point Cadets Celebrate 12-mile 'March Back.'" [Middletown, NY] *Times Herald-Record.* August 14, 2017. https://tinyurl.com/7knzf65k.

Ravenscraft, Eric. "Use the Combat Breathing Technique to Help Control Nervous Shaking." *Lifehacker.* September 16, 2015. https://tinyurl.com/y5rmhj8c.

Reimer, Allison. "My Magical Northern Lights, Have You Seen Them?" September 21, 2017. https://tinyurl.com/vhyj9pwx.

Reinold, Mike. "The Problem with the Kinetic Chain Concept." MikeReinold.com. November 29, 2011. https://tinyurl.com/3usj97zc.

Ricard, Matthieu, Antoine Lutz, and Richard J. Davidson. "Mind of the Meditator." *Scientific American* 311:5 (November 2014), 38–45. doi:10.1038/scientificamerican1114-38.

Rielly, Robin L. *Karate Training: The Samurai Legacy and Modern Practice.* Tokyo: Tuttle, 1985.

Rist, Curtis. "The Physics of ... Karate." *Discover.* January 19, 2000. https://tinyurl.com/crzhk5un.

Ritterbusch, Erin. "The Importance of the Lumbopelvic Hip Complex." Athletic Lab. April 12, 2017. https://tinyurl.com/y6654h2b.

Roberts, R. Edward, Elaine J. Anderson, and Masud Husain. "White Matter Microstructure and Cognitive Function." *Neuroscientist* 19:1 (2013), 8–15. doi:10.1177/1073858411421218.

Roberts, Thomas J., and Nicolai Konow. "How Tendons Buffer Energy Dissipation by Muscles." *Exercise and Sport Sciences Reviews* 41:4 (2013), 186–93. doi:10.1097/JES.0b013e3182a4e6d5.

Robinson, H. G. R. "United States Military and Martial Arts: A History of the Strategic Air Command (SAC) and Its Combative Measures Program." *International Armed Services Judo and Jujitsu Academy.* 2018. https://asjja.com.

Robinson, S. "The Seven Biomechanical Principles." Exercise Science Portfolio. 2019.

Ross, Christina L. "Energy Medicine: Current Status and Future Perspectives." *Global Advances in Health and Medicine* 8 (2019), 1–10. doi:10.1177/2164956119831221.

Rowell, Richard E. "One-Inch Distance: Life and Death in the Thickness of Paper." *Budo Theory: Exploring Martial Arts Principles.* Nanton, Canada: Richard E. Rowell, 2011.

Rubik, Beverly, David Muehsam, Richard Hammerschlag, and Shamini Jain. "Biofield Science and Healing: History, Terminology, and Concepts." *Global Advances in Health and Medicine* 4 (2015), 8–14. doi:10.7453/gahmj.2015.038.suppl.

Russell, Daniel A. "Bat Weight, Swing Speed and Ball Velocity." *Physics and Acoustics of Baseball and Softball Bats.* March 27, 2008. https://tinyurl.com/y6en38bb.

Sanders, Lisa. "Mysterious Psychosis." *New York Times Magazine.* March 10, 2009. https://tinyurl.com/55wdfw4t.

Santona, Ottica. "Posture and Vision." n.d. Santona.it. https://tinyurl.com/wvtz8287.

Schafer, Matthew. "Shaolin vs. Wudang vs. History." Schafer's Self-Defense Corner. Blog. December 24, 2008. https://tinyurl.com/mzbnf77j.

Schildmeier, Leigh Ann. "What Is Kata?" Park Avenue Solutions. n.d. https://tinyurl.com/thezrtx7.

Schlimm, Karl. "Channeling Adrenaline: Upset Prevention and Recovery Training." *Skies.* 2019. https://tinyurl.com/2c5v7cdd.

Scholkmann, Felix, Daniel Fels, and Michal Cifra. "Non-Chemical and Non-Contact Cell-to-Cell Communication: A Short Review." *American Journal of Translational Research* 5:6 (2013), 586–93.

Schonbrun, Zach. *The Performance Cortex: How Neuroscience is Redefining Athletic Genius.* New York: Penguin, 2018.

Schwartz, Mel. *The Possibility Principle: How Quantum Physics Can Improve the Way You Think, Live, and Love.* Boulder, CO: Sounds True, 2017.

Sekine, M. "Takeshi Oishi, All Japan Champion." *Samurai* 1 (1971), 37.

Sellers, William Irvin, Todd C. Pataky, Paolo Caravaggi, and Robin Huw Crompton. "Evolutionary Robotic Approaches in Primate Gait Analysis." *International Journal of Primatology* 31 (2010), 321–38. doi:10.1007/s10764-010-9396-4.

Selye, Hans. "Stress Without Distress." In *Psychopathology of Human Adaptation.* Edited by George Serban. 137–46. Philadelphia: J. B. Lippincott, 1974.

Seroyer, Shane T., Shane J. Nho, Bernard R. Bach, Charles A. Bush-Joseph, Gregory P. Nicholson, and Anthony A. Romeo. "The Kinetic Chain in Overhand Pitching: Its Potential Role for Performance Enhancement and Injury Prevention." *Sports Health* 2:2 (2010), 135–46. doi:10.1177/1941738110362656.

Shan, Chen Qing. *A Scholar's Path: An Anthology of Classical Chinese Poems and Prose of Chen Qing Shan.* Translated by Peter Chen and Michael Chan. Hackensack, NJ: World Scientific, 2010. doi:10.1142/7840.

Sharrock, Chris, Jarrod Cropper, Joel Mostad, Matt Johnson, and Terry Malone. "A Pilot Study of Core Stability and Athletic Performance: Is There a Relationship?" *International Journal of Sports Physical Therapy* 6:2 (2011), 63–74.

Shiah, Yung-Jong. "From Self to Nonself: The Nonself Theory." *Frontiers in Psychology* 4:7 (February 2016), 124. doi:10.3389/fpsyg.2016.00124.

Shiffrin, Richard M., and Walter Schneider. "Controlled and Automatic Human Information Processing: II. Perceptual Learning, Automatic Attending, and a General Theory." *Psychological Review* 84 (1977), 127–90. doi:10.1037/0033-295X.84.2.127.

Shirtcliff, Elizabeth A., Michael J. Vitacco, Alexander R. Graf, Andrew J. Gostisha, Jenna L. Merz, and Carolyn Zahn-Waxler. "Neurobiology of Empathy and Callousness: Implications for the Development of Antisocial Behavior." *Behavioral Sciences and the Law* 27:2 (March–April 2009), 137–71. doi:10.1002/bsl.862.

Shreeve, James. "New Skeleton Gives Path from Trees to Ground an Odd Turn." *Science* 272:5262 (1996), 654. doi:10.1126/science.272.5262.654.

Simon, Leslie V., Richard A. Lopez, and Kevin C. King. "Blunt Force Trauma." *StatPearls.* January 2021. https://tinyurl.com/3jkh6csp.

Simons, Daniel. "Selective Attention Test." YouTube. Video. March 10, 2010. https://youtu.be/vJG698U2Mvo.

Sluyter, Tomas. "Waza Explained." Renshinjuku Kendo. October 4, 2013. https://tinyurl.com/2r2we9pr.

Smith, Detric. "Heavier Isn't Always Better: How the Force-Velocity Curve Impacts Your Training." Stack. July 8, 2019. https://tinyurl.com/4ecj97xh.

Smith, Philip. "Moving Zen: Seido Karate in Pictures and Words." *Tricycle.* Spring 2000. https://tinyurl.com/386jbw7a.

Spagna, Alfredo, Melissa-Ann Mackie, and Jin Fan. "Supramodal Executive Control of Attention." *Frontiers in Psychology.* February 24, 2015. doi:10.3389/fpsyg.2015.00065.

Spivey, Michael. *The Continuity of Mind.* New York: Oxford University Press, 2007.

Srado, Antonio. "The Biomechanics of a Boxer's Cross Punch." Antonio Srado. Blog. October 21, 2013. https://tinyurl.com/425tf6a2.

Srinivasan, T. M. "Biophotons as Subtle Energy Carriers." *International Journal of Yoga* 10:2 (May–August 2017), 57–58. doi:10.4103/ijoy.IJOY_18_17.

St. Laurent, Stephen. "Torajiro Mori: The Essence of Kata." *Samurai* 1:3 (March 1975).

Staal, Mark A. "Stress, Cognition, and Human Performance: A Literature Review and Conceptual Framework." NASA Ames Research Center. August 2004. https://tinyurl.com/nufrpvkx.

Stasulli, Dominique. "Reaction Time in Track and Field Athletes." Freelap. January 13, 2015. https://tinyurl.com/phcy6bx3.

Statista. "Number of Participants in Martial Arts in the United States from 2006–2017 (in Millions)." Statista.com. 2019.

Stieger, Allison. "Living with Apollo and Dionysus." Mythic Stories. Blog. May 2, 2013. https://tinyurl.com/adbv72ub.

Sugiyama, Shojiro. *25 Shoto-kan Kata.* Chicago: Shojiro Sugiyama, 1984.

Suzuki, Daisetz Teitaro. *Manual of Zen Buddhism.* 1935. Reprinted Baghdad, Iraq: M. G. Sheet, 2005. https://tinyurl.com/z32hhf2c.

Suzuki, Daisetz Teitaro. *The Zen Doctrine of No Mind.* York Beach, ME: Weiser Books, 1991.

Swanson, Aaron. "Basic Biomechanics: Moment Arm and Torque." AaronSwansonPT.com. July 3, 2011. https://tinyurl.com/26jt9772.

Swinton, Paul A., Ray Lloyd, Justin W. L. Keogh, Ioannis Agouris, and Arthur D. Stewart. "A Bio-mechanical Comparison of the Traditional Squat, Powerlifting Squat, and Box Squat." *Journal of Strength and Conditioning Research* 26:7 (July 2012), 1805–16. doi:10.1519/JSC.0b013e3182577067.

Szczepanski, Kallie. "About Seppuku (or Harakiri)." ThoughtCo. March 1, 2019. https://tinyurl.com/chv5dwh7.

Tan, Sor-Hoon. "The Concept of Yi (义) in the Mencius and the Problems of Distributive Justice." *Australasian Journal of Philosophy* 92:3 (2014),489–505. doi:10.1080/00048402.2014.882961.

Tang, Yi-Yuan, and Michael I. Posner. "Attention Training and Attention State Training." *Trends in Cognitive Sciences* 13:5 (May 2009), 222–27. doi:10.1016/j.tics.2009.01.009.

Tang, Yi-Yuan, Yinghua Ma, Junhong Wang, Yaxin Fan, Shigang Feng, Qilin Lu, Qingbao Yu, Danni Sui, Mary K. Rothbart, Ming Fan, and Michael I. Posner. "Short-Term Meditation Training Improves Attention and Self-Regulation." *Proceedings of the National Academy of Sciences USA* 104 (2007), 17152–56. doi:10.1073/pnas.0707678104.

Taylor, Jim. "What Mental Training for Sports Is Really All About." *Psychology Today.* November 12, 2018. https://tinyurl.com/pfpw4hxr.

Theiss, F. B. "Karate: A Scientific View." *Samurai,* summer 1971, 42–47.

Tims, Catherine. "Body, Soul, and Spirit." Autumn Damask. September 7, 2019. https://tinyurl.com/cm4rn76s.

Todor, John I. "Sequential Motor Ability of Left-handed Inverted and Non-inverted Writers." *Acta Psychologica* 44 (1980), 165–73. doi:10.1016/0001-6918(80)90065-7.

Tokeshi, Jinichi. *Kendo: Elements, Rules, and Philosophy.* Honolulu: University of Hawaii Press, 2003.

Tomboc, Kyjean. "What Happens When You Neglect Your Lower Body Composition." InBodyUSA.com. November 10, 2016. https://tinyurl.com/zr2szped.

Trautwein, Fynn-Mathis, Philipp Kanske, Anne Böckler, and Tania Singer. "Differential Benefits of Mental Training Types for Attention, Compassion, and Theory of Mind." *Cognition* 194 (January 2020), 104039. doi:10.1016/j.cognition.2019.104039.

U.S. Centers for Disease Control and Prevention. "1918 Pandemic (H1N1 Virus)." CDC National Center for Immunization and Respiratory Diseases. March 20, 2019. https://tinyurl.com/4spdhwvn.

Uddén, Julia, Vasiliki Folia, and Karl Magnus Petersson. "The Neuropharmacology of Implicit Learning." *Current Neuropharmacology* 8:4 (2010), 367–81. doi:10.2174/157015910793358178.

University of Helsinki. "What Is Time?" *Science Daily.* April 15, 2005. https://tinyurl.com/dvxh6kz8.

Vallortigara, Giorgio, and Lesley J. Rogers. "Survival with an Asymmetrical Brain: Advantages and Disadvantages of Cerebral Lateralization." *Behavioral and Brain Sciences* 28:4 (2005), 575–89. doi:10.1017/S0140525X05000105.

van der Stockt, Tarina, Evan Thomas, Kim Jackson, Olajumoke Ogunleye, and Rewan Elsayed Elkanafany. "Kinetic Chain." Physiopedia.com. 2012. https://tinyurl.com/aussdts.

Vera, Amir, and Jamiel Lynch. "Deadly California Wildfires Scorch More Than 1 Million Acres with No End in Sight." CNN. August 23, 2020. https://tinyurl.com/3pcv9d74.

Vogt, B. A., and O. Devinsky. "Topography and Relationship of Mind and Brain." *Progress in Brain Research* 122 (2000), 11–22. doi:10.1016/s0079-6123(08)62127-5.

Vonk, Kathleen. "Police Performance Under Stress." *Law and Order* 56:10 (October 2008), 86–90. https://tinyurl.com/t77wrm26.

Waaijman, Kees. *Spirituality: Forms, Foundations, Methods.* Leuven, Belgium: Peeters, 2003.

Wager, Tor D., and Edward E. Smith. "Neuroimaging Studies of Working Memory: A Meta-Analysis." *Cognitive, Affective and Behavioral Neuroscience* 3:4 (December 2003), 256–74. doi:10.3758/cabn.3.4.255.

Wahbeh, H., A. Sagher, W. Back, P. Pundhir, and F. Travis. "A Systematic Review of Transcendent States Across Medication and Contemplative Traditions." *Explore (NY)* 14:1 (2014), 19–35.

Walker, Jearl D. "Karate Strikes." *American Journal of Physics* 43:10 (October 1, 1975), 845. doi:10.1119 /1.9966.

Walton, Alice G. "7 Ways Meditation Can Actually Change the Brain." *Forbes.* February 9, 2015. https:// tinyurl.com/5kzzz35n.

Warner, Jennifer. "Catcher's Mitts Strike Out at Hand Protection." WebMD. July 1, 2005. https://tinyurl .com/4289c46d.

Watson, Christopher, and Roy Suenaka. *Complete Aikido: Aikido Kyohan: The Definitive Guide to the Way of Harmony.* North Clarendon, VT: Tuttle, 1997.

Wayman, Erin. "Becoming Human: The Evolution of Walking Upright." *Smithsonian Magazine,* August 6, 2012. https://tinyurl.com/4p4t3svd.

Weaver, Janelle. "Motor Learning Unfolds Over Different Timescales in Distinct Neural Systems." *PLoS Biology* 13:12 (December 8, 2015), e1002313. doi:10.1371/journal.pbio.1002313.

Webber, James T., and David A. Raichlen. "The Role of Plantigrady and Heel-Strike in the Mechanics and Energetics of Human Walking with Implications for the Evolution of the Human Foot." *Journal of Experimental Biology* 219:23 (December 2016), 3729–37. doi:10.1242/jeb.138610.

Wenger, M. A., and B. K. Bagchi. "Studies of Autonomic Functions in Practitioners of Yoga in India." *Behavioral Science* (October 1961), 312–23. doi:10.1002/bs.3830060407.

Wien, Aikinomichi. "Aikido: Yoko Okamoto Sensei Berlin 2017." YouTube. Video. February 5, 2018. https://youtu.be/KrpFhe6jFyI.

Wilder, Kris. "Sanchin Kata—Ancient Wisdom." YMAA. March 8, 2010. https://tinyurl.com/rjcj36eb.

Wilkerson, Gary B., Jessica L. Giles, Dustin K. Seibel. "Prediction of Core and Lower Extremity Strains and Sprains in Collegiate Football Players: A Preliminary Study." *Journal of Athletic Training* 47:3 (2012), 264–72. doi:10.4085/1062-6050-47.3.17.

Williams, A. Mark, and David Elliott. "Anxiety, Expertise and Visual Search Strategy in Karate." *Journal of Sport and Exercise Psychology* 21 (1999), 362–75. doi:10.1123/jsep.21.4.362.

Wilson, Wendell E. "Mushin and Zanshin." Essays on the Martial Arts. 2010. https://tinyurl.com /yt52teav.

Wilson, William Scott. *Ideals of the Samurai: Writings of Japanese Warriors.* 1982. Reprinted Valencia CA: Black Belt Books, 2014.

Winerman, Lea. "What Sets High Achievers Apart?" *Monitor on Psychology* 44:11 (2013), 28. https:// tinyurl.com/f9w33dae.

Winter, David A. *Biomechanics and Motor Control of Human Movement.* 4th ed. Hoboken, NJ: Wiley & Sons, 2009.

Wise, T. "I Thought It'd Be Easy to Fix My Bad Posture—I Was Wrong." Yahoo Life. August 7, 2019. https://tinyurl.com/yhjd9hjf.

Wong Kiew Kit. *The Art of Shaolin Kung Fu.* Rockport, MD: Element Books, 1996.

Wong Kiew Kit. *The Complete Book of Tai Chi Chuan: A Comprehensive Guide to the Principles and Practice.* Revised edition. Boston: Tuttle, 2016.

Wong, Nathan Colin. "The 10,000 Hour Rule." *Journal of the Canadian Urological Association* 9:9–10 (Sept.–Oct. 2015), 299. doi:10.5489/cuaj.3267.

Yamaguchi, Gogen. *Gōjū-ryū: Karate Dō Kyohan.* Hamilton, Canada: Masters Publication, 2006.

Yamamoto Tsunetomo. *Hagakure: The Book of the Samurai.* Translated by William Scott Wilson. Boston: Shambhala, 2012.

Yang, Jwing-Ming. *The Essence of Shaolin White Crane: Martial Power and Qigong.* Jamaica Plain, MA: YMAA Publication Center, 1996.

Yang, Jwing-Ming. *The Root of Chinese Chi Kung: The Secrets of Chi Kung Training.* Jamaica Plain, MA: YMAA Publication Center, 1989.

Yang, Jwing-Ming. "What Is Qi and What Is Qigong?" YMAA. March 14, 2016. https://tinyurl.com /hyvy8crz.

Yarrow, Kielan, Peter Brown, and John W Krakauer. "Inside the Brain of an Elite Athlete: The Neural Processes That Support High Achievement in Sports." *Nature Review Neuroscience* 10 (2009), 585–96. doi:10.1038/nrn2672.

Yee, Jackson. "Improving Your Self-Discipline with Your Training." EliteFTS.com. August 29, 2011. https://tinyurl.com/8aw34vn8.

Young, Richard W. "Evolution of the Human Hand: The Role of Throwing and Clubbing." *Journal of Anatomy* 202:1 (2003), 165–74. doi:10.1046/j.1469-7580.2003.00144.x.

Zakrzewski, Vicki. "How Humility Will Make You the Greatest Person Ever." *Greater Good Magazine.* January 12, 2016. https://tinyurl.com/2ktkhr28.

Zebrowitz, Leslie A. "First Impressions from Faces." *Current Directions in Psychological Science* 26:3 (2018), 237–42. doi:10.1177/0963721416683996.

Zhou, Xuan-Yun. "Daoist Breathing Techniques." YMAA. May 20, 2009. https://tinyurl.com/2s4xfj5z.

IMAGE CREDITS

Figure 10.5: © Artwork by Caitlyn Tong.

Figure 10.6: © Photography by Madeleine Tong.

Figure 10.7: © Photography by Madeleine Tong.

Figure 10.8: Public domain, historic image, circa 1924.

Figure 11.1: NASA line drawing, Wikimedia Commons GNU FDL 1.2 or later, modified.

Figure 11.2: © Shutterstock.com 1068897173.

Figure 11.3: © Shutterstock.com 1460617385, modified.

Figure 11.4: © Shutterstock.com 1622243914, modified.

Figure 11.5: © Photography by Madeleine Tong, modified.

Figure 11.6: © Left: Author's computer artwork; right, Shutterstock.com 1724092468, modified.

Figure 12.1: © Artwork by Caitlyn Tong.

Figure 12.2: © Author's computer artwork.

Figure 12.3: © Artwork by Caitlyn Tong.

Figure 13.1: © Author's computer artwork, modified from © Shutterstock.com 440006554.

Figure 13.2: © Author's computer artwork.

Figure 14.1: © Artwork by Caitlyn Tong.

Figure 14.2: ©Photography by Madeleine Tong.

Figure 14.3: © left: JKA Dallas archive, photography by Madeleine Tong; right: © Wikimedia Commons GNU FDL 1.2 or later, Eri seoi otoshi by Ari Kyllönen.

Figure 14.4: © Artwork by Caitlyn Tong.

Figure 15.1: © AAKF.org archive.

Figure 16.1: © JKA Dallas archive.

Figure 17.1: © Author's computer artwork.

Figure 17.2: © Shutterstock.com 723403525.

Figure 18.1: © Photography by Caitlyn Tong (Kankū Dai, Bassai Dai, Empi, Nijūshiho, Gojūshiho Dai) and Madeleine Tong (Sōchin, Jion, Chinte, Unsū, Gankaku).

Figure B.1: © Author's computer artwork.

Front Cover Design: © Photography and Illustration by Caitlyn Tong.

Author Portrait: © Photo by Beau Bumpus Photography, Dallas (https://beaubumpas.com).

INDEX

ABOUT THE AUTHOR

ALEX W. TONG, PhD, grew up in Hong Kong, immersed in its heritage of Chinese and European cultures. He started formal karate training while an undergraduate at the University of Oregon. A career opportunity brought Tong to Texas, where he founded the Japan Traditional Karate Association of Dallas, while leading a highly productive Cancer Immunology research program at the Baylor Sammons Cancer Center. JTKA Dallas celebrates its fortieth anniversary in 2022, having graduated more than thirty black belts and producing many world-class competitors. As Tong traverses the fascinating worlds of martial arts and biomedicine, he is constantly amazed by the intricate link of physical prowess, mental aptitude, and spirituality. The monograph is a synthesis of these ideas, presented against the backdrop of his enduring martial arts journey. Tong has practiced Nishiyama Shōtōkan Karate for fifty years. He holds the rank of *hachidan* (eighth-degree black belt) by the World Traditional Karate-dō Federation based in Geneva, Switzerland. Tong is current President of the American Amateur Karate Federation, the oldest non-governmental organization for karate competition in the United States.

About North Atlantic Books

North Atlantic Books (NAB) is a 501(c)(3) nonprofit publisher committed to a bold exploration of the relationships between mind, body, spirit, culture, and nature. Founded in 1974, NAB aims to nurture a holistic view of the arts, sciences, humanities, and healing. To make a donation or to learn more about our books, authors, events, and newsletter, please visit www.northatlanticbooks.com.